Multivariate Analysis of Ecological Data using Canoco 5

This revised and updated edition focuses on constrained ordination (RDA, CCA), variation partitioning and the use of permutation tests of statistical hypotheses about multivariate data. Both classification and modern regression methods (GLM, GAM, loess) are reviewed and species functional traits and spatial structures are analysed.

Nine case studies of varying difficulty help to illustrate the suggested analytical methods, using the latest version of Canoco 5. All studies utilise descriptive and manipulative approaches, and are supported by data sets and project files available from the book website: http://regent.prf.jcu.cz/maed2/.

Written primarily for community ecologists needing to analyse data resulting from field observations and experiments, this book is a valuable resource for students and researchers dealing with both simple and complex ecological problems, such as the variation of biotic communities with environmental conditions or their response to experimental manipulation.

Petr Šmilauer is Associate Professor of Ecology in the Department of Ecosystem Biology, at the University of South Bohemia. His main research interests are: multivariate statistical analysis, modern regression methods, as well as the role of arbuscular mycorrhizal symbiosis in plant communities. He is co-author of the multivariate analysis software Canoco 5, CANOCO for Windows 4.5, CanoDraw, and TWINSPAN for Windows.

Jan Lepš is Professor of Ecology in the Department of Botany, at the University of South Bohemia, and in the Institute of Entomology at the Czech Academy of Sciences. His main research interests include: plant community biology, statistical analysis in the field of ecology, as well as the studies of species diversity, and the role of functional traits in plant community ecology and ecology of hemiparasitic plants. Together with P. Šmilauer, he regularly offers international courses on multivariate statistics.

Multivariate Analysis of Ecological Data using Canoco 5

Second Edition

PETR ŠMILAUER
University of South Bohemia, Czech Republic

JAN LEPŠ
University of South Bohemia, Czech Republic

CAMBRIDGE
UNIVERSITY PRESS

University Printing House, Cambridge CB2 8BS, United Kingdom

Published in the United States of America by Cambridge University Press, New York

Cambridge University Press is part of the University of Cambridge.

It furthers the University's mission by disseminating knowledge in the pursuit of education, learning and research at the highest international levels of excellence.

www.cambridge.org
Information on this title: www.cambridge.org/9781107694408

© Cambridge University Press 2014

This publication is in copyright. Subject to statutory exception and to the provisions of relevant collective licensing agreements, no reproduction of any part may take place without the written permission of Cambridge University Press.

First published 2003
Second Edition 2014

A catalogue record for this publication is available from the British Library

Library of Congress Cataloguing in Publication data
Šmilauer, Petr, 1967–
Multivariate analysis of ecological data using CANOCO 5 / Petr Šmilauer, University of South Bohemia, Czech Republic, Jan Lepš, University of South Bohemia, Czech Republic. – Second edition.
   pages   cm
Includes bibliographical references and index.
ISBN 978-1-107-69440-8 (pbk.)
1. Ecology – Statistical methods.   2. Multivariate analysis.   3. Ecology – Statistical methods – Case studies.   4. Multivariate analysis – Case studies.   I. Lepš, Jan, 1953–   II. Title.
QH541.15.S72S63   2014
577.01'5195 – dc23   2013048954

ISBN 978-1-107-69440-8 Paperback

Additional resources for this publication at http://regent.prf.jcu.cz/maed2/

Cambridge University Press has no responsibility for the persistence or accuracy of URLs for external or third-party internet websites referred to in this publication, and does not guarantee that any content on such websites is, or will remain, accurate or appropriate.

Contents

	Preface		*page* x
1	**Introduction and data types**		1
	1.1	Why ordination?	1
	1.2	Data types	4
	1.3	Data transformation and standardisation	7
	1.4	Missing values	11
	1.5	Types of analyses	12
2	**Using Canoco 5**		15
	2.1	Philosophy of Canoco 5	15
	2.2	Data import and editing	17
	2.3	Defining analyses	24
	2.4	Visualising results	33
	2.5	Beware, CANOCO 4.x users!	36
3	**Experimental design**		39
	3.1	Completely randomised design	39
	3.2	Randomised complete blocks	40
	3.3	Latin square design	41
	3.4	Pseudoreplicates	42
	3.5	Combining more than one factor	44
	3.6	Following the development of objects in time: repeated observations	45
	3.7	Experimental and observational data	48
4	**Basics of gradient analysis**		50
	4.1	Techniques of gradient analysis	51
	4.2	Models of response to gradients	51
	4.3	Estimating species optima by weighted averaging	53
	4.4	Calibration	56
	4.5	Unconstrained ordination	57

	4.6	Constrained ordination	60
	4.7	Basic ordination techniques	61
	4.8	Ordination axes as optimal predictors	62
	4.9	Ordination diagrams	64
	4.10	Two approaches	66
	4.11	Testing significance of the relation with explanatory variables	66
	4.12	Monte Carlo permutation tests for the significance of regression	67
	4.13	Relating two biotic communities	68
	4.14	Community composition as a cause: using reverse analysis	69
5	**Permutation tests and variation partitioning**	71
	5.1	Permutation tests: the philosophy	71
	5.2	Pseudo-F statistics and significance	72
	5.3	Testing individual constrained axes	74
	5.4	Tests with spatial or temporal constraints	75
	5.5	Tests with hierarchical constraints	79
	5.6	Simple versus conditional effects and stepwise selection	83
	5.7	Variation partitioning	88
	5.8	Significance adjustment for multiple tests	91
6	**Similarity measures and distance-based methods**	92
	6.1	Similarity measures for presence–absence data	93
	6.2	Similarity measures for quantitative data	96
	6.3	Similarity of cases versus similarity of communities	101
	6.4	Similarity between species in trait values	102
	6.5	Principal coordinates analysis	103
	6.6	Constrained principal coordinates analysis (db–RDA)	106
	6.7	Non-metric multidimensional scaling	107
	6.8	Mantel test	108
7	**Classification methods**	112
	7.1	Example data set properties	112
	7.2	Non-hierarchical classification (K-means clustering)	113
	7.3	Hierarchical classification	116
	7.4	TWINSPAN	121
8	**Regression methods**	129
	8.1	Regression models in general	129
	8.2	General linear model: terms	131
	8.3	Generalized linear models (GLM)	133
	8.4	Loess smoother	135

	8.5 Generalized additive models (GAM)	136
	8.6 Mixed-effect models (LMM, GLMM and GAMM)	137
	8.7 Classification and regression trees (CART)	139
	8.8 Modelling species response curves with Canoco	140
9	**Interpreting community composition with functional traits**	**151**
	9.1 Required data	152
	9.2 Two approaches in traits – environment studies	154
	9.3 Community-based approach	158
	9.4 Species-based approach	162
10	**Advanced use of ordination**	**167**
	10.1 Principal response curves (PRC)	167
	10.2 Separating spatial variation	169
	10.3 Linear discriminant analysis	173
	10.4 Hierarchical analysis of community variation	174
	10.5 Partitioning diversity indices into alpha and beta components	177
	10.6 Predicting community composition	182
11	**Visualising multivariate data**	**184**
	11.1 Reading ordination diagrams of linear methods	186
	11.2 Reading ordination diagrams of unimodal methods	195
	11.3 Attribute plots	199
	11.4 Visualising classification, groups, and sequences	202
	11.5 T-value biplot	205
12	**Case study 1: Variation in forest bird assemblages**	**208**
	12.1 Unconstrained ordination: portraying variation in bird community	209
	12.2 Simple constrained ordination: the effect of altitude on bird community	215
	12.3 Partial constrained ordination: additional effect of other habitat characteristics	218
	12.4 Separating and testing alpha and beta diversity	221
13	**Case study 2: Search for community composition patterns and their environmental correlates: vegetation of spring meadows**	**226**
	13.1 Unconstrained ordination	227
	13.2 Constrained ordination	230
	13.3 Classification	237
	13.4 Suggestions for additional analyses	238
	13.5 Comparing two communities	239

14	**Case study 3: Separating the effects of explanatory variables**	**246**
	14.1 Introduction	246
	14.2 Data	247
	14.3 Changes in species richness and composition	247
	14.4 Changes in species traits	255
15	**Case study 4: Evaluation of experiments in randomised complete blocks**	**258**
	15.1 Introduction	258
	15.2 Data	258
	15.3 Analysis	259
	15.4 Calculating ANOVA using constrained ordination	265
16	**Case study 5: Analysis of repeated observations of species composition from a factorial experiment**	**267**
	16.1 Introduction	267
	16.2 Experimental design	267
	16.3 Data coding and use	268
	16.4 Univariate analyses	270
	16.5 Constrained ordinations	270
	16.6 Principal response curves	275
	16.7 Temporal changes across treatments	280
	16.8 Changes in composition of functional traits	285
17	**Case study 6: Hierarchical analysis of crayfish community variation**	**301**
	17.1 Data and design	301
	17.2 Differences among sampling locations	302
	17.3 Hierarchical decomposition of community variation	305
18	**Case study 7: Analysis of taxonomic data with discriminant analysis and distance-based ordination**	**309**
	18.1 Data	309
	18.2 Summarising morphological data with PCA	310
	18.3 Linear discriminant analysis of morphological data	313
	18.4 Principal coordinates analysis of AFLP data	317
	18.5 Testing taxon differences in AFLP data using db-RDA	320
	18.6 Taking populations into account	322
19	**Case study 8: Separating effects of space and environment on oribatid community with PCNM**	**324**
	19.1 Ignoring the space	324
	19.2 Detecting spatial trends	326

	19.3 All-scale spatial variation of community and environment	328
	19.4 Variation partitioning with spatial predictors	332
	19.5 Visualising spatial variation	333
20	**Case study 9: Performing linear regression with redundancy analysis**	**337**
	20.1 Data	337
	20.2 Linear regression using program R	337
	20.3 Linear regression with redundancy analysis	340
	20.4 Fitting generalized linear models in Canoco	342

Appendix A Glossary 343
Appendix B Sample data sets and projects 346
Appendix C Access to Canoco and overview of other software 347
Appendix D Working with R 350

References 351
Index to useful tasks in Canoco 5 359
Subject index 360

Preface

The multidimensional data on community composition, properties of individual populations, or properties of environment are the bread and butter of an ecologist's life. Such data need to be analysed taking into account their multidimensionality. A reductionist approach of looking at the properties of each variable separately does not work in most cases. The methods for statistical analysis of such data sets fit under the umbrella of 'multivariate statistical methods'.

In this book, we present a consistent set of approaches to answer many of the questions that an ecologist might have about the studied systems. Nevertheless, we happily admit that other quantitative ecologists may approach the same set of questions with a toolbox of methods (partly) different from those presented here. We pay only a limited attention to other, less parametric methods, such as the family of non-metric multidimensional scaling (NMDS) algorithms or the group of methods similar to the Mantel test. We do not want to fuel the sometimes seen controversy between proponents of various approaches to analysing multivariate data. We simply claim that the solutions presented here are not the only ones possible, but they worked for us, as well as for many other researchers.

We also give greater emphasis to ordination methods compared to classification approaches, but we do not imply that the classification methods are not useful. Our description of multivariate methods is extended by a short overview of regression analysis, including some of the more recent developments such as the generalized additive models or CART models, because the regression models often complement the results of multivariate analyses.

We assume the reader has knowledge of testing statistical hypotheses, of linear regression and ANOVA in the range covered by introductory statistical courses for undergraduates and we make no attempt here to explain corresponding terms or principles.

Our intention is to provide the reader with both the basic understanding of principles of multivariate methods and the skills needed to use those methods in his/her own work. Consequently, the methods are illustrated by examples. For all of them, we provide the data on our web page (see Appendix B), and for all the analyses carried out by the Canoco[1] program, we also provide Canoco 5 project files, containing analyses with required settings and precomputed results.

---

[1] Although the CANOCO name was originally an acronym, it became a recognized entity over the years and this is reflected in the change from upper-case letters, as also made in the Canoco 5 manual.

The nine case studies which conclude the book contain tutorials, where the project/analysis options are explained and the software use is described. The individual case studies differ intentionally in the depth of explanation of the necessary steps. In the first two case studies, the tutorial is more in a 'cookbook' form, whereas a detailed description of individual steps in the subsequent case studies is only provided for the more complicated and advanced methods that are not described in the preceding tutorial chapters. You can work with offered case studies in any order, but if you are new to ordination methods, we recommend you to read at least Chapters 1 to 4 first, and if you are an experienced user of ordination methods (including constrained ones), but not yet friendly with Canoco software, read at least Chapter 2 that introduces the Canoco program[2] and then start with Case studies 1 and 2. The 'Index to useful tasks in Canoco 5', which is located before the standard Index, allows the reader to quickly find solutions to many common technical tasks in work with Canoco 5 software, which are described in this book.

In the second edition, we have tried to cater for both beginning and more advanced users. We have therefore put many of the more advanced comments or suggestions into an extensive set of footnotes. The main text of this book can be understood while ignoring the footnotes, but they provide greater insight for the advanced topics or explain technical details.

The methods discussed in this book are widely used among plant, animal and soil biologists, as well as in freshwater and marine biology or in landscape ecology. In most of these fields, the methods are now routinely applied also to the data sets obtained with molecular biology techniques.

We hope that this book provides an easy-to-read supplement to the more exact and detailed publications such as the collection of Cajo ter Braak's papers, the Canoco 5 manual, or the Legendre and Legendre (2012) textbook. The Reference manual and user's guide to the new version of Canoco 5 is, in fact, so often referred to that instead of citing Ter Braak and Šmilauer (2012), we use 'Canoco 5 manual' throughout this book.

In some case studies, we needed to compare multivariate methods with their univariate counterparts. The univariate methods are demonstrated using the freely available R software (R Core Team 2013), which can be also used to work with multivariate methods described here, but not available in Canoco 5 software (cluster analysis, Mantel test). See Appendix D for a link to a brief tutorial on working with R (in the context of tasks present in this book). But these methods are also available in other statistical packages so the readers can hopefully use their favourite software, if different from R (see Appendix C with an overview of alternative software for multivariate statistical analysis).

Please note that we have omitted the trademark and registered trademark symbols when referring to commercial software products.

We would like to thank John Birks, Robert Pillsbury, and Samara Hamzé for correcting our English in the first edition. We are grateful to all who read drafts of the first edition and gave us many useful comments: Cajo ter Braak, John Birks, Mike Palmer, and Marek Rejmánek; additional useful comments on the text and the language were provided by

---

[2] Reading Chapter 2 is recommended even to experienced users of CANOCO version 4.0 or 4.5.

the students of Oklahoma State University: Jerad Linneman, Jerry Husak, Kris Karsten, Raelene Crandall, and Krysten Schuler.

We are much indebted to Mike Palmer for his meticulous reading of the whole manuscript of the second edition and many great suggestions concerning the language, presented theory and book usability, vastly improving the book quality. We are thankful to Cajo ter Braak for his numerous comments on the second edition. Francesco de Bello provided multiple suggestions on the text concerning the analysis of species traits.

Camille Flinders, Michal Hájek, and Milan Štech kindly provided data sets, respectively, for case studies 6, 2, and 7.

PŠ wants to thank his wife Marie for her continuous support (and drawing Figure 10–3) and to his daughters Marie and Tereza for their patience with him. He also insisted on stating that the ordering of authorship is based purely on the alphabetical order of their names, this time reversed – for a change from the first edition.

JL wants to thank his parents and Olina for support, his daughters Anna and Tereza for patience and his grandchildren Anna, Eliška, and Matěj for keeping him in a good mood.

1 Introduction and data types

1.1 Why ordination?

When you investigate the variation of plant or animal communities across a range of different environmental conditions, you typically find not only large differences in species composition of the studied communities, but also a certain consistency or predictability of this variation. For example, if you look at the variation of grassland vegetation in a landscape and describe the plant community composition using vegetation plots, then the individual plots can be usually ordered along one, two or three imaginary axes. The change in the vegetation composition is often small as you move your focus from one plot to those nearby on such a hypothetical axis.

This gradual change in the community composition can often be related to differing, but partially overlapping demands of individual species for environmental factors such as the average soil moisture, its fluctuations throughout the season, the ability of species to compete with other ones for the available nutrients and light, etc. If the axes along which you originally ordered the plots can be identified with a particular environmental factor (such as moisture or richness of soil nutrients), you can call them a soil moisture gradient, or a nutrient availability gradient. Occasionally, such gradients can be identified in a real landscape, e.g. as a spatial gradient along a slope from a riverbank, with gradually decreasing soil moisture. But more often you can identify such axes along which the plant or animal communities vary in a more or less smooth, predictable way, yet you cannot find them in nature as a visible spatial gradient and neither can you identify them uniquely with a particular measurable environmental factor. In such cases, we speak about **gradients of species composition change**.

The variation in biotic communities can be summarised using one of a wide range of statistical methods, but if we stress the continuity of change in community composition, the so-called **ordination methods** are the tools of the trade. They have been used by ecologists since the early 1950s, and during their evolution these methods have radiated into a rich and sometimes confusing mixture of various techniques. Their simplest use can be illustrated by the example introduced above. When you collect data (cases) representing the species composition of selected quadrats in a vegetation stand, you can arrange the cases into a table where individual species are represented by columns and individual cases by rows. When you analyse such data with an ordination method (using the approaches described in this book), you can obtain a fairly representative summary

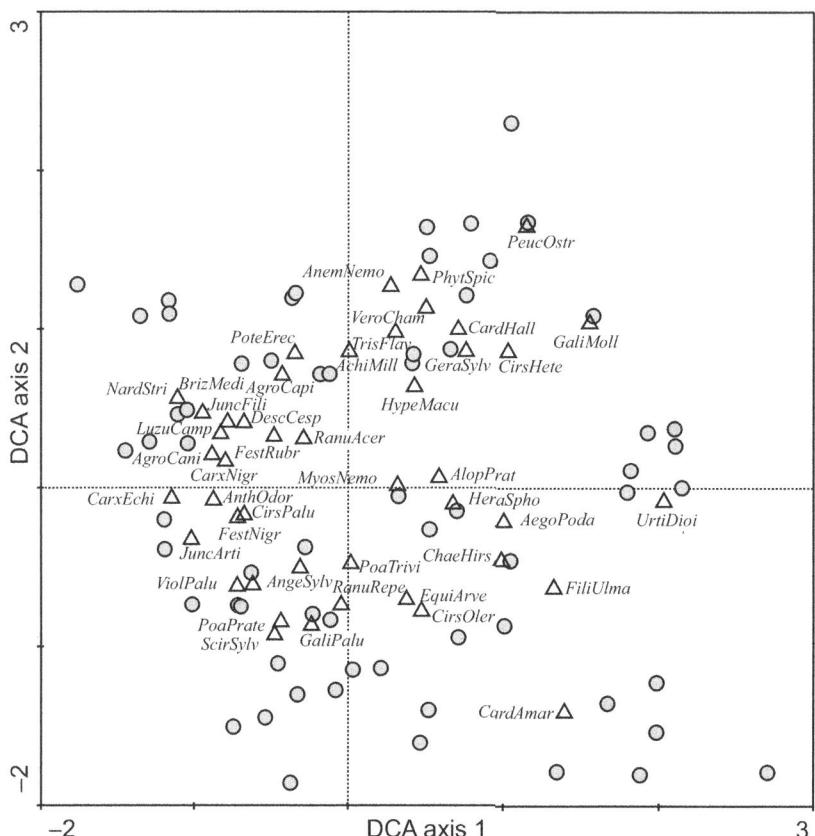

Figure 1-1 Summarising grassland vegetation composition with ordination: ordination diagram from detrended correspondence analysis displaying first two axes, explaining respectively 13% and 8% of the total variation in community composition.

of the grassland vegetation using an ordination diagram, such as the one displayed in Figure 1-1.[1]

The rules for reading such ordination diagrams will be discussed thoroughly later on (in Chapter 11), but even without their knowledge you can read much from the diagram, using the idea of continuous change of composition along the gradients (suggested here by the diagram axes) and the idea that **proximity implies similarity**. The individual cases are represented in Figure 1-1 by grey circles. You can expect that two cases that lay near to each other will be much more similar in their lists of occurring species and

---

[1] In a research paper, it is appropriate to describe the identity of ordination axes either directly using axes labels (as we do in Figure 1-1) or in figure caption (also illustrated in the caption of Figure 1-1). This book, however, contains a multitude of ordination diagrams and to save some ink, we often omit the labelling of their axes.

1.1 Why ordination?

even in the relative importance of individual species populations, compared to cases far apart in the diagram.

The triangle symbols represent the individual plant species occurring in the studied type of vegetation (not all species present in the data are shown in the diagram). In this example, our knowledge of the ecological properties of the displayed species[2] can aid us in an **ecological interpretation of the gradients** represented by the diagram axes. The species preferring nutrient rich soils (such as *Urtica dioica*, *Aegopodium podagraria*, or *Filipendula ulmaria*) are located at the right side of the diagram, while the species occurring mostly on soils poor in available nutrients are on the left side (*Viola palustris*, *Carex echinata*, or *Nardus stricta*). The horizontal axis can therefore be informally interpreted as a gradient of nutrient availability, increasing from the left to the right side. Similarly, the species with their points at the bottom of the diagram are from the wetter stands (*Galium palustre*, *Scirpus sylvaticus*, or *Ranunculus repens*) than the species in the upper part of the diagram (such as *Achillea millefolium*, *Trisetum flavescens*, or *Veronica chamaedrys*). The second axis, therefore, represents a gradient of soil moisture.

As you probably already guessed, the proximity of species symbols (triangles) with respect to a particular case symbol (a circle) indicates that these species are likely to occur there more often and/or with a higher (relative) abundance than the species with symbols more distant from the case.

Our example illustrates a frequent way of using ordination methods in community ecology. You can use such an analysis to visually summarise community patterns in an intuitive way and compare the suggested gradients with your independent knowledge of environmental conditions. But you can also test statistically the predictive power of such knowledge; i.e. address the questions such as 'Does the community composition change with the soil moisture or are the identified patterns just a matter of chance?' Such analyses can be done with the help of **constrained ordination methods** and their use will be illustrated later in this book.

However, you do not need to stop with such exploratory or simple confirmatory analyses and this is the focus of the rest of the book. The rich toolbox of various types of regression and analysis of variance, including analysis of repeated measurements on permanent sites, analysis of spatially structured data, various types of hierarchical analysis of variance (ANOVA), etc., allows ecologists to address more complex, and often more realistic questions. Given the fact that the populations of different species occupying the same environment often share similar strategies in relation to the environmental factors, it would be very profitable if one could ask similar complex questions for the whole biotic community. In this book, we demonstrate that this can be done and we show you how to do it.

Unlike the statistical models with a single response variable, the constrained ordination methods enable you to simply compare all the response variables (species) present in

---

[2] The knowledge of habitat preferences of many plant species is a traditional asset of plant ecologists in Europe. It might be, however, lacking for other groups of organisms or other parts of the world.

the data in terms of their relation to predictors (e.g. environmental variables or human impacts), but also to interpret their similarity and differences using known properties, often representing so-called functional traits of biotic species. And this allows you to relate the functional traits (directly or indirectly) to the properties of environment, generalising your findings beyond the context of the particular area and particular group of organisms you have studied. In this book, we demonstrate the methods working with species traits in sufficient depth for their practical use with Canoco 5.

And yet another type of question arises when you start to compare different kinds of biotic communities. Imagine, for example, that you were able to extend your data set with records on grassland plant community composition with another one, where the community of leaf-eating insect herbivores was quantified for each recorded vegetation plot. How does the compositional variation within the plant and insect communities relate and can we find some gradients, summarising in some optimal way their relation (co-variation)? We will demonstrate a useful method addressing such questions.

1.2 Data types

The terminology for multivariate statistical methods is quite complicated. There are at least two different sets of terms. One, more general and abstract, contains purely statistical terms applicable across the whole field of science. In this chapter we give the terms from this set in italics. The other set contains terms coming from the application domain. As an example, if you study marine phytoplankton, you think about the data in terms of phytoplankton species, sampling station, and environmental characteristics. Starting with version 5, Canoco expects you to define these domain-specific terms (called **item terms** in Canoco 5) for each data table and it uses them afterwards throughout its user interface whenever possible. As this second set varies among projects, we will use the statistical terms in this book, except in the case study chapters or where the discussed concepts are strictly bound to the notion of biological species.

In all multivariate statistical methods we have one data table that can be labelled as the **response data**. This data table contains a collection of observations – *cases*. Each *case* comprises values for multiple *response variables*. The response data can be represented by a rectangular matrix (table), where the rows represent individual *cases* and the columns represent individual *response variables* (species, chemical or physical properties of the water or soil, etc.).[3]

If the response data represent the species composition of a community, we describe the composition using various abundance measures, including counts, frequency estimates, or biomass estimates. Alternatively, we might have information only on the presence or

---

[3] Note that this arrangement is transposed in comparison with the tables used, for example, in traditional vegetation analyses (phytosociological studies). The classical vegetation tables have individual taxa represented by rows and the columns represent individual records or community types.

1.2 Data types

absence of species in individual *cases*; such data essentially correspond to the list of species present in each of the cases.

> An important feature of the data types introduced in the preceding paragraph is that summing up the values of individual *response variables* (species) within each *case* results in a meaningful characteristic: for species abundances, this is the total abundance in a case, for presence–absence data (recorded using 1 and 0 values), this is the total number of species in each case. Data tables with this type of values are called **compositional** in Canoco 5 – as opposed to the **general** type – and this is an important attribute you must set correctly for each created data table to obtain useful advice for your analyses.

In some cases, we estimate the values for the response data on a simple, semi-quantitative scale. Good examples are the various semi-quantitative scales used in recording the composition of plant communities (e.g. original Braun-Blanquet scale or its various modifications).[4]

If our response variables represent the properties of the chemical or physical environment (e.g. concentrations of ions or more complicated compounds in the water, soil acidity, water temperature, etc.), we usually get quantitative values for them, but with an additional limitation: these characteristics do not share the same units of measurement and cannot be meaningfully added, even if they share the units (such as $\mu g \cdot l^{-1}$ for various ion types[5]). In other words, these are non-compositional, *general* data as discussed above. This fact precludes the use of some of the ordination methods[6] and dictates the way the variables are standardised if used in the other ordinations (see Section 1.3).

Very often the response data table is accompanied by another one containing predictor variables that we want to use to understand the response data table contents. If our response data represent community composition, then the predictor data set typically contains measurements of the soil or water properties (for the terrestrial or aquatic ecosystems, respectively), a semi-quantitative or categorical scoring of human impact, etc. When using these variables in a model to predict the response data (like community composition), we might divide them into two different types. The first type is called the *explanatory variables* and refers to the variables that are of the prime interest (in the role of predictors) in our particular analysis. The other type represents the *covariates* which are also variables with an acknowledged (or hypothesised) influence on the

---

[4] And although the sums of such values are sometimes not fully intuitive entities, they still represent rough estimates of the more precise abundance estimates and so we should treat them as compositional data type too.

[5] We admit that in some context, adding concentrations of various ions makes sense, but in ecological studies, these (additive) concentrations are usually supplemented with other chemical or physical measures that are not additive.

[6] Namely correspondence analysis (CA), detrended correspondence analysis (DCA), or canonical correspondence analysis (CCA) and related partial versions. But such general variables can be used in these methods as supplementary or explanatory variables or as covariates – they only cannot be used as the response data for them.

response variables. We want to account for (factor-out or partial-out) such an influence **before** focusing on the influence of the variables of prime interest (i.e. on the effect of *explanatory* variables).

As an example, imagine a situation where you study the effects of soil properties and type of management (hay cutting or pasturing) on the species composition of grassland in a particular area. In one analysis, you might be interested in the effect of soil properties, paying no attention to the management regime. In this analysis, you use the grassland composition as the *response data* (with individual plant species as individual *response variables*) and the measured soil properties as the *explanatory variables*. Based on the results, you can make conclusions about the relation of plant species populations to particular environmental gradients, which are described (more or less appropriately) by the measured soil properties. Similarly, you can ask how the management type influences plant composition. In the corresponding analysis, the variables describing the management regime act as *explanatory variables*. Further, you might expect that the management also influences the soil properties and this is probably one of the ways in which management acts upon the community composition. Based on such expectation, you may ask about the influence of management regime **beyond** that mediated through the changes of soil properties. To address such a question, you must use the variables describing the management regime as the *explanatory variables* and the measured soil properties as the *covariates*.[7] Of course, there might also exist unique effects of soil properties not related to management, and to test and explore them you need to define another analysis, where the management descriptors act as covariates and the soil characteristics as explanatory variables.

Another typical example of *covariate* use is for an experimental design where *cases* are grouped into logical or physical blocks. The values of *response variables* (e.g. species composition) for a group of *cases* might be similar due to their (spatial) proximity, so we need to model this influence and account for it in our data. The differences in *response variables* that are due to the membership of *cases* in different blocks can be removed (i.e. 'partialled-out') from the model by using a factor identifying experimental blocks as a *covariate*.

Beside *explanatory variables* and *covariates*, we recognise yet another kind of predictor variables, called *supplementary variables*. These are used in unconstrained ordination (also called indirect gradient analysis), defined in the following section, to interpret its results.[8]

Predictors can be quantitative variables (concentration of nitrate ions in soil), semi-quantitative estimates (degree of human influence estimated on a 0–3 scale) or factors (nominal or categorical – also categorial – variables). The simplest predictor form is a binary variable, where the presence or absence of a certain feature or event (e.g. vegetation was mown, trap is located near a road, etc.) is indicated, respectively, by a 1 or 0 value.

---

[7] This particular example is discussed in Canoco 5 manual, section 6.3.1.
[8] Supplementary variables are projected *post hoc* into an ordination space already computed for cases and response variables.

Factors with multiple values (levels) are the natural way of expressing the classification of our *cases* or subjects: for example, classes of management type for meadows or the type of stream for a study of pollution impact on rivers.

1.3 Data transformation and standardisation

1.3.1 Transformation

As will be shown in Chapter 4, ordination methods find the axes representing regression predictors that are optimal for predicting the values of *response variables*, i.e. the values in the response data table. Therefore, the problem of selecting a transformation for the *response variables* is rather similar to the problem one needs to solve when using any of the variables in the (multiple) regression. The one additional restriction is the need to specify an identical data transformation for all the *response variables* when working with so-called compositional data, see the preceding section, because such variables are often measured on the same scale. In the unimodal (weighted averaging) ordination methods (see Section 4.2), the data values cannot be negative and this imposes a further restriction on the outcome of any potential transformation.

This restriction is particularly important in the case of the **log transformation**. The logarithm of 1.0 is zero and logarithms of values between 0 and 1 are negative and hence non-acceptable for unimodal ordination. Therefore, Canoco provides a flexible log-transformation formula

$$y' = \log(A \cdot y + B)$$

You should specify the values of A and B so that before the transformation is applied to your data values, the result $A \cdot y + B$ is always greater than zero. The default values of both A and B are 1.0, which neatly map the zero values again to zero, and positive y values remain positive. Nevertheless, if your original values are small (say, in the range 0.0 to 0.1), the shift caused by adding the relatively large value of 1.0 dominates the resulting structure of the data matrix. You can adjust the transformation in this case by increasing the value of A to 10.[9] The default log transformation (i.e. $\log(y + 1)$) works well with the percentage data on the 0 to 100 scale, or with the ordinary counts of objects (e.g. caught individuals of each species).

Whether to use a log transformation or keep the original scale is a difficult question, with different answers from different statisticians. We suggest that you do not consider the variable distribution (at least not in the sense of testing its difference from a Normal distribution, as routinely and often incorrectly done[10]), but you base your decision on how you phrase the hypothesis standing behind your research, as described in the following paragraph.

---

[9] Such change is automatically done by Canoco when you, for example, specify A = 1 and B = 0.1, changing A to 10.0 and B to 1.0.

[10] Many users forget that only the residuals of a statistical model are expected to have a Normal distribution and they test the response variable values instead.

As stated above, ordination methods can be viewed as an extension of multiple regression methods, so this **semantic-based approach** will be explained in the simpler regression context. You might try to predict the abundance of a particular biotic species in cases, based on the values of one or more predictors (environmental variables, or ordination axes in the context of ordination methods). One can formulate the question addressed by such a regression model (assuming just a single predictor variable for simplicity) as 'How does the average value of species Y change with a change in the environmental variable X by one unit?' If neither the response variable nor the predictors are log transformed, the answer coming from a regression model can take the form: 'The value of species Y increases by B if the value of environmental variable X increases by one measurement unit'. Of course, B is then the regression coefficient of the linear model equation $Y = B_0 + B \cdot X + E$. But often you can feel that the answer should have a different form, such as 'If the value of environmental variable X increases by one unit, the average abundance of the species increases by 10%.' Alternatively, you can say, 'the abundance increases 1.10 times'. In both cases, you are thinking on a multiplicative scale, which is not the scale assumed by the linear regression model. In such a situation, you should **log-transform the response variable**. Similarly, if the effect of a predictor (environmental) variable changes in a multiplicative way, **the predictor variable should be log-transformed**.[11]

Plant community composition data are often collected on a semi-quantitative estimation scale and the Braun-Blanquet scale with seven levels (r, $+$, 1, 2, 3, 4, 5) is a typical example. Such a scale is then quantified in the spreadsheets using corresponding ordinal levels (from 1 to 7 in the above case). This coding, called **ordinal transformation**, already implies a log-like transformation because the actual cover/abundance differences between the successive levels are generally increasing. An alternative approach to using such estimates in data analysis is to replace them by the assumed centres of the corresponding range of percentage cover. But doing so, however, you find a problem with the r and $+$ levels because these are based more on the abundance (number of individuals) of the species than on their estimated cover. Nevertheless, using very rough replacements, such as 0.1 for r and 0.5 for $+$, rarely harms the analysis (compared to the alternative solutions).

Another useful transformation of the response data available in Canoco is the square-root transformation. This might be the best transformation to apply to count data, such as the number of specimens of individual species collected in a soil trap, number of individuals of various ant species passing over a marked 'count line', etc., but the log transformation also handles well such data. Further transformations available in Canoco 5 analyses are two variants of arcsine transformation (one for fractional data on a 0–1 scale, another one for percentage data on the 0–100 scale) and binarising transformation, turning any positive value into 1.0 and other values into 0.0. Additionally,

---

[11] Strictly speaking, the log-transformation turns a multiplicative relation into an additive one only if you can use it without the B constant (see the formula above). But even if the presence of zero values in your data requires you to add a positive B, the bias is small if B is small compared with the range of transformed variable values.

if you need any kind of transformation that is not provided by the Canoco software, you might do it in your spreadsheet software and import the transformed data into Canoco project.

When you work with binary (0/1) data, any of the transformations discussed above do not have any real effect, so it is best to keep such data untransformed.

The transformation you choose for a data table when it is used as the response data for the first time is remembered and offered during the setup of following analyses. You can also set or change the default transformation directly using the *Data | Default transformation and standardization* menu command, when the particular table is in foreground. Shared transformation can be set only for data tables of the *compositional* type (see Section 1.2), but for the *general* table type you can set the implicit transformation for individual variables in the table.

1.3.2 Standardisation

In this book, we treat the transformation and standardisation processes separately, even though both 'transform' (change) the original data in the usual meaning. In our view, the **transformation** can be represented by an algebraic function $Y'_{ik} = f(Y_{ik})$ which is applied to each value independently of the other values. **Standardisation** is done, on the other hand, with respect to either the values of other variables measured for the same case (standardisation by cases) or the values of the same variable measured for the other cases (standardisation by variables).

In fact, even the term standardisation can be understood more broadly, as we do here, or in a more narrow sense, as used in the Canoco software: *standardisation* means there the adjustment of values affecting their variability, while so-called **centring**[12] changes mean value. The most common type of centring leads to zero average of variables (or – more rarely used – of cases), while the most common type of standardisation (in the narrower sense) is the standardisation to unit norm (a square root of the sum of squared variable/case values).[13] For variables, the standardisation to unit norm must be almost always combined with the centring, so that the resulting variables have not only a unit norm, but also a unit variance.

Canoco centres and standardises any *explanatory* or *supplementary variables* and any *covariates*, to bring their means to zero and their variances to one,[14] but for the *response variables*, use of standardisation (and of centring, to a lesser extent) is an important choice in the **linear** ordination methods.[15] For constrained linear methods (i.e. redundancy analysis), the centring by variables is required (and enforced in Canoco 5),

---

[12] Note that the Canoco 5 user interface uses US spelling ('center'/'centering') and so we do the same whenever referring directly to program user interface elements.

[13] Alternatively, the Analysis Setup Wizard offers the standardisation of cases to unit sum, but see Table 6–1 and Section 6.2.2 for a discussion of existing issues with this standardisation.

[14] This treatment of predictors makes their effects, as seen in ordination diagrams and in numerical summaries, comparable, but it also assures numerical stability of the calculations.

[15] Not so in unimodal methods, where a special form of double standardisation (both by rows and by variables) is implied by the weighted averaging algorithm and the standardisation cannot be therefore selectively applied by the user.

while the standardisation is optional (also in unconstrained linear ordination), at least for compositional data tables.

> With compositional data tables, you should be extremely careful with standardisation by variables (typically species). The intention of this procedure is to give all the species (response variables) the same weight. But the result is often counter-productive, because a species with a low frequency of occurrence might become very influential. If the species is found in one case only, then all of its quantity is in this case alone, which makes this case very different from the others. On the other hand, the species that are found in many cases do not attain, after standardisation, a high share in any of them and their effect is relatively small.

For general (non-compositional) data tables where each variable has its own scale, it is necessary to centre and standardise the variables (this is also often referred to as calculating the *z-scores*). A typical example of this comes from classical taxonomy: each object (individual, population) is described by several characteristics, measured in different units (e.g. number of petals, density of hairs, weight of seeds, etc.). When a similarity among measured individuals or among populations is calculated from the rough data, the weight of individual variables changes when you change their units – and the final result is completely meaningless.

The difference of running principal components analysis (PCA) on response data that were only centred or both centred and standardised by variables is reflected in the traditional names for these two variants: the former one is called 'PCA on variance-covariance matrix', while the latter one is called 'PCA on correlation matrix'.[16]

For the redundancy analysis (RDA, constrained linear ordination), Canoco also offers another kind of standardisation by response variables, called the **standardisation by error variance**. In this case, Canoco proceeds as if the standard centring and standardisation by variables was chosen, but in addition it calculates, separately for each response variable, how much of its variance was not explained by the explanatory variables (and covariates, for partial RDA). The inverse of that *error variance* is then used as the relative weight of each response variable. Therefore, the better a response variable is described by the explanatory variables, the greater impact it has on the analysis results.

> For response data representing biotic communities, the standardisation by cases (either by case norm or by the total) has a clear ecological meaning. If you use it, you are interested only in proportions of species (both for the standardisation by totals and by the case norm). With standardisation, two cases containing three species, in the first case with 50, 20 and 10 individuals, and in the second case with 5, 2 and 1 individual, will be found identical.[17] Standardisation by the total (i.e. to percentages)

---

[16] PCA on correlation matrix is the most common type of PCA outside the field of ecology.
[17] The differences in ecological interpretations of analyses with and without standardisation by case norm are discussed in Section 15.3.

> is more intuitive: it produces relative abundance values (called 'dominance' by zoologists). The standardisation by case norm is particularly suitable when applied before calculating the Euclidean distance (see Section 6.2). The standardisation by cases should be routinely used[18] when the abundance total in a case is affected by the sampling effort or other factors that you are not able to control, but which should not be reflected in the analysis results.[19]

More details about the various centring and standardisation options available for the response data in linear ordination methods can be found in the Canoco 5 manual, pp. 121–123.

1.4 Missing values

Whatever precautions you take, you are often not able to collect all the data values you need: a soil sample sent to a regional lab gets lost, you forget to fill in a particular slot in your data collection sheet, etc.

Most often, you cannot go back and fill in the empty slots, usually because the subjects you study change in time. You can attempt to leave those slots empty, but this is often not the best decision. For example, when recording sparse community data (you might have a pool of, say, 300 species, but the average number of species per plot is much lower), you interpret the empty cells in a spreadsheet as absences, i.e. zero values. But the absence of a species is very different from the situation where you simply forgot to look for this species! Like the other statistical software, Canoco 5 provides a notion of missing values (it might be represented as a word 'NA'), but this is only a notational convenience. The actual analysis done on such data must deal further with the fact that there are missing values. Here are few options you might consider:

(1) You can remove the cases in which the missing values occur. This works well if the missing values are concentrated in a few cases. If you have, for example, a data set with 30 variables and 500 cases and there are 20 missing values from only 3 cases, it might be wise to ignore these 3 cases in the analysis. This strategy (called 'case-wise deletion') is often used by general statistical packages and it represents the default handling of missing values in Canoco 5 as well.
(2) On the other hand, if the missing values are concentrated in a few variables that are not deemed critical, you might remove the variables from your data set. Such a situation often occurs when you are dealing with data representing chemical analyses. If 'every thinkable' cation concentration was measured, there is usually a strong correlation among them. For example, if you know the values of cadmium

---

[18] With linear methods – alternatively, for compositional data, you can use the unimodal methods where a standardisation is implied.

[19] For example, the number of grasshoppers caught in a net depends on the sampling effort and also on the weather at the time of sampling. If we are not sure that both factors were constant during all the sampling, it is better to standardise the data.

concentration in air deposits, you can usually predict the concentration of mercury with reasonable precision (although this depends on the type of pollution source). Strong correlation between these two characteristics implies that you can make good predictions with only one of these variables. So, if you have a lot of missing values in Cd concentrations, it might be best to drop this variable from your data. Removal of variables containing missing values is also available in Canoco 5. The removal of variables and the removal of rows can be also combined or applied as alternatives, with subsequent comparison of the analysis' conclusions.

(3) The two methods of handling missing values described above might seem rather crude, because you lose so much of your data that you often collected at considerable expense. Indeed, there are various **imputation** methods. The simplest one is to take the average (or median) value of the variable (calculated, of course, only from the *cases* where the value is not missing) and replace the missing values with it. Another, more sophisticated one, is to build a (multiple) regression model, using the *cases* with no missing values, to predict the missing value of a variable for *cases* where the values of the other variables (predictors in the regression model) are not missing. This way, you might fill in all the holes in your data table, without deleting any *cases* or variables.

Yet, you are deceiving yourself, as you only duplicate the information you have. The degrees of freedom you lost initially cannot be recovered. If you then use such imputed data in a statistical test, this test makes an erroneous assumption about the number of degrees of freedom (number of independent observations in your data) that support the conclusion made. Therefore, the significance estimates are not quite correct (they are 'over-optimistic').[20] You can alleviate this problem partially by decreasing the statistical weight for the *cases* where missing values were estimated using one or another method. The calculation can be quite simple: in a data set with 20 variables, a *case* with missing values replaced for five variables gets a weight 0.75 ($=1.0-5/20$). Nevertheless, this solution is not perfect. If you work only with a subset of the variables (for example, during a stepwise selection of explanatory variables), the *cases* with any variable being imputed carry the penalty even if (some of) the imputed variables are not used.

Canoco 5 implements the removal of variables with missing values and the replacement of missing values with variable average or median, but these options are not active by default. You can change the handling of missing values for any data table using the *Data | Set handling of missing values* menu command. The methods of handling missing data values are treated in more detail e.g. in the book by Little and Rubin (1987).

1.5 Types of analyses

When we try to describe the variation in values of one or more *response variables*, the appropriate statistical modelling methodology depends on whether we study each of the

---

[20] Same concern applies also to permutation tests, such as those used in Canoco, even though they do not work explicitly with the degrees of freedom.

Table 1–1 The types of statistical models.

Response variable(s) ...	Predictor variable(s)	
	Absent	Present
... is one	distribution summary	regression models *sensu lato*
... are many	unconstrained ordination (PCA, CA, DCA, NMDS) cluster analysis	constrained ordination unconstrained ordination with supplementary variables discriminant analysis (CVA)

response variables separately (or many variables at the same time) and whether we have any *predictor variables* available when we build the model.

Table 1–1 summarises the most important statistical methodologies used in these different situations (you should also check Section 4.1 for more detailed treatment).

If we look at a single response variable and there are no predictor variables available, then we can summarise only the distributional properties of that variable (e.g. by a histogram, median, standard deviation, inter-quartile range, etc.). In the case of multivariate data, we might use the ordination approach represented by the methods of **unconstrained ordination** (also called indirect gradient analysis). Most prominent are principal components analysis (PCA), correspondence analysis (CA), detrended correspondence analysis (DCA) and non-metric multidimensional scaling (NMDS). Even with unconstrained ordination methods, we can interpret their results by projecting available descriptors (called *supplementary variables*) into the ordination space. Alternatively to ordination methods, we can try to (hierarchically) divide our set of cases into compact distinct groups based on the similarity of the values of *response variables* (methods of cluster analysis, see Chapter 7).

If we have one or more *predictor variables* available and we describe values of a single *response variable*, then we can use **regression models** in the broad sense, i.e. including both traditional regression methods and methods of analysis of variance (ANOVA) and analysis of covariance (ANCOVA). This group of methods is unified under the so-called **general linear model** and was later extended and enhanced by the methodology of **generalized linear models (GLM)** and **generalized additive models (GAM)** in one direction, and by adding provision for random effects[21] in another direction, leading to (generalized) linear/additive mixed effects models (i.e. (G)LMM and GAMM). Further information on regression models is provided in Chapter 8.

If we have *explanatory variables* for a set of *response variables*, we can summarise relations between multiple *response variables* (typically biological species) and one or several *explanatory variables* using the methods of **constrained ordination** (also called direct gradient analysis, although this term has variable meaning in statistical literature).

---

[21] Random effects can be (roughly) understood as categorical predictors (factors, e.g. block or individual identity) with actually observed realisations being just a small subset of all possible. For example, there can be infinitively many experimental blocks established and I am not particularly interested in the identity of those present in my experiment, only in the extent of difference among them.

Most prominent are redundancy analysis (RDA) and canonical correspondence analysis (CCA), but there are several other methods in this category.

When we have two data sets of similar nature, such as the records of two types of biotic communities, measured at the same set of locations, we can compare them with methods studying co-variation among both data sets. In this book, we focus on the method of symmetric co-correspondence analysis (CoCA), but co-inertia analysis or linear canonical correlation analysis represent other possibilities. The CoCA is introduced in Section 4.13 and an example of its use is provided in Section 13.5.

In this book, we discuss the methods and demonstrate related analyses for all the types of statistical models outlined in this section, but first you must become familiar with the primary tool used in the case study chapters, the Canoco 5 software.[22] This is the subject of Chapter 2.

---

[22] See Appendix D for a link to the information about how to work with R software, used in this book for 'univariate' analyses and cluster analysis.

2   Using Canoco 5

In this chapter you will learn about the style of work which Canoco 5 supports and encourages and about the basic techniques of importing and editing data tables, defining and executing statistical analyses, and visualising their results. This introduction to the use of Canoco 5 is necessarily brief and cannot cover all its aspects. You are advised to obtain additional information from the Canoco 5 manual, starting with its tutorial (Chapter 2).

2.1   Philosophy of Canoco 5

Canoco 5 is implemented around a way of working with data, statistical analyses and their results, which is quite different from earlier versions. All the work with data and analyses is now more tightly integrated and when properly understood, Canoco 5 aids users not only in routine work, but also in making correct decisions about the statistical analyses. The relations among the most important entities of work in Canoco 5 are summarised in Figure 2–1.

The central stage is taken by projects and analyses. A **project** contains one or multiple **data tables**, as well as analyses of the data. You can open only one project in the Canoco 5 program at any time. But you can also open multiple Canoco 5 instances (by repeatedly double-clicking the Canoco 5 icon at the computer desktop) and work with different projects in each. Each **analysis** in a project represents a single task, addressing one or multiple research questions you have about the data. When an analysis is accomplished (by choosing an appropriate analysis template, specifying its options, and executing the analysis), analysis results can be inspected in an **analysis notebook**, which is automatically opened after the analysis is performed and contains one to many pages. When working with ordination methods, visual representation of their results plays a key role and Canoco 5 makes it easy to quickly create appropriate ordination diagrams using the Graph Wizard. With the help of commands located in the Graph menu a large variety of additional ordination diagrams and attribute plots can be created. All created graphs are also contained, as separate pages, in the analysis notebook.

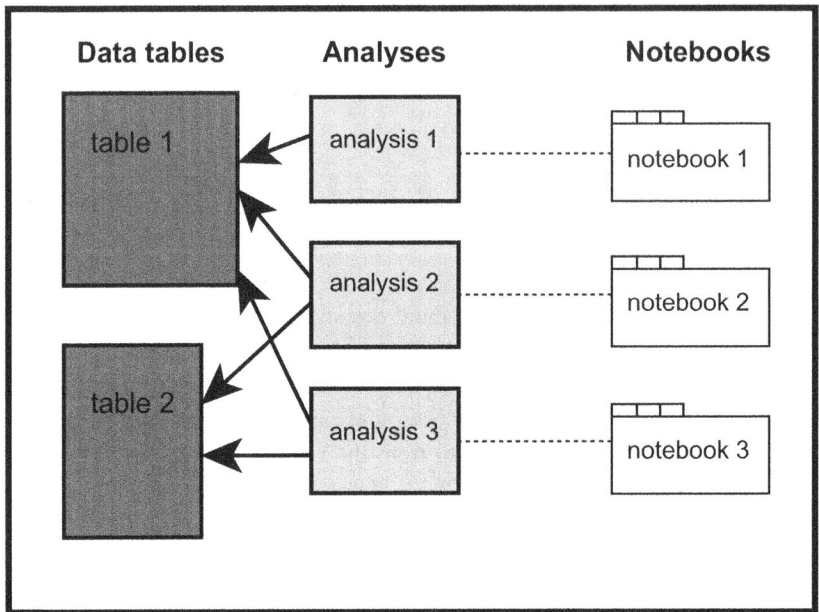

Figure 2–1 Canoco 5 project components.

Canoco Adviser and the Wizards

Canoco Adviser is a mysterious character of the Canoco 5 user interface: you read a lot about him in the software documentation, but you never meet him in person. He works with you through multiple wizards. Wizards are specific sequences of dialog pages, which help you to accomplish complex tasks by dividing them into simpler sub-tasks on individual pages. You proceed through a page sequence using the *Next* button, until you come to the end, where the *Next* button is replaced by another one, labelled *Finish*. In addition, each wizard page has a *Back* button to return to an earlier page and a *Help* button, displaying a help window explaining the contents of the current page.

The page sequence shown in each wizard is not fully fixed, the number and identity of displayed pages depends on the context. Beside displaying a preset range of options, wizard pages sometimes provide advice based on the type and statistical properties of the analysed data or on the type of chosen analysis. Such advice is assembled by an expert system built into Canoco 5. The entire set of such advisory features is named the Canoco Adviser.

Import Wizards help you select the data present in Excel spreadsheet files or in legacy data files formatted for older versions of CANOCO, describe their properties, and transform them into project data tables.

> **New Analysis Wizard** suggests various types of analyses (analysis templates) that are appropriate for the data available in your project and allows you to select one of them, matching your task. Based on your choice, a new analysis is then created. New analyses can be created also without the wizard as so-called customised analyses (including imported CANOCO 4.x project files), but this is not recommended if you can achieve your task using the wizard.
>
> **Analysis Setup Wizard** takes a newly created or existing analysis and allows you to adjust its settings (variables to use, number of ordination axes to compute, the type of test to apply, etc.). It corresponds to the only wizard present in CANOCO 4.5.
>
> **Graph Wizard** is available for most analyses created with the New Analysis Wizard and it is shown automatically the first time the analysis was executed. It provides quick access to a subset of graphing utilities of Canoco 5.

2.2 Data import and editing

The preparation of input data for multivariate analyses has always been one of the big obstacles to their effective use. In earlier versions of the CANOCO program, one had to understand the overly complicated and unforgiving format of the data files, which was based on the requirements of the FORTRAN programming language on which CANOCO was based. Version 4.x of CANOCO alleviated this problem by an easy method of transforming data stored in spreadsheets into CANOCO format files via the export utility called WCanoImp.

Canoco 5 now offers direct import of Excel spreadsheet files into Canoco project. Alternatively, you can import – in a similar way – the legacy CANOCO 4.x data files or you can enter the data directly within Canoco 5 workspace, using its integrated spreadsheet editor. This introduction describes only the import from Excel spreadsheet files.

Let us start with the data in your spreadsheet files. While the majority of users will work with Microsoft Excel, Excel formats can be also exported from open-source spreadsheet software running under Microsoft Windows. If the data are stored in a relational database (Oracle, FoxBASE, Access, etc.), you can use the facilities of your spreadsheet program to first import the data into it. In the spreadsheet, you must arrange your data into a rectangular structure, as laid out by the spreadsheet grid. In the default layout, the individual *cases* correspond to the rows, while the individual spreadsheet columns represent the variables.[1]

In addition, you should have at least a simple heading for both rows and columns: the first row (except the empty upper left corner cell) contains the names of variables, while the first column contains the names of individual cases. Unlike the earlier versions of CANOCO 4.x, the length of the row and column names is not too limited (it should not exceed 255 characters). Nevertheless, **brief names** as defined in the earlier CANOCO

---

[1] This is the layout always used in data tables of Canoco 5 projects, but Excel spreadsheets to be imported may have this arrangement rotated by 90 degrees.

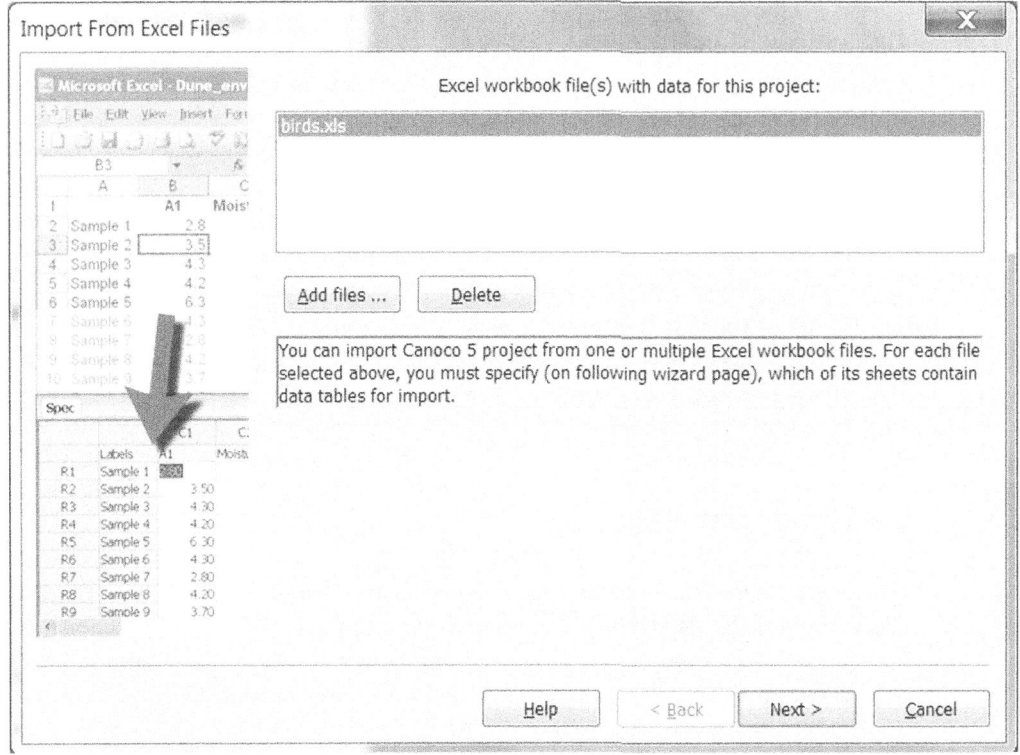

Figure 2–2 First page of the Excel Import Wizard after one input file was added.

versions are still very useful for creating informative ordination diagrams. Canoco 5 therefore generates them from the **full names** while importing the data. Alternatively, you can have the first two rows of the source spreadsheet filled with full and brief names (in this order) of columns and/or the first two columns filled with full and brief names of rows (cases). The brief names cannot have more than eight characters. When you let Canoco 5 generate the brief names during import, you can still adjust their contents later in the Canoco 5 spreadsheet editor.[2]

The remaining cells of the spreadsheet must be numbers (whole or decimal) for numeric variables or text values for factors or they can be empty. An empty cell in imported data always implied a zero value in CANOCO 4.x, but in Canoco 5, empty cells can be used to code either zero values or the missing values and this choice can be set for each data table before its import.[3] Missing values can be always specified using *NA* text.

Data can be imported from an Excel spreadsheet file either as a part of creating a new Canoco 5 project (using the *File | Import project | from Excel* menu command) or you can add one or more tables to an existing project (using the *Data | Add new table(s) |*

---

[2] Note that the Canoco 5 spreadsheet shows, at any time, only the brief or only the full names and you can change the current display (separately for table rows and columns) e.g. using the *Data | Short Labels* submenu.

[3] For a factor variable, empty cell always means a missing value, because zero value makes sense only for numeric variables.

2.2 Data import and editing

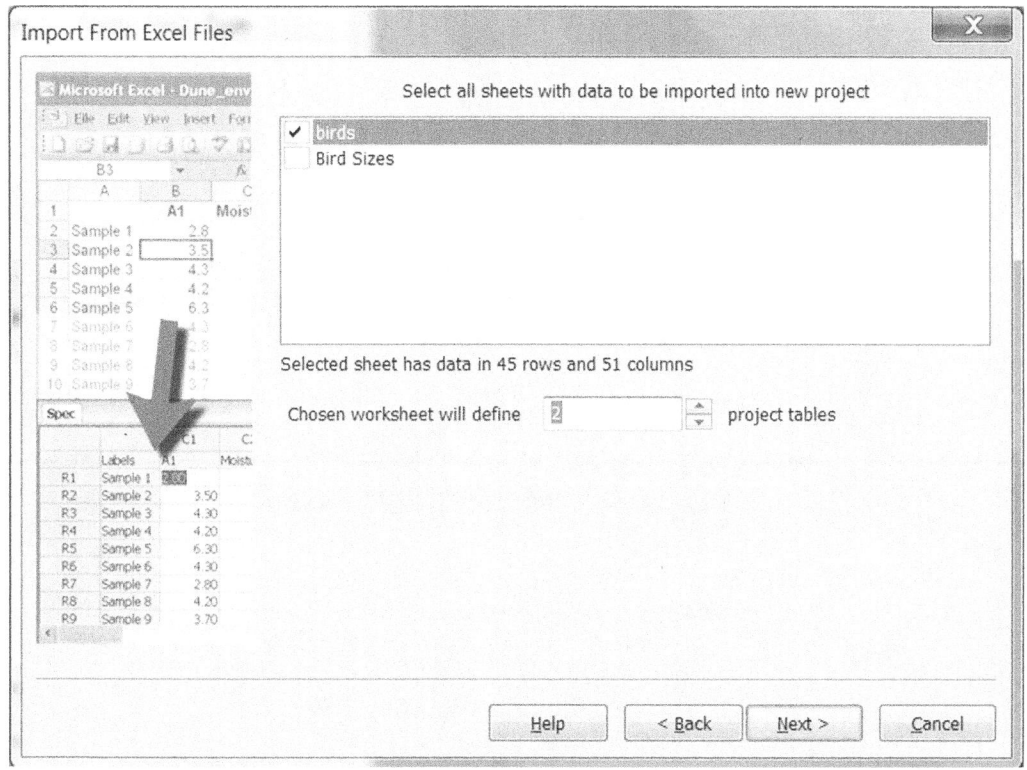

Figure 2–3 Second page of the Excel Import Wizard when only one input file was specified on preceding page.

Import from Excel menu command). Either way, the user interface of the Excel import wizard is essentially identical and we describe it briefly here.

The first page of the wizard (as illustrated in Figure 2–2) allows you to choose one or multiple Excel files containing your data. Canoco 5 is able to import either from *.xls* files with the format used up to Microsoft Office 2003 or from *.xlsx* files used in the newer Office versions.[4]

On the following wizard page (illustrated in Figure 2–3), you must identify the sheets containing the data and also specify how many tables shall be created. Note that this count does not follow from the number of chosen sheets, because one data table can be created by combining multiple rectangular areas possibly distributed across multiple sheets, and one sheet can also contain multiple separate data tables (as in the case illustrated here).

The number of data tables specified at this wizard page is then reflected in the following wizard pages: for each table, the wizard displays two consecutive pages: the first page in each pair is used to set table options, while the second one is used to specify the exact range(s) of spreadsheet cells containing the data to be imported.

---

[4] These two Excel formats cannot be combined in the same import wizard session, but this limitation can be circumvented with the ability of Canoco 5 to add tables to an existing project.

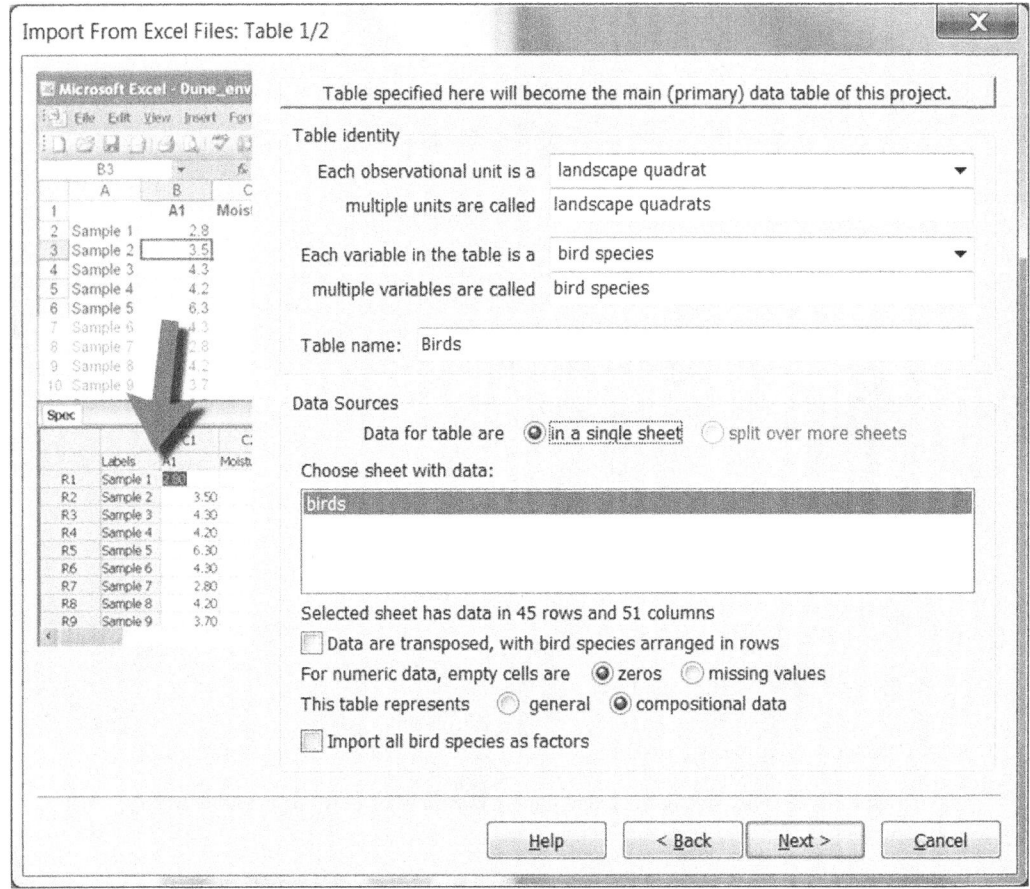

Figure 2–4 First of the two pages defining particular data table in the Excel Import Wizard.

Figure 2–4 illustrates the first of the two pages as shown for the first imported data table. Details for individual options can be found for example by clicking the *Help* button, but here we emphasise the necessity to specify carefully the terms to be used for each table's rows and columns. For both kinds of terms, you must specify their singular and plural. Also, the interpretation of empty cells is set here, in the lower part of the page, and the table content is characterised as *compositional data* in the snapshot (see the explanation at the end of Section 1.2).

> It is advisable to import the table containing response data as the first data table in the project. This will make your work with New Analysis Wizard easier, and the first imported data table is also specified as compositional type by default, corresponding to the fact that in many applications, the response data describe community composition, often explained by the environmental properties that would then naturally come in a second data table of the project.
>
> **You should never merge the response and explanatory variables into a single data table, always keep the response data as a separate table!**

2.2 Data import and editing

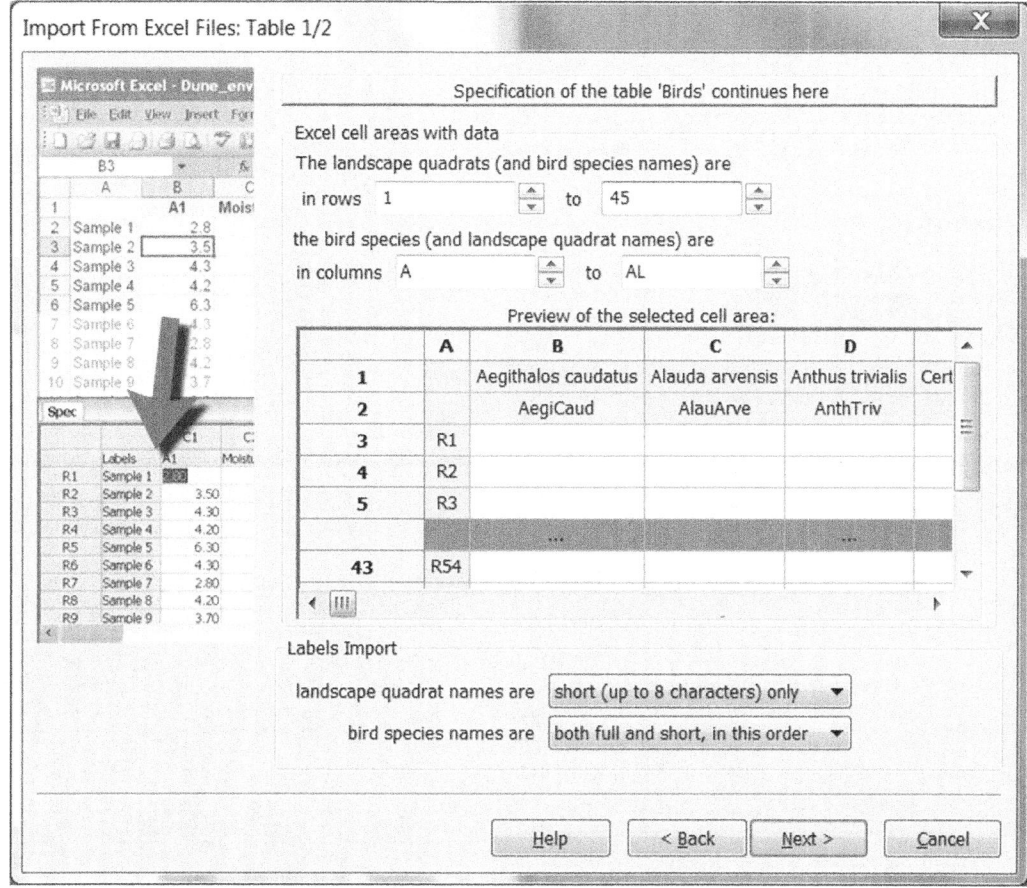

Figure 2–5 Second of the two pages defining particular data table in the Excel Import Wizard.

For the second and later data tables of a project, the terms for their rows cannot be specified ad hoc. Instead, they must be chosen from the terms used for rows (or columns) of the tables defined earlier. This simple constraint guarantees coherence among multiple tables of a project and this coherence is also reflected in the synchronisation of the number of rows (*cases*) across project tables. When you insert or delete rows in one of the tables, the same number of rows is inserted or removed in corresponding positions of the other tables sharing the same row identity. In many projects, row identity is the same for all data tables, but another possibility is to base the identity of the rows of a table on the columns of an earlier defined data table. Such a table can then represent, for example, species traits for the species defined by columns of an earlier data table.

The second of the two wizard pages shown for each imported data table (illustrated in Figure 2–5) displays in its central part a preview of the content of the sheet(s) selected on the preceding page. The Excel import wizard attempts to guess the position of the rectangle with input data, as well as the presence and type of row and column labels. Both options can be adjusted, however.

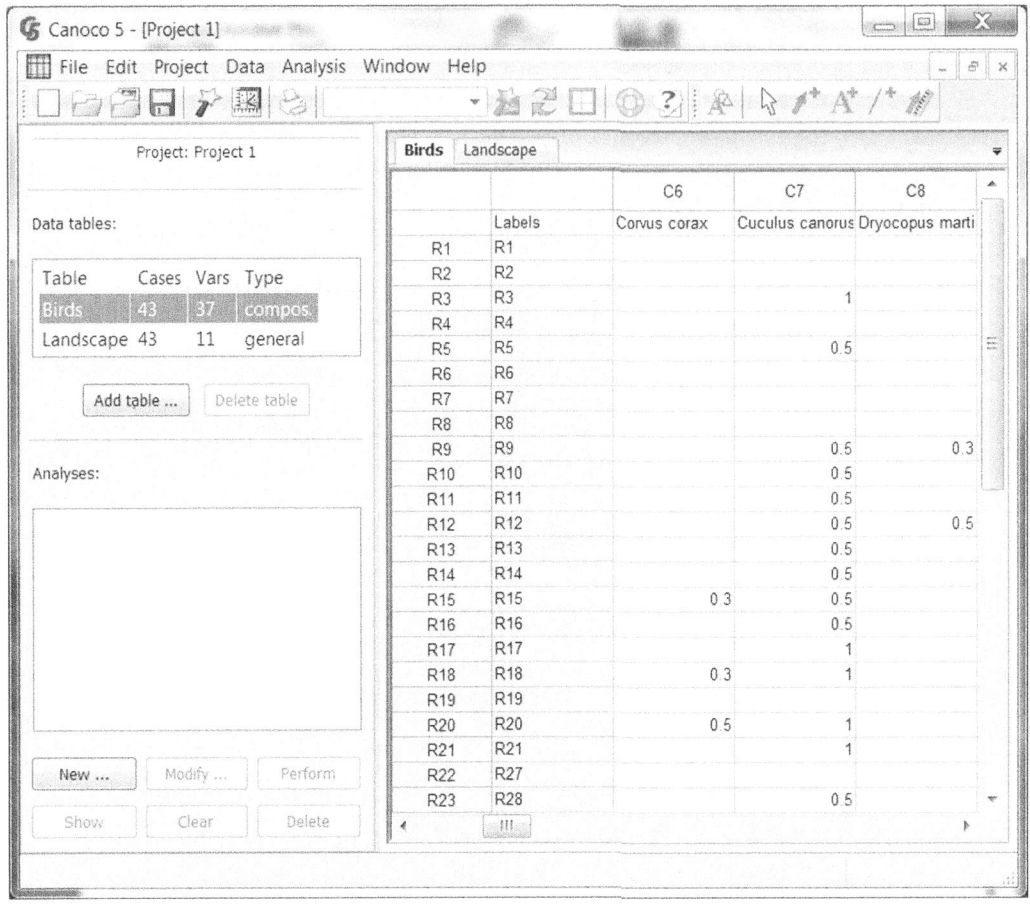

Figure 2–6 Canoco 5 workspace with integrated spreadsheet window in the foreground.

After you have specified import options for all data tables, the wizard imports the data into a new (or existing) Canoco 5 project and it is shown in a Canoco 5 spreadsheet,[5] illustrated in Figure 2–6.

This spreadsheet looks similar to the Microsoft Excel user interface and you can indeed enter new (or change existing) values there. But there are also differences: the Canoco 5 spreadsheet offers some functionality specific for working with multivariate data (e.g. recoding numeric variables into factors and vice versa) and some of the Excel functionality (such as using formulas) is not available. The labels are treated specifically: you can always see just one of the two label sets (full or brief labels, see above) and they always stay at the left and top edges of the window, even when you scroll the

---

[5] Before that, you might also see a dialog box offering 'Introductory Analysis' and if you select the *Yes* button in it, analysis setup is offered. After you execute the analysis, its notebook comes into foreground. But the data spreadsheet is still available and can be displayed e.g. using the *Data | Show project data* menu command.

2.2 Data import and editing

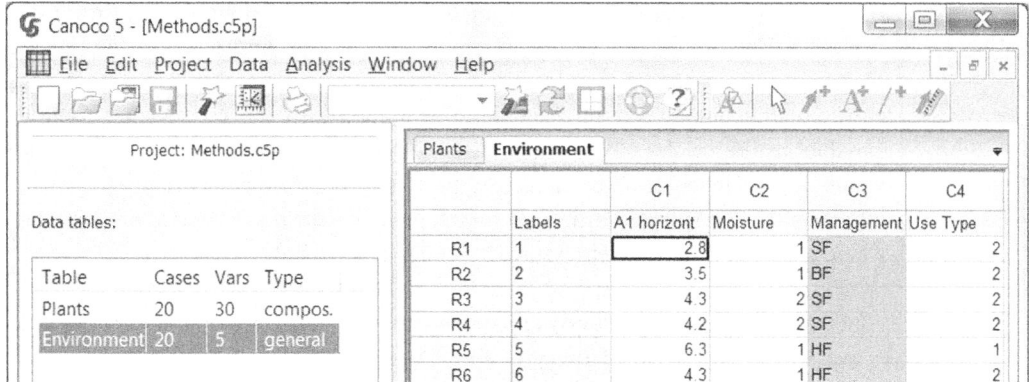

Figure 2–7 Canoco workspace with visible data table spreadsheet containing three numeric variables and one factor (*Management*).

spreadsheet contents. Each data table is represented by one page in the spreadsheet window and you can choose the currently visible table using the tabs shown at the top of the spreadsheet.

You can create a new project also by importing data from files used by older CANOCO versions (4.0 and 4.5) using the *File | Import project | from Canoco 4.x files* menu command. The corresponding import wizard is simpler, as the individual imported files match the newly created data tables one-to-one. Using this wizard, you can also import CANOCO 4.x 'project' files (with *.con* file extension), turning them into analyses in Canoco 5 project. After importing legacy data containing factor predictors, you must turn the dummy variables (used in older versions to code factor values) into real factor columns, as described below.

After you create a new project by import or even after you defined some analyses in the project, you can still add new data tables to it. This is accomplished through the commands in the *Data | Add new table(s)* submenu. These commands allow you to import additional data tables from Excel or CANOCO 4.x data files or to create empty ones or, to support advanced analyses, to define data tables in a specialised way, e.g. as summary statistics of other data tables or as averages of functional trait values per individual cases, or as a distance matrix computed from a phylogenetic tree.

You can also create a new Canoco 5 project with empty data tables and gradually fill them by entering data directly in the spreadsheet visible in Canoco workspace. The spreadsheet supports automatic data table growth, as you enter new values into its border zone (data cells with light orange background colour) at the bottom and right edges of the spreadsheet.

Variables present in Canoco 5 data tables can be either numeric ones or factors, which are (under default settings) recognisable by their light blue background colour, as illustrated in Figure 2–7.[6]

---

[6] Semi-quantitative variables, particularly those on ordinal scale, are not treated in any special way in Canoco and we recommend using them as numeric ones, with necessarily more cautious interpretation of the results.

> In CANOCO 4.x versions, each factor had to be coded using so-called **dummy variables**. There was one separate dummy variable for each different value (level) of a factor. If a case (observation) had a particular value of the factor, then the corresponding dummy variable had the value 1.0 for this case and the other dummy variables had zero values for the same case. This explicit coding of factors by dummy variables is no longer used in Canoco 5, but it is still performed internally (as in the other statistical packages using factors), and it is reflected in ordination diagrams, where each factor level is represented by a separate symbol.[7]

But the explicit decomposition of factors into dummy variables is still required, when you want to create so-called **fuzzy coding**. As an example, you might be recording vegetation and you record the type of agricultural management at each site, e.g. as a factor with three levels: *pasture*, *meadow*, and *abandoned*. To record such data in Canoco 5, a single column with a factor variable is sufficient. But imagine that you might have in your data set a site that had been used as a hay-cut meadow until the previous year, but it is used as a pasture in the current year. You can reasonably expect that both types of management influenced the present community composition. Therefore, you can define three 'dummy' (fuzzy-coding) variables named, say, *Meadow*, *Pasture* and *Abandoned* and for such a 'mixed' case give values larger than 0.0 and less than 1.0 for both the first and second variable. The important restriction here is that the values of all the variables coding such a fuzzy factor must sum to 1.0. Unless you can quantify the relative importance of the two management types acting on this site, your best guess is to use values 0.5, 0.5, and 0.0 (respectively for the *Meadow*, *Pasture*, and *Abandoned* variables).

2.3 Defining analyses

With data present in a Canoco 5 project, you can start addressing your research questions using analyses. There are multiple ways of defining an analysis and here we focus on their creation with the New Analysis Wizard. It can be started using the *New...* button below the list of analyses (in the lower left corner of the Canoco 5 workspace). This wizard is relatively simple, with one to three pages (the first two pages are not shown if your project contains just a single data table). On the first page, all data tables present in your project are initially selected, but if you do not want to use some of them in the new analysis, uncheck their boxes. The second page displays selected data tables and asks you to choose the **focal table** for your analysis. Typically, this is the table with the response data that you try to explain using the ordination model.[8] Most important is the last page of the New Analysis Wizard, illustrated in Figure 2–8.

---

[7] Canoco 5 supports replacement of a set of dummy variables (e.g. imported from a CANOCO 4.x data file) with a single factor variable. To do so, select all the dummy variables forming the factor, right-click the header of the first variable and select the *Aggregate dummy variables* menu command. Reverse change is also possible by selecting one factor variable, right-clicking its header and choosing the *Expand into dummy variables* command.

[8] In more advanced analyses relating functional traits and environmental descriptors, the choice made on this page is less trivial, as described for corresponding analysis templates in the Canoco 5 manual (section 4.3.4.5).

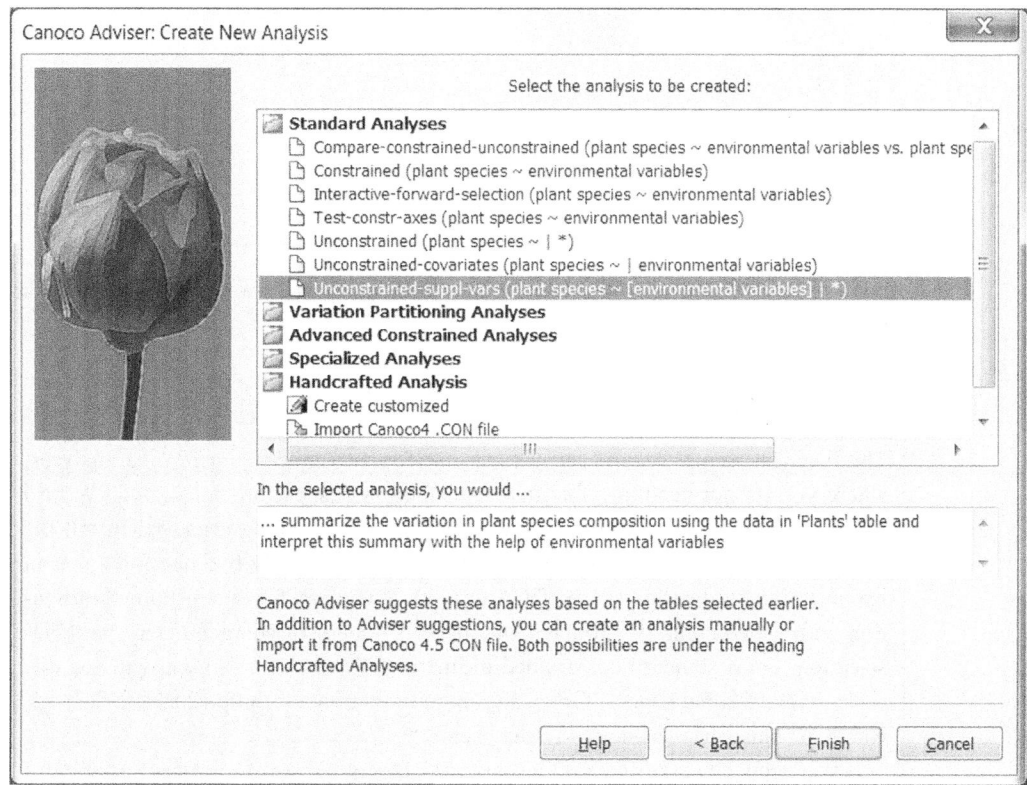

Figure 2–8 Last page of New Analysis Wizard, with suggested analysis templates.

This page offers multiple analysis templates, arranged into a set of categories (represented by the bold text in the largest dialog field). Categories like *Variation Partitioning Analyses* in Figure 2–8 are **not** empty, the list of their members can be seen when you unfold the category by double-clicking its title.

Each analysis template represents a specific type of analytical task, with its roles assigned to data tables in your project. In the lower part of the dialog, a text field describes (using project-specific terms – 'plant species' and 'environmental variables' in the above figure) what you can achieve with the particular template. Standard ordination analyses of constrained and unconstrained ordination are offered in the first *Standard Analyses* category and the Canoco Adviser initially suggest one of these templates for you (depending on the state of your project).

After you select an analysis template and click the *Finish* button, the new analysis is created in your project based on that template and it is placed into the list of analyses. But before you can execute the analysis, its settings must be adjusted or at least reviewed. Therefore, another wizard window is shown, called the Analysis Setup Wizard.[9]

---

[9] You can also display this wizard for already defined (and even executed) analysis by selecting the analysis and clicking the *Modify* button below the list of analyses. Canoco 5 gives you a choice of either modifying the existing analysis or first creating its copy and making the changes to it.

Figure 2–9 QuickWizard mode is active: the QuickWizard toggle button, fifth from the left in the toolbar, looks pushed-in.

Quick Wizard and Slow Wizard

The Analysis Setup Wizard can present itself in two modes: for routine analyses, where you do not need any fine-tuning of the options, the so-called QuickWizard mode is an appropriate choice. It is, in fact, a default setting after you install the software. With QuickWizard mode set, the wizard shows just two pages for a plain unconstrained ordination (PCA, DCA or CA). If you ask for unconstrained ordination with supplementary variables, one page is added for supplementary variables selection. For a standard constrained ordination (RDA or CCA), you will see one (or more) additional page(s) for specifying permutation test options, beside the page where explanatory variables can be chosen.

If you need, however, to select a subset of cases or subset of response variables or if you want to perform explicit weighting of particular cases or to create interaction terms for your predictors, you must switch to 'Slow Wizard' mode, before you define the new analysis. To do so, click the QuickWizard button in the Canoco 5 main toolbar (illustrated in Figure 2–9). It is a toggle button – to restore the QuickWizard mode, click it again.

Detailed information about the options available in setup wizard pages can be found in the Canoco 5 manual (pp. 111–143) and the work with this wizard is also illustrated by some of the tutorials in this book. Here we will discuss just a single page, named *Ordination Options* and representing the core setting page for unconstrained and constrained ordination. Its look (for a compositional response data) is illustrated in Figure 2–10.

At the page top, four main types of ordination methods supported by Canoco 5 (see Section 4.7) are shown.[10] Partial analyses (with covariates) or analyses with supplementary variables are not explicitly distinguished here, just by the presence of the corresponding variable type in the analysis. Similarly the hybrid constrained analyses, explicitly identified in earlier CANOCO versions, are now distinguished by a non-default choice in the *Hybrid analysis* field in the middle of the page. The two choices for

---

[10] Specialised multivariate methods newly available in Canoco 5 (e.g. principal coordinates analysis, non-metric multidimensional scaling, or co-correspondence analysis) do not show this page in their Analysis Setup Wizard sequence.

2.3 Defining analyses

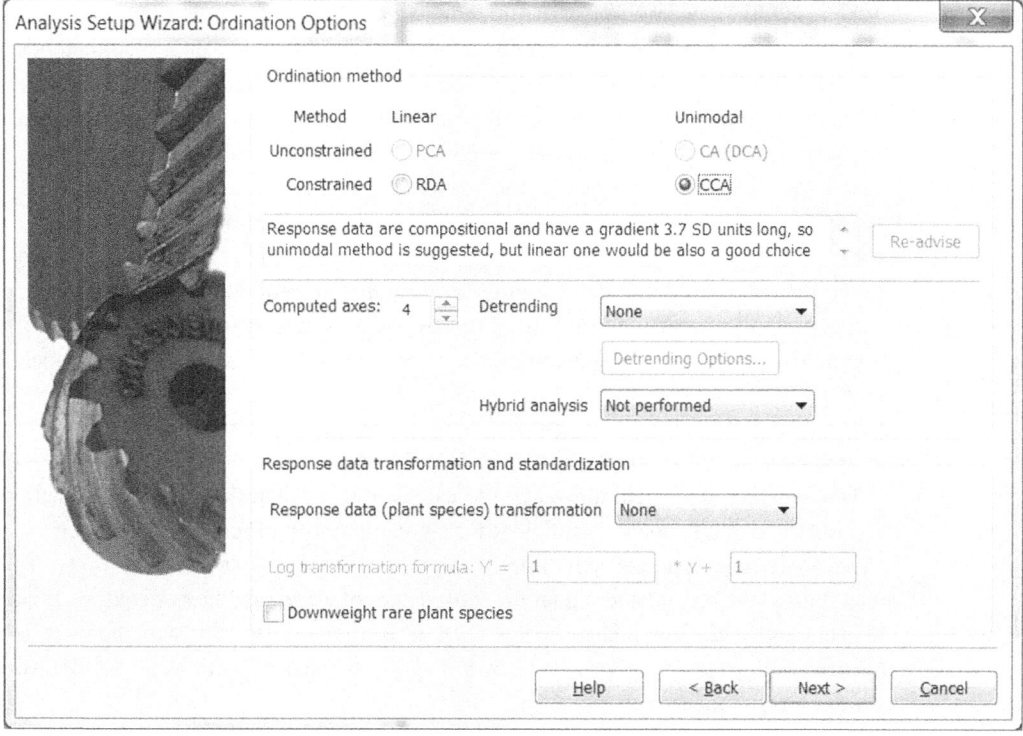

Figure 2–10 Ordination Options page of the Analysis Setup Wizard.

unconstrained ordination are disabled in the above snapshot, as the use of constrained ordination was fixed in the New Analysis Wizard by selecting the *Constrained* analysis template.

Canoco 5 allows you to choose between the linear and the unimodal ordination model,[11] but the choice shown for your analysis is not a dumb default (as it used to be in earlier CANOCO versions). Instead, the Canoco Adviser performed, in a background check, an unconstrained unimodal ordination (DCA) and measured the length of its ordination axes (see the report shown right below the analysis choice). In a DCA with detrending by segments, this length is measured in so-called **turnover** (or **SD**) **units** and for community composition data represents a measure of its beta diversity. The Canoco Adviser applies here a traditional heuristic rule and for this data set (with the longest axis of 3.7 turnover units) recommends a unimodal method, i.e. canonical correspondence analysis (CCA). For data with this length smaller than three turnover units, the linear method is offered as a preferable choice, while for a length of more than four turnover units, the linear method is no longer recommended.[12]

---

[11] Except when the response data table is marked as general, because the unimodal (weighted averaging) ordination methods are appropriate only for compositional data.

[12] Less technically, using an example of biotic community as your response data table, you can decide based on your a priori (or field-based) knowledge of the extent of community change along the major gradients

> When deciding whether to use the linear or unimodal type of ordination, you must take into account another important difference among them. The unimodal methods always implicitly standardise the data. CCA, CA or DCA methods summarise the variation in relative frequencies of response variables (e.g. species). An important implication of this fact is that these methods cannot work with 'empty' cases, i.e. records in which no variable has non-zero value (e.g. no species is present). Also, the unimodal ordination cannot be used when the response variables do not share measurement units, but this mistake is now largely prevented by the Canoco Adviser.
>
> On the other hand you can usefully apply unimodal ordination methods to data sets with low compositional variation (where DCA would produce axes with small length in turnover units), because they also have a 'linear face' (see Canoco 5 manual, p. 89).

The *Computed axes* field allows you to calculate more than the first four ordination axes, but the default choice is sufficient for a great majority of real-world data sets.

The *Detrending* options are offered only for unimodal ordination methods. The detrending of the second and higher ordination axes of a correspondence analysis (leading to detrended correspondence analysis, DCA) is often used to cope with the so-called arch effect, illustrated in Figure 2–11 by a diagram with case positions on the first two axes of correspondence analysis (CA).

The position of cases on the second (vertical) axis is here strongly – but not linearly – dependent on their position on the first (horizontal) axis. This effect can be interpreted as a limitation of the method, because the consecutive axes are made mutually independent but only a linear independence is required or, alternatively, as a consequence of the projection of the non-linear relations of response variables to the underlying gradients into a linear Euclidean drawing space (see Legendre & Legendre 2012, pp. 482–485 for more detailed discussion). **Detrending by segments** (Hill & Gauch 1980), while lacking a convincing theoretical basis and considered as inappropriate by some authors (e.g. Wartenberg et al. 1987 or Knox 1989), is the most frequently used way to making the recovered compositional gradient straight (linear). When you select a unimodal ordination model for your analysis in the *Ordination Options* page of the Analysis Setup Wizard, Canoco offers the following choices for the *Detrending* field, as illustrated in Figure 2–12.

Use of detrending by segments is not recommended for unimodal ordination methods where either covariates or explanatory variables are present (i.e. for partial and/or constrained methods). In such cases, if the detrending is needed, **detrending by polynomials** is the recommended choice. You are advised to check the Canoco 5 manual

---

reflected in the data. If species mostly change only their proportions on these gradients, then the linear approximation is appropriate and so are the linear techniques. If you expect qualitative changes in the community composition (many species appearing and disappearing along the gradient), then unimodal (weighted-averaging) techniques are a better choice.

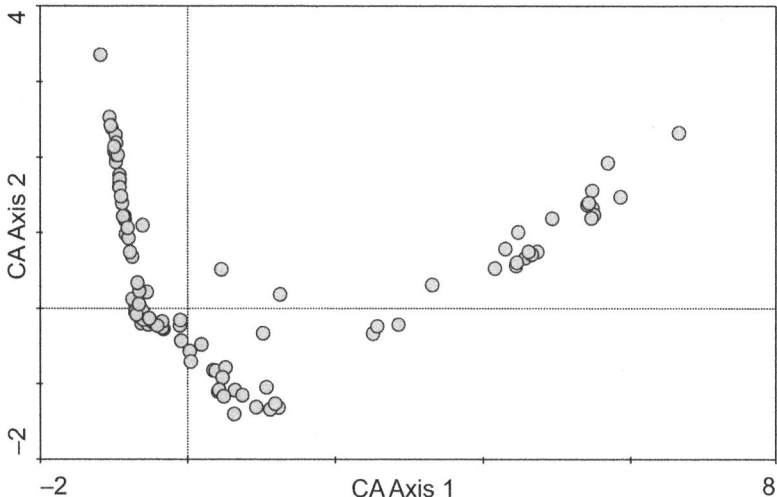

Figure 2–11 Scatter of cases from correspondence analysis, featuring so-called arch effect. You can see the 'arch' by rotating your book by 180 degrees.

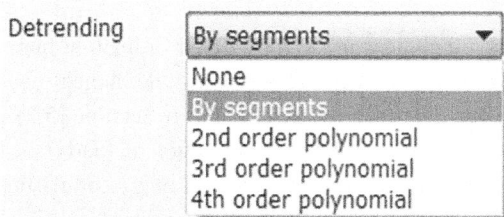

Figure 2–12 Detrending method selection in the Analysis Setup Wizard.

(section 4.4.2.7) for more details on how to decide among the use of polynomials of second, third, or fourth degree.

But the **detrending** procedure is often **not needed** for a constrained unimodal ordination, when analysing the same data table that required detrending for unconstrained ordination. If an arch effect occurs in CCA, this is often a sign of some unnecessary explanatory variables being present. For example, there may be two or more explanatory variables in your set, strongly correlated (positively or negatively and not necessarily in a linear way) with each other. If you retain only one variable from such a group and remove also the variables that are not strongly correlated with the other ones, but do not have an important relation with response data, the arch effect often disappears. The selection of a subset of explanatory variables can be performed using the forward selection of explanatory variables in Canoco 5 – see Section 5.6.

In the *Response data transformation and standardization* section, present in the lower part of the Ordination Options page, you can choose one of the standardisations discussed in Section 1.3. But even here, the Canoco Adviser tries to help and automatically selects

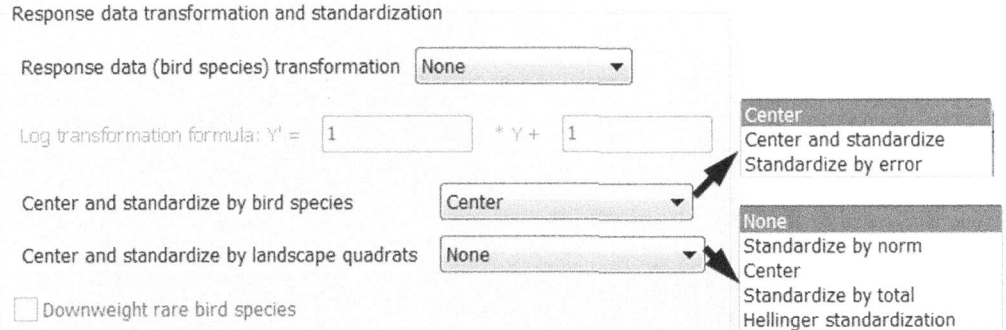

Figure 2–13 Centring and standardisation options in the Ordination Options page.

for you either no transformation or log transformation (with its two parameters properly selected, see Section 1.3), depending on the range and distribution of data values in the response data table. You will note that there are no centring or standardisation options (discussed in Section 1.3) offered in the snapshot in Figure 2–10. This is because the unimodal methods imply a double standardisation by both cases and response variables, as explained in Section 4.5.

Centring and standardisation choices made for a linear ordination may substantially change the outcome of the analysis. In fact, by varying the choice, you can address different questions about your data (see the discussion in Section 15.3). The choices available for centring and standardisation in linear ordination methods (see Figure 2–13) are discussed in detail in Section 1.3. Here we provide only a brief summary of the consequences of individual centring and standardisation options.

Centring by cases (referred to as 'landscape quadrats' in Figure 2–13) results in a zero average for each table row. Similarly, centring by response variable (named 'bird species' in the above figure) results in a zero average for each column. Centring by response variables is obligatory for a constrained linear ordination (RDA, to which Figure 2–13 refers) or for any **partial** linear ordination (i.e. where covariates are used).

Standardisation by response variables (or by cases) results in the norm of each column (or each row) being equal to one. The **norm** is the square root of the sum of squares of the column (row) values. If you apply both centring and standardisation, the centring is done first.[13] Therefore, after centring **and** standardising by response variables, the columns represent variables with zero average and unit variance.

The *Downweight rare* <response-variables> option is available only for unimodal methods and it targets the sensitivity of chi-square distances (implicitly used by these methods) to (rare) species with a low total abundance (see Section 6.2.2). We recommend to choose it particularly when working with a large, heterogeneous set of cases, with many response variables (e.g. species) having only a few non-zero values (occurrences).

---

[13] And if you centre/standardise both by rows and columns, the centring/standardisation is done first by rows, then by columns.

2.3 Defining analyses

An alternative (but not fully equivalent) approach is the omission of such variables. See Canoco 5 manual, p. 124, for additional details.

Whatever pages the setup wizard sequence contains, the last page is always the *Finish* page. On it, there is only one option you can change, namely a check-box labelled *Execute this analysis after Finish*.[14] By default, it is always checked and this implies analysis execution immediately after you click the *Finish* button.

After you executed the analysis (either by leaving the option, described in the preceding paragraph, checked, or by clicking the *Perform* button below the list of analyses), analysis results are shown in an analysis notebook.[15] The analysis notebook almost always contains a Summary page and the pages with individual graphs created for the analysis, and – depending on its presentation mode (see the *Brief and non-brief view of analysis notebooks* box below) – also additional pages, namely those with computed ordination scores and analysis log. Figure 2–14 illustrates the content of *Summary* page for a constrained ordination set up in our example above.

The upper part of the Summary page provides an overview of the size and type of the data used in the analysis (including the real number of model degrees of freedom for chosen explanatory variables), followed by a summary of the explained variation and additional statistics for each of the computed ordination axes. Interpretation of these values is illustrated in the first two case studies (Chapters 12 and 13) and full details can be found in section 5.2 of the Canoco 5 manual. Finally, results of permutation tests performed on the constrained ordination model are shown.

For other kinds of analyses, some of this information may be missing and other information can be present, namely the results of stepwise selection of explanatory variables or the evaluation of their simple and conditional effects.

Brief and non-brief view of analysis notebooks

In the same way you can switch on and off the comprehensive content of Analysis Setup Wizard using the QuickWizard button, you can also decide whether the analysis notebooks are shown in a brief (default) or non-brief mode. To do so, select the *Edit | Settings | Canoco5 options* menu command and in the *General* page uncheck or check the *Show brief version of notebooks with analysis results* box. The effect of such change is not immediately visible in already opened notebooks: you must close the notebook(s) using the *Hide* button in the lower left corner of Canoco 5 workspace and then reopen it again using the same button, now with changed text *Show*. Note also that when you close an analysis notebook, Canoco 5 usually switches to another analysis (when it exists), so you must possibly activate the original analysis before re-displaying its notebook.

---

[14] This check-box control is not shown when you invoke the Analysis Setup Wizard on an already executed analysis and do not change any of its settings.
[15] For a newly created analysis, the notebook is shown automatically. Alternatively, you can display it by selecting the analysis name in the list of analyses and clicking the *Show* button. When the notebook is visible, text of the button changes to *Hide*, with corresponding change in functionality.

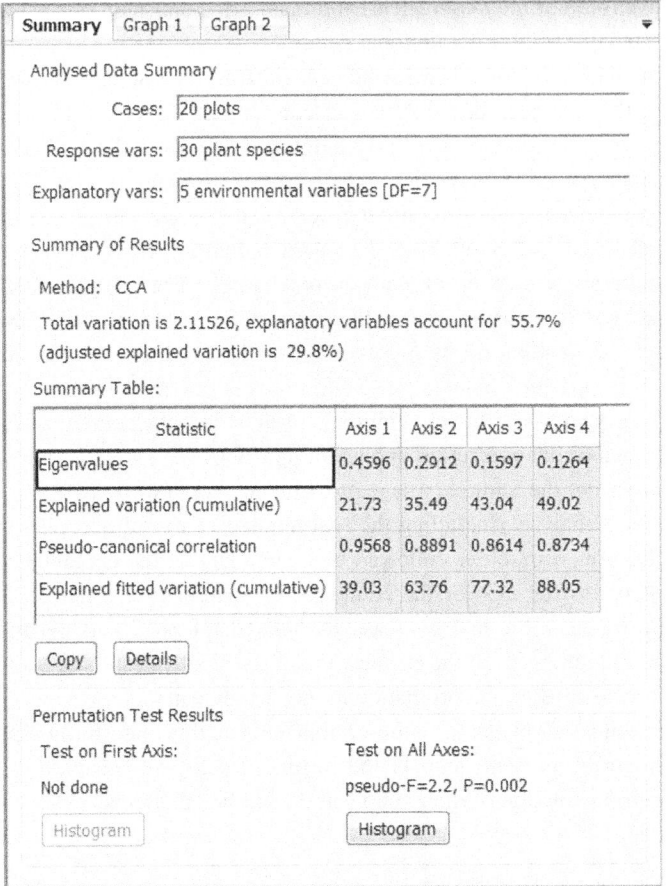

Figure 2–14 Summary page of an analysis notebook for a constrained ordination.

Unlike earlier versions, Canoco 5 now offers even very complex tasks as a single analysis. For example, variation partitioning for two groups of predictors did require in CANOCO 4.x at least three separately executed analyses (called 'projects' by then) plus additional calculations on their results. In Canoco 5, this is now a single analysis (with optional testing and stepwise selection for each group of predictors) that immediately provides a variation partitioning table, including the results of permutation tests. To perform the partitioning, however, Canoco 5 must still fit multiple (partial) constrained ordination models, corresponding to individual executions of the ordination algorithm, and these components of complex analysis are called **analysis steps**. With multiple steps in an analysis, the Summary page of the analysis notebook always refers just to one of the steps, but you can use the scroll control in the page's left-hand corner to review the results of other steps, when needed. Standard ordination analyses have just a single step and its existence is hidden from the user.

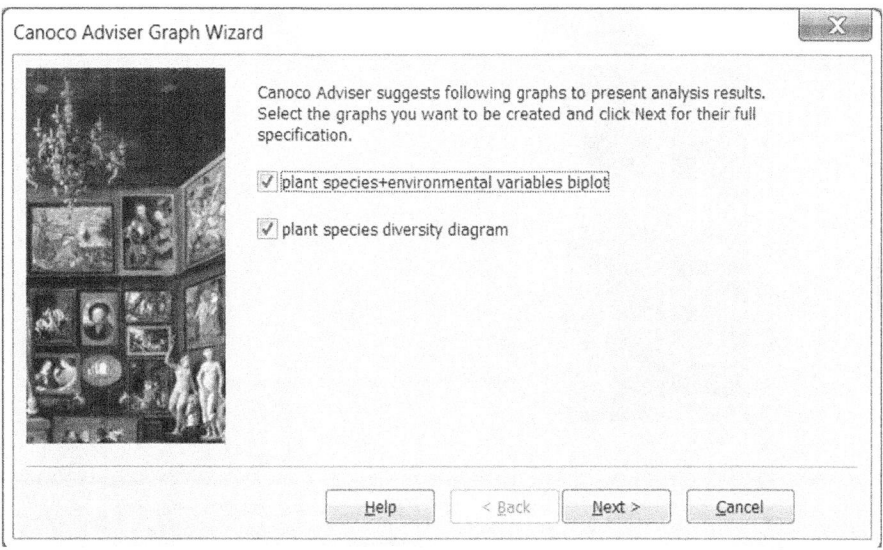

Figure 2–15 First page of the Graph Wizard for a constrained ordination.

2.4 Visualising results

In this section, we introduce the basic steps of graph creation in Canoco 5. The diagram types available in Canoco 5 and their interpretation are treated in more detail in Chapter 11.

The results provided by project analyses can be visualised in two ways. The full toolbox of visualisation methods is available from the *Graph* menu shown for an active analysis with existing results. But for the analyses created using the New Analysis Wizard, there is an easy and powerful way to create an initial set of diagrams to inspect analysis results. After a newly created analysis is executed, Canoco 5 displays the Graph Wizard, where a set of graphs appropriate for the analysis template is offered.

In our example (see Figure 2–15), Graph Wizard suggests creating a biplot with response variables (plant species) and explanatory variables (environmental variables), as well as an attribute plot illustrating the change of plant species diversity across the ordination space defined by the environmental descriptors. In this first page, you should select the graph types you are interested in, and for the types you select, Graph Wizard displays additional pages, where the graph content can be further fine-tuned, as illustrated in Figures 2–16 and 2–17.

Figure 2–16 shows a Graph Wizard page displayed for the above-selected biplot of plant species and environmental variables.

Some of the options are disabled, as they are not appropriate for current analysis or graph. The Graph Wizard considers both plotted components (plant species scores and the scores of environmental variables) as required contents, so they cannot be removed from the plot. For response variables (here plant species), however, you can select only

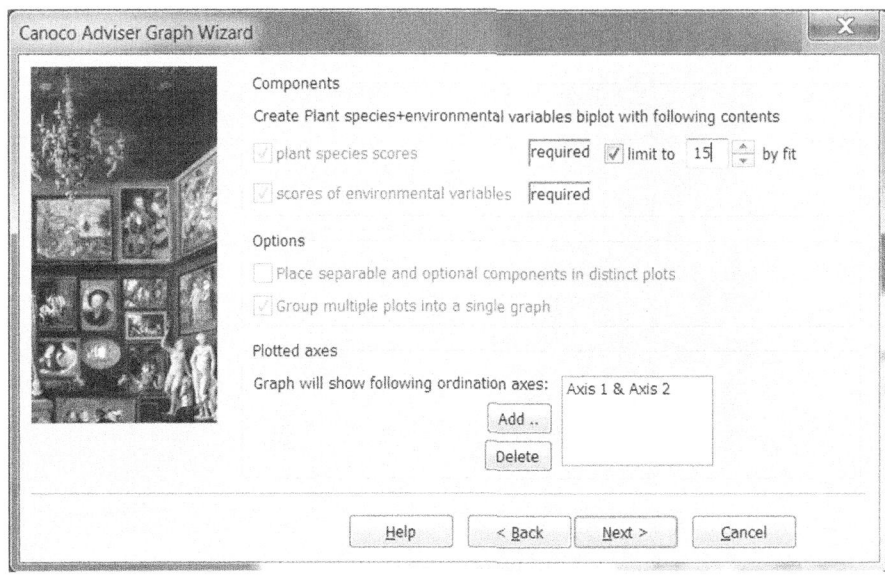

Figure 2–16 Graph Wizard page for standard ordination diagrams.

a subset of their total number, those best predicted (fitted) by the constrained ordination axes. Another option available in this page is the possibility to create this ordination diagram not only with the two most important ordination axes (first and second), but also with other ones, for example the first and third axis.

Figure 2–17 shows a wizard page that allows easy creation of another useful graph, displaying community diversity in its relation to environmental variables.

As in the earlier page, case scores and environmental variables are obligatory parts of the diagram and cannot be avoided, although the positions of individual cases will not be plotted in the diagram, if you choose the 'contour plot' option in the page line introduced by the 'and presented using a' label. In such case, a loess smoother is fitted to the original data and the resulting surface presented with isolines. Alternatively, 'symbol plot' displays case symbols with varying diameter, proportional to the diversity value of a particular case, and the 'colour-coded plot' displays symbols for all cases with the same size, but with the fill colour varying along a colour scale (gradient) reflecting the case diversity. Another important choice concerns the type of diversity measure visualised in the diagram. Standard species richness (number of present species) is the default choice, but other diversity measures, such as Shannon–Wiener H or N_2 diversity index (inverse of the Simpson index), are also available.

The Graph Wizard also offers other kinds of diagrams, not present in our example analysis – two types of diagrams useful when the explanatory variables represent a classification (i.e. a single factor), a specialised diagram for principal response curves (PRC) method (see Section 16.6 for a practical example), as well as a specialised diagram for co-correspondence analysis (see Section 13.5 for an example). Unlike the other diagrams, the PRC and co-correspondence analysis diagrams cannot be created

2.4 Visualising results

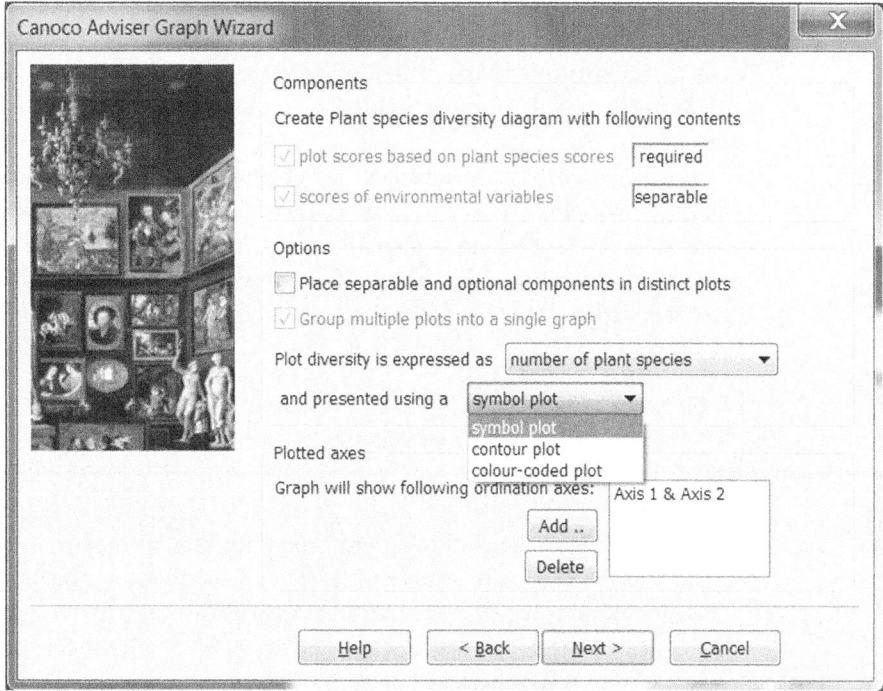

Figure 2–17 Graph Wizard page for attribute plot visualising alpha diversity changes.

using the commands in the *Graph* menu, only with the Graph Wizard. The wizard can be invoked at any time using either its button in the graphing toolbar or the *Graph | Advise on graphs* menu command.

When working with the commands present in the *Graph* menu's submenus (*Scatterplots*, *Biplots*, *Triplots*, and *Attribute plots*), you can create some graphs not available through the Graph Wizard, but this flexibility comes at the price of larger complexity. For example, in the case of multi-step analyses (see the end of Section 2.3 above), you must select the step that calculates the ordination scores for graphing with the *Analysis | Select graphing step* menu command.

When creating diagrams using the commands in the submenus of the *Graph* menu (and to a limited extent also when working with the Graph Wizard), the look and content of the created diagrams are affected by the options chosen at three different levels in the Canoco 5 user interface:

1. At the level of Canoco 5 application, the choices made through the four commands in the *Edit | Settings* submenu (most importantly the *Graphing options* and *Visual attributes*) affect created graphs in all projects and analyses opened in Canoco 5. Dialog boxes shown by the commands in the *Settings* submenu are described in Canoco 5 manual, section 7.3.3.
2. At the level of a particular Canoco 5 project, the changes made through the commands in the *Project* menu affect all graphs created (after the settings change) in the analyses

of the currently open project. To store them across different Canoco 5 sessions, you must save the project before closing the application. The most important changes at this level affect the classification of data items (cases, variables), definitions of groups and of spatial or temporal series, as well as the suppression of a subset of data items.
3. For a particular analysis, you can change the graphing options using the *Analysis | Plot creation options* menu command. This includes settings like the identity of plotted axes or the selection of plotted data items based on the information provided in the analysis results (such as the fit of response variables into the ordination space illustrated earlier). You can find additional information about this important dialog box in Canoco 5 manual, section 7.6.9.

To apply new settings to an already created graph, you must recreate it using either the green double-arrow in the graphing toolbar or the *Recreate graph* command in the *Graph* menu.

> If you prefer to have the axes of an ordination diagram labelled with the name of ordination method and axis number (or even with additional information), you can achieve this easily by checking the *Show transformation and axis labels dialog* box in the *General* page of the dialog shown by the *Edit | Settings | Graphing options* menu command. When checked, Canoco 5 displays – before a diagram is created – a dialog box where default axis labels are suggested and the label text can be adjusted. For general XY(Z) diagrams, axis labels are plotted by default and the dialog offers additional on-fly transformations of plotted variables.

You can save the current state of any Canoco 5 graph in a separate file with a .c5g extension. To do so, the selected graph must be first copied from its analysis notebook using the *Graph | Copy to standalone graph* menu command and then saved. Both the standalone graphs and those included in the analysis notebooks can be printed directly from the program, copied to other applications through the Windows Clipboard, or exported in a variety of graph formats: BMP, PNG, JPEG, TIFF, Adobe Illustrator, PDF, and Enhanced Metafile Format.

2.5   Beware, CANOCO 4.x users!

Long-term users of the older versions of Canoco software have the advantage of greater expertise with the choice of ordination methods and interpretation of their results, but there are also some dangerous waters for them, when they set sail with Canoco 5. Unlike the novice users, they have working habits to unlearn, because otherwise these habits will stand in their way towards an enjoyable experience with the new version. Here we summarise the nine main differences of Canoco 5 from the previous version CANOCO 4.5 and suggest that users of older versions learn them by heart.

2.5 Beware, CANOCO 4.x users!

1. Make sure that for each data table created in a Canoco 5 project you specify correctly whether it is of **compositional** or **general** type (see the end of Section 1.2, Section 2.2, and Canoco 5 manual, p. 48). When specified incorrectly, you will not be able to use the full range of the Canoco Adviser functionality or to find required analysis templates in the New Analysis Wizard. If you get it wrong during project definition, you can always correct it later (for a currently selected data table) using the *Data | Change table kind to...* menu command. Data table type affects which analysis templates the Canoco Adviser offers and if your intended analysis is not among the offered templates, then perhaps the data table type must be changed.
2. When your data layout fits the widespread pattern of having a single compositional data table (typically describing community composition for a set of cases) and one or multiple data tables with explanatory variables, make sure the compositional data table is imported first into the project. In this way, the data type (compositional vs. general, see above) is preset correctly in the import wizard and your use of the New Analysis Wizard is more straightforward, as you can rely on its default settings.
3. In CANOCO 4.x, the response variables were always called 'species', your cases were 'samples' and your predictors 'environmental variables'. This was perhaps a good choice for community ecologists, but not for everyone. In Canoco 5, the choice of proper item terms is in your hands and if you play along, the expert advice provided by the Canoco Adviser during your work with Canoco 5 will be more understandable. Do not rely on default item terms offered when you define (import) the data tables, but choose the terms appropriate for your research field. You do not have cases, but 'traps', 'quadrats', 'stations'; perhaps you do not have just 'species' but 'plant species', or 'gene loci' or 'land-use categories'; and your 'environmental variables' might be in fact better called 'experimental treatments' or 'socio-economic impacts'. These terms are used not only in the menu commands and in the Analysis Setup and Graphing Wizards, but also in the Describe Graph Contents tool that aids you when interpreting ordination diagrams.
4. Data tables referring to the same set of cases must have identical number of rows, with matching identity: Canoco 5 now works with missing (NA) values, so if you do not have environmental descriptors for a particular record of community composition, the corresponding data table row must contain NA values. Canoco 5 spreadsheet editor takes care of maintaining the consistency across data tables (e.g. the case labels are shared), so just go along with it.
5. Dummy (0/1) variables are no longer the appropriate representation of factors. If you import old data files, get rid of dummy variables (convert them to factors) as soon as possible. Neither principal response curves (PRC), nor the linear discriminant analysis (LDA) can be set up properly with dummy variables; these methods are only offered if there are appropriate factors in the data tables to be analysed.
6. You no longer need to transform numeric predictors (explanatory and supplementary variables or covariates) **before** you import them into Canoco 5 project. You can set the transformation for individual predictors using the *Data | Default transformation and standardization* menu command. In fact, when you choose this command, the Canoco

Adviser may suggest using log transformation for some variables.[16] Transformation settings you specify in this dialog are not reflected in the data tables themselves (they show the non-transformed data values), but they are applied to the variables when you execute your analyses.

7. Canoco 5 data tables have both the traditional 'short' labels for cases and variables as well as full labels without practical limitation on their length. If possible, try to define (import) primarily the full labels, the short ones will be automatically generated from them and you can correct the result when needed (see section 2.9.3 of the Canoco 5 manual). Brief labels are also used in ordination diagrams by default, but this can be changed for any kind of data items using the *Project | Visibility and labelling* menu command.

8. The choice of scaling for ordination scores is no longer a part of the analysis setup. For an executed analysis, you can change its scaling type 'on the fly', using the *Edit scaling options . . .* tool in the main toolbar (sixth from the left, next to the QuickWizard mode button). See Sections 11.1 and 11.2 for additional information on how the chosen scaling affects the interpretation of ordination diagrams. For specialised analyses requiring particular score scaling (such as principal response curves or discriminant analysis), the scaling is fixed within the analysis by the Canoco Adviser.

9. You might eventually find that there is a possibility of creating new analyses using the *Analysis | Add new analysis | Customized* menu command, which feels very similar to the way the analyses were created in CANOCO 4.5. But we urge you to resist this 'temptation' and create your new analyses using the *New . . .* button below the list of analyses. This is because the customised analyses are second-class citizens in Canoco 5 projects and should be used only for very advanced tasks, not supported by the analysis templates of the Canoco Adviser, or when you need to validate Canoco 5 analysis template results with those done earlier in CANOCO 4.5 (in this case, the customised analyses result from importing CANOCO 4.5 projects).

---

[16] You should always check their suitability for your variables, of course.

3 Experimental design

Some researchers still believe that multivariate methods are restricted to the exploration of data and to the generation of new hypotheses, but this has not been the case for several decades. In particular, constrained ordination is a powerful tool for analysing data from manipulative experiments. In this chapter, we review the basic types of experimental design, with an emphasis on manipulative field experiments.[1]

Generally, we expect that the aim of an experiment is to compare the response of studied objects (e.g. an ecological community) to several treatments (treatment levels). One of the treatment levels is usually a control treatment, although in real ecological studies, it might be often difficult to decide what is the control (for example, when we compare several types of grassland management, which of the management types is the control one?). Detailed treatment of the topics handled in this chapter can be found for example in Underwood (1997).

If the response is univariate (e.g. number of species, total biomass), then the most common analytical tools are ANOVA, general linear models (which include both ANOVA, linear regression and their combinations), or generalized linear models. Generalized linear models are an extension of general linear models for the cases where the distribution of the response variable cannot be approximated by the normal distribution. These types of statistical models are further discussed in Chapter 8.

3.1 Completely randomised design

The simplest design is the completely randomised one (Figure 3–1). We first select the plots, and then randomly assign treatment levels to individual plots. This design is correct, but not always the best, as it does not control for environmental heterogeneity. This heterogeneity is always present as an unexplained variability. If the among-plot heterogeneity is large, use of this design might decrease the power of the test[2] of the null hypothesis that the treatment does not influence the response.

---

[1] In exploratory field studies, the question of sampling design is less crucial. Nevertheless, the ideas of randomisation and of taking any internal sampling structure (e.g. hierarchical sampling or repeated measures) into account when performing tests on our data are still relevant.

[2] Power of a test is the probability that the test will reject the null hypothesis if it is not correct.

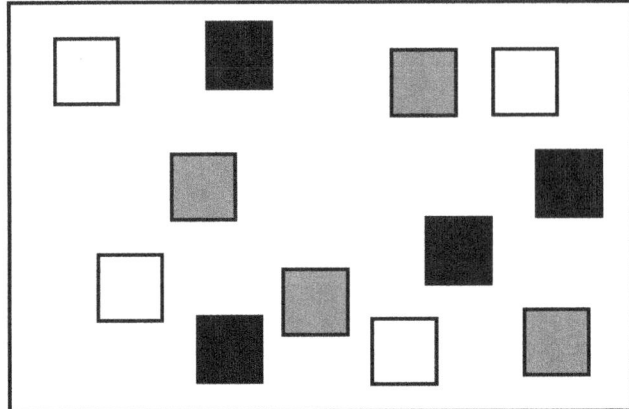

Figure 3–1 Completely randomised design, with three treatment levels (white, grey, black) and four replicates (or replications) per treatment.

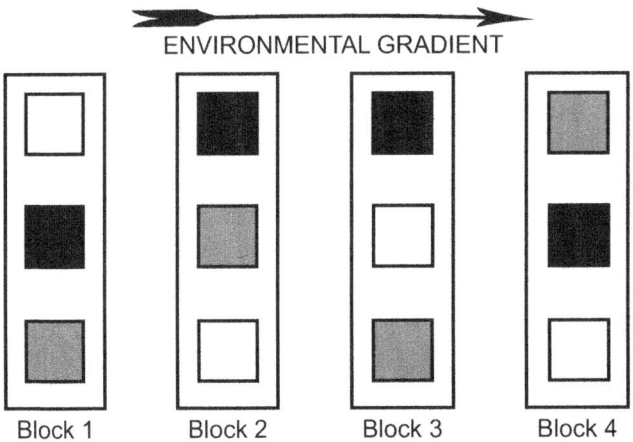

Figure 3–2 The randomised complete block design for three treatment levels.

3.2 Randomised complete blocks

There are several ways to control for environmental heterogeneity. Probably the most popular one in ecology is the randomised complete blocks design. Here, we first try to select the blocks so that they are internally as homogeneous as possible (e.g. rectangles with the longer side perpendicular to the environmental gradient, Figure 3–2). Each block contains just one plot for each treatment level, and their spatial position within a block is randomised. The number of blocks is equal to the number of replicates of each treatment level.

The usefulness of randomised complete blocks design is by no means limited to a situation with a single gradient indicated in Figure 3–2. It is useful e.g. for repeating experiments at multiple disjunctive and distant locations and generally in cases where we

3.3 Latin square design

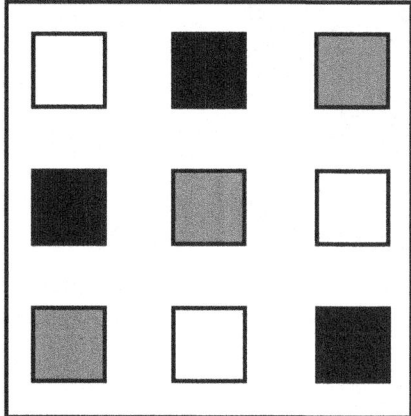

Figure 3-3 Latin square design for three treatment levels.

expect the heterogeneity of background conditions will grow with the distance between experimental units.

For a field experiment, we can place the required number of cases in randomly chosen locations (preventing their spatial overlap) or in a rectangular grid arrangement and screen the variation of community composition before the treatments are applied. Assignment of individual cases to a particular block can be then based on their compositional similarity rather than spatial proximity (with random assignment of cases to treatments, within each block). We can, for example, define the blocks by partitioning the ordination space of unconstrained ordination of the pre-treatment data.

If there are differences among the blocks,[3] the randomised complete blocks design provides a more powerful test than the completely randomised design. On the other hand, when applied in a situation where there are no differences among blocks, the power of the test will be lower in comparison to the completely randomised design, because the number of degrees of freedom is reduced. This is particularly true for designs with a low number of replicates and/or a low number of levels of the experimental treatment. There is no consensus among statisticians about when the block structure can be ignored if it appears that it does not explain anything.

3.3 Latin square design

Latin square design (see Figure 3-3) assumes that there are gradients, both in the direction of the rows and the columns of a square. The square is constructed in such a way that each column and each row contains just one instance of each treatment level. Consequently, the number of replicates is equal to the number of treatments. This might be an unwanted restriction, but more than one Latin square can be used in the same

---

[3] And often there are: the spatial proximity alone usually implies that the plots within a block are more similar to each other than to the plots from different blocks.

42   **Experimental design**

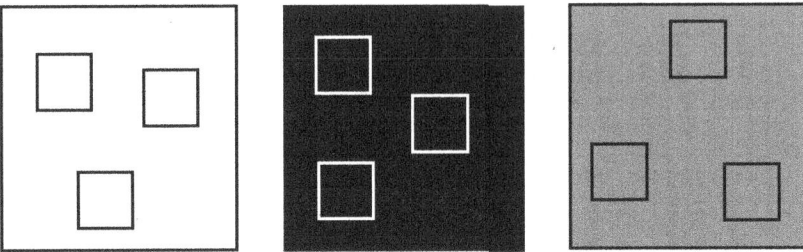

Figure 3–4 Pseudoreplicates: the plots (large squares) have been assigned different treatments, but the treatments are not replicated. Instead, several subsamples are taken within each plot.

experiment. Latin squares are more popular in agricultural research than in ecology. As with randomised complete blocks, the environmental variability is powerfully controlled. However, when there is no such variability (i.e. variability that can be explained by the rows and the columns), the test is weaker than a test for a completely randomised design, because of the loss of degrees of freedom.

3.4   Pseudoreplicates

Pseudoreplication (illustrated in Figure 3–4) is among the most frequent errors in ecological research (Hurlbert 1984). A possible test, performed on the data collected using such a design (nine observations in Figure 3–4), evaluates differences among plot means, not the differences among treatments.

Distances of subsamples within a plot are smaller than the distances between subsamples from different plots. In most cases, this means that it can be reasonably expected that the means of subsamples coming from different plots are different: just the proximity of subsamples suggests that there are some differences between plots, regardless of the treatment. Consequently, a significant result of the statistical test does not prove (in a statistical sense) that the treatment has any effect on the measured response.

Avoiding pseudoreplication is usually easy, when we start a new, relatively small-scale field experiment with, say, quadrats of the size 2m × 2m as basic experimental units. A paper reporting such an experiment carried out using the pseudoreplication design should be rejected. But in many instances, the smallest possible size of experimental unit is determined by the scale of ecological processes – for example, regulation of water regime in a peat bog must be done in the entire peat bog, and we will probably not have any chance to control water regime in multiple independent peat bogs.

As another example, there are historical non-replicated experiments, sometimes running for a long time. The question is then, whether it is better to use, for example, a short-term management experiment (where the long-term consequences of management need not be sufficiently detectable), but which is executed in a correct statistical design, or to rely on a pseudoreplicated long-term experiment (or experiment-like arrangement in an observation study), where we can study directly the long-term consequences of the management. As the resources for carrying out a research are always limited, we often face a dilemma whether to sacrifice relevant spatial and temporal scales to achieve a correct statistical design.

This situation was nicely illustrated by Oksanen (2001): 'Hurlbert divides experimental ecologists into "those who do not see any need for dispersion (of replicated treatments and controls), and those who do recognise its importance and take whatever measures are necessary to achieve a good dose of it." Experimental ecologists could also be divided into those who do not see any problems with sacrificing spatial and temporal scales in order to obtain replication, and those who understand that appropriate scale must always have priority over replication.' As can be expected, Oksanen's paper raised a discussion, which was not clearly resolved (and probably cannot be resolved). Nevertheless, it seems that it is still much easier to publish a correctly replicated paper based on experiments using inappropriate scales, than to publish an interesting paper using appropriate scales, but without true replicates.

In many non-manipulative studies, we are interested in the relationship between environmental characteristics (in the widest sense, including e.g. area management, etc.) and community composition, with the community composition considered as response and the environmental characteristics as predictors. Strictly speaking, the 'pseudoreplication' term does not apply here (it was coined for manipulative studies), but we should be aware that the spatial autocorrelation might have similar adverse effects on the interpretability of results. We should therefore accommodate our sampling design in a way that minimises possible effects of spatial autocorrelation.

Spatial autocorrelation is typical for the majority of field data – the similarity of sampling units (both in terms of their environment and the community composition) increases with their proximity. The sampling units that are very close to each other are thus not truly independent. Whereas the random placement of sampling units is generally recommended, it does not avoid the problems with spatial autocorrelation. Sampling in a regular grid tends to maximise the distance between neighbouring units and, moreover, some permutation tests, reflecting the spatial sampling design including spatial correlation, require this type of sampling (Section 5.4). Other popular arrangement of sampling units is in transects, where the units are located at equidistant positions along a line, and also for them there are permutation tests available that reflect the spatial autocorrelation (Section 5.4). In some cases, the transects are positioned to follow an (expected) spatial gradient in the field.[4] These are sometimes called gradsects (Wessels et al. 1998), and the position on such a transect can be considered as an explanatory variable. At the other extreme of spatial arrangement for sampling units, their placement in spatial clusters, which might be logistically easier in many cases, increases the effect of spatial autocorrelation.

Recently, several methods separating the spatial dependence from the dependence on measured environmental factors (provided that the location of individual sampling units is known) were suggested (see Section 10.2). It is therefore of vital importance that

---

[4] In this case, the spatial autocorrelation is not a nuisance, but a part of the phenomenon we study.

whatever sampling plan we use, we record the exact location of each sampling point. Even in a completely random location plan, the effect of unit proximity can be accounted for in the analyses.

3.5 Combining more than one factor

Often, we want to test several factors (treatments) in a single experiment. For example, in ecological research, we might want to test the effects of fertilisation and mowing.

3.5.1 Factorial designs

The most common way to combine two factors is through a factorial design. This means that each level of one factor is combined with each level of the second factor. If we consider fertilisation (with two levels, fertilised and non-fertilised) and mowing (mown and non-mown), we get four possible combinations. Those four combinations have to be distributed in space either randomly, or they can be arranged in randomised complete blocks or in a Latin square design.

3.5.2 Hierarchical designs

In hierarchical designs, each plot contains several subplots. The difference from the pseudoreplication design (Figure 3–4) is that treatment levels are now replicated in different plots. So in the situation of Figure 3–4 with three treatment levels, each of the treatment levels must be replicated. If we accept that three replicates are sufficient, then we will need 9 *main plots*, with several subplots in each.

For example, we can study the effect of fertilisation on soil organisms. For practical reasons (e.g. edge effects), the plots should have, say, a minimum size of 5 m × 5 m. This limits the number of replicates, given the space available for the experiment. Nevertheless, the soil organisms are sampled using soil cores of diameter 3 cm. Common sense suggests that more than one core can be taken from each of the plots. This is correct, but the individual cores are not independent observations. Here we have one more level of variability – the plots (in ANOVA terminology, the plot is a random factor).

The cores in our example are said to be nested within the plots (sometimes hierarchical design is called **nested design**). In the example above, the effect of fertilisation is tested against the variability among plots with the same fertilisation treatment level, **not** against the variability among soil cores. By taking more soil cores from a plot we do not increase the error degrees of freedom, but we decrease the variability among the plots within the treatment level, and increase the power of the test in this way.

Sometimes, this design is called a **split-plot design**, particularly when the subplots within the plot are treated with a second experimental factor as shown in Figure 3–5 for the factors *bedrock* (limestone versus granite) and *fertilisation* (with levels control, nitrogen and phosphorus).

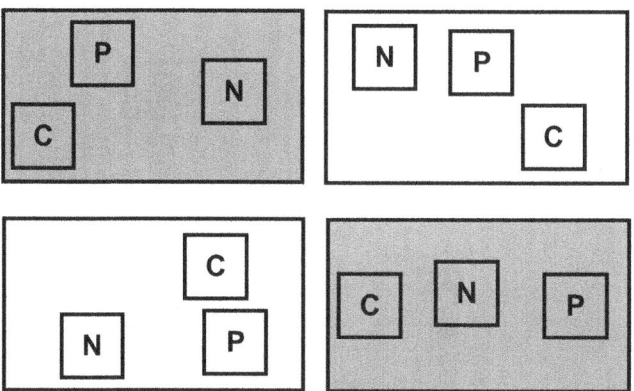

Figure 3–5 The split-plot design. In this example, the effect of fertilisation was studied on four plots, two of them on limestone (shaded) and two on granite (empty). The following treatments were established in each plot: control (C), fertilised by nitrogen (N), and fertilised by phosphorus (P).

In this case, we have the **main plots** (with two replicates for each level) and the **split-plots**. The main plots are alternatively called whole plots.

The effect of the bedrock must be tested against the variability among the main plots within the bedrock type. Note that this design is different from a factorial design with two factors, bedrock and fertilisation and consequently has to be analysed differently: there is one additional level of variability, that of the plots, which is nested within the bedrock type.

3.6 Following the development of objects in time: repeated observations

The biological/ecological objects (whether individual organisms or communities studied at permanent plots) develop in time. We are usually interested not only in their static state, but also in the dynamics of their development, and we can investigate the effect of experimental manipulations on their dynamics. In all the cases, the inference is much stronger if the **baseline data** (the data recorded on the experimental objects before the manipulation was imposed) are available. In this case, we can apply a BACI (before-after control-impact) design (Green 1979).

In designed manipulative experiments, the objects should be part of a correct statistical design; we then speak about replicated BACI designs. As has been pointed out by Hurlbert (1984), the non-replicated BACI is not a statistically correct design, but it is often the best possibility in environmental impact studies where the 'experiment' is not designed in order to enable testing of the effect.

An example is given in Figure 3–6. In this case, we want to assess the effect of a newly built factory on the quality of water in a river (and, consequently, we do not have replicated sites). We can reasonably expect that upstream of the factory, the water

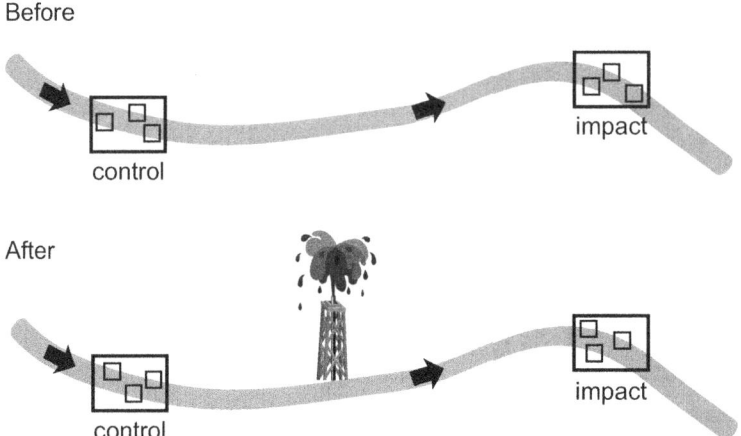

Figure 3–6 The non-replicated BACI (Before-After-Control-Impact) design.

should not be affected, so we need a control site above the factory and an impact site below the factory (nevertheless, there might be differences in water quality along the river course). We also need to know the state before the factory starts operating and after the start, but we should be aware that there might be temporal changes independent of the factory's presence. We then consider the changes that happen on the impact site but not on the control site[5] to be a proof of the factory's impact.

Nevertheless, the observations within a single cell (e.g. on the control site before the impact) are just pseudoreplicates. Even so, the test performed in this way is much more reliable than a simple demonstration of either temporal changes on the impact site itself or the differences between the impact and control sites in the time after the impact was imposed.[6]

Even if we miss the 'before' situation, we can usually demonstrate a lot by a combination of common sense and some design. For example, in our example case, we can have a series of sites along the river and observe an abrupt change just between the two sites closest to the factory (one above and one below it), with the differences between the other sites being gradual.[7] This is often a suggestive demonstration of the factory effect (see Reckhow 1990 for additional discussion).

Another possibility is to repeat the sampling over time and use time points as (possibly correlated) replicates (Stewart-Oaten et al. 1986; Canoco 5 manual, section 6.5.7).

In designed experiments, the replicated BACI (as in Figure 3–7) is probably the best solution. In this design, the individual sampling units are arranged in some correct design (e.g. completely randomised or in randomised complete blocks). The first measurement is done before the experimental treatment is imposed (the baseline measurement) and

---

[5] In ANOVA terminology, this is the interaction between the time and site factors.
[6] The only possibility of having a replicated BACI design here would be to build several factories of the same kind on several rivers, which is clearly not a workable suggestion! But we can have multiple control sites.
[7] Or similarly, time series from the impacted site, with an abrupt change temporarily coincident with the impact-causing event.

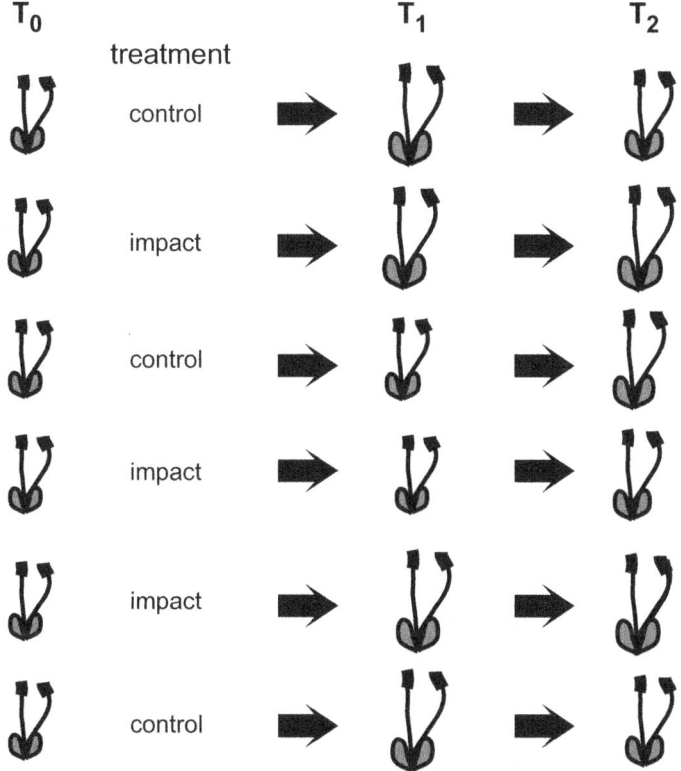

Figure 3–7 Replicated BACI design. The objects (individual plants) are completely randomised and measured at the first time T_0. Thereafter, the impact (fertilisation) is imposed and the development is followed (with two sampling times T_1 and T_2 in our example). The null hypothesis we are interested in is: *The growth is the same in the control and impact groups*.

then the development (dynamics) of the objects is followed. In such a design, there should be no differences among the experimental groups in the baseline measurement. Again, the interaction of treatment and time is of greatest interest, and usually provides a stronger test than a comparison of the state at the end of experiment. Nevertheless, even here we have various possible analyses, with differing emphasis on the various aspects that we are interested in (see Lindsey 1993).

One should be aware that, even if we correctly randomise the sampling units at the beginning, we can find (when testing at $\alpha = 0.05$) significant differences between treatments in the baseline data with 5 per cent probability. It is therefore advisable to test the baseline data before imposing the treatments and if significant differences are found re-assign the treatments to the units (with the same random process as used originally).

With continuing development of statistical methods, the designs are becoming more and more flexible. Often the design of our experiment is a compromise between the statistical requirements and the feasibility of the implied manipulations in the field. The presence of pseudoreplication in designed experiments is, nevertheless, one of the most common mistakes and can lead to serious errors in our conclusions. Each researcher is advised to have a clear idea how the data will be analysed before performing the experiment. This will prevent some disappointment later on.

3.7 Experimental and observational data

Manipulative experiments in ecology are limited both in space and time (Diamond 1986). For larger spatial or temporal scales we have to rely on observational data. Also, each manipulation has inevitably some side-effects and, consequently, the combination of observational data and manipulative experiments is necessary. Generally, the observational data are more often used in hypotheses generation, whereas the data from manipulative experiments are used for hypotheses testing. But this need not be always so. In manipulations on the community level, we usually affect a whole chain of causally interconnected variables. For example in the fertilisation experiment in Case study 3 (Chapter 14), fertilisation affected the cover of the crop, which, together with the direct effect of the nutrients, affected the weed community.

In manipulative experiments on community level replicated in space and time, there are many factors that we are not able to control, but which still affect the results (weather in a year, bedrock at a site). When the response is species community composition, we often know the biological traits of some species (which we are not able to manipulate). We primarily test the hypothesis concerning change of species composition, but the results very often suggest new hypotheses on the relationship between species traits and species response to the manipulation (see Case study 5, Chapter 16, for an example). All those situations call for careful exploration of data, even in the analysis of manipulative experiments (Hallgren et al. 1999).

Also in non-manipulative studies, we often collect data to test a specific hypothesis. It is useful to collect data in a design that resembles the manipulative experiment. Diamond (1986) even coined the term 'natural experiment', for the situations in nature that resemble experimental designs and can be analysed accordingly (for example, to compare the abundance of spiders on islands with and without lizards). A similar approach can be used when studying long-term effects of management practices that we have no chance to affect. According to Diamond, such situations are under-utilised: 'While field experimentalists are laboriously manipulating species abundances on tiny plots, analogous ongoing manipulations on a gigantic scale are receiving little attention (e.g. expansion of parasitic cowbirds in North America and the West Indies, decimation of native fish by introduced piscivores, and on-going elimination of American elms by disease).'

In non-manipulative situations, however, one can never exclude the possibility of confounding effects. In this case, our sampling design should aim to minimise the spatial autocorrelation of units of the same category, and we should attempt to avoid any confounding effects. We should aim that our design resembles as much as possible a correct experimental design, although this is often not possible. For example, if we want to study the effect of ground rock (limestone versus granite), the individual units should be intermingled – sampling just one limestone and one granite area resembles too closely pseudoreplication, and the possible (statistically significant) differences might be easily due to physical proximity of units of the same category rather than an effect of ground rock.

When studying the effect of management (e.g. long-term abandonment of meadows versus traditional mowing), we should carefully consider whether the management decision to continue or discontinue mowing, made many years ago, was not based on meadow fertility and whether we, therefore, do not incorrectly ascribe an environmental effect to management differences. This might be particularly tricky as the management probably changes the soil properties that we can measure. There are no unequivocal rules in this case, we need to use all our ecological knowledge to separate the considered effect from possible confounding effects and at the same time to avoid spatial aggregation of sampling units with the same or similar values of intended explanatory variables (Diamond 1986).

In the presence of environmental heterogeneity, various forms of stratified sampling are used. For example, stratified random sampling design divides first the area of interest into subareas, based on the (expected) most influential environmental variable, and then performs random sampling in each of the subareas. For example, if we expect that mowing versus grazing has the highest impact on grassland species composition, we might first divide the area into grazed and mown patches (according to some preliminary mapping) and then take random samples from each of these two groups.

Recently, more and more community composition records are available in various databases, most typically the databases of phytosociological records. On one hand, they are an invaluable source of information not available elsewhere (e.g. the Czech National Phytosociological Database grew recently to more than 100 000 records and covers last ca 100 years of phytosociological research). On the other hand, most of the phytosociological records stored in such databases were placed intentionally to various 'typical' places, which is a serious obstacle to their use in rigorous statistical analysis. We still believe that they can (and should) be statistically analysed, provided that the type of sampling design is taken into account in interpretations and various methods minimising the adverse effects are used (Lepš & Šmilauer 2007). The ever growing number of publications based on such data supports this view.

4 Basics of gradient analysis

The methods for analysing community composition or similar kinds of multivariate ecological data are usually divided into gradient analysis and classification. The term **gradient analysis** is used here in a broad sense, for any method attempting to relate community composition to the (measured or hypothetical) environmental gradients.

Traditionally, the classification methods, when used in plant community ecology, were connected with the **discontinuum approach** (or vegetation unit approach) or sometimes even with the Clementsian superorganismal approach, whereas the methods of gradient analysis were connected with the **continuum concept** or with the Gleasonian individualistic concept of communities (Whittaker 1975). While this might reflect the history of the methods, this distinction is no longer valid. The methods are complementary and their choice depends mainly on the purpose of a study.

For example, in vegetation mapping some classification is usually needed. Even if there are no distinct boundaries between adjacent vegetation types, we have to cut the continuum and create distinct vegetation units for mapping purposes.[1] Ordination methods can help find repeatable vegetation patterns and discontinuities in species composition, and show any transitional types, etc. These methods are now accepted even in phytosociology. Also, the methods are no longer restricted to plant community ecology. They became widespread in most studies of ecological communities with major emphasis on species composition and its relationship with the underlying factors, and they have found their way also into research fields unrelated to natural sciences (archaeology, social sciences). In fact, it seems to us that the advanced applications of gradient analysis are nowadays found outside the vegetation sciences and the methods become more frequently used in freshwater and marine studies.[2]

In Sections 4.1 to 4.4, we have preferred to replace the general statistical terms (response variables and explanatory variables) by the terms *species* and *environmental variables*, representing the common usage of the general terms in the field of community ecology. We hope this makes the introductory part more easily understandable to the majority of readers.

---

[1] It is possible and straightforward to map a quantitative variable (e.g. a record position on an ordination axis) and hence prepare vegetation maps without classification. But this approach is seldom used and the maps based on community typology are generally preferred.

[2] For example, among more than 5600 citations of CANOCO 4.x manual in research papers published between 2003 and 2012 (recorded in Web of Science), 39% studies belong to fields of marine and freshwater science, followed by 30% belonging to terrestrial plant ecology (including vegetation science), 13% to terrestrial animal ecology, and 10% to soil and microbial research (Šmilauer, unpublished data).

4.1 Techniques of gradient analysis

Table 4–1 provides an overview of the problems solved by gradient analysis and related methods (the methods are categorised according to Ter Braak & Prentice 1988). The methods are selected mainly according to the data that are available for the analysis and according to the desired result (which is determined by the question we are asking).

The goal of **regression** is to find the dependence of a univariate response (e.g. the quantity of a species, or some synthetic characteristic of a community, such as diversity or biomass) on explanatory (e.g. environmental) variables. As the explanatory variables can be represented by a mixture of quantitative and categorical ones, this group of methods also includes all types of general or generalized linear models (see Chapter 8). By **calibration** we mean the estimation of values of environmental characteristics based on the species composition of a community. Typical examples are the estimates based on Ellenberg indicator values ('Zeigerwerte', Ellenberg 1991), or the estimates of water acidity based on the species composition of diatom communities (Batterbee 1984, Birks 2012). To use calibration procedures, we need to know a priori the species' responses to the environmental gradients being estimated.[3]

The goal of **unconstrained ordination** is to find axes of the greatest variability in community composition (the ordination axes) for a set of cases and to visualise (using an ordination diagram) the patterns of similarity among the cases and among response variables (e.g. species). It is often expected that the ordination axes will coincide with some measurable explanatory variables (e.g. properties of environment), and when such variables are measured we usually correlate them with the ordination axes. The aim of **constrained ordination** is to find the variability in species composition data that can be explained by the measured environmental (explanatory) variables. Measured variables useful for interpretation can be also used with the unconstrained ordination, to correlate them with the ordination axes as so-called **supplementary variables**. Unconstrained and constrained ordinations are also available in **partial** versions (not shown in Table 4–1), as **partial ordination** and **partial constrained ordination**. In partial analyses, we first remove the variability in the species composition data, explainable by the **covariates** and then perform a (constrained) ordination on the residual variability. In **hybrid ordination** analyses, first x (x is usually 1, 2, or 3) ordination axes are constrained and the remaining axes are unconstrained.

The explanatory or supplementary variables and the covariates can be both quantitative and categorical.

4.2 Models of response to gradients

The change of a species abundance with the values of a continuous environmental variable can be described by a rich variety of shapes, implied by the chosen regression model.

---

[3] Canoco 5 supports only the basic calibration methods, see section 6.3.3 of Canoco 5 manual for an example.

Table 4-1 Methods of gradient analysis.

Data used in calculations		A priori knowledge of species–environment relationships	Method	Result
No of envir. variables	No of species			
1, n	1	no	Regression	Relation of species to environment
None	n	yes	Calibration	Estimates of environmental values
None	n	no	Ordination	Axes of variation in species composition
1, n	n	no	Constrained ordination	Variation in species composition explained by environmental variables. Relationship of environmental variables to ordination axes

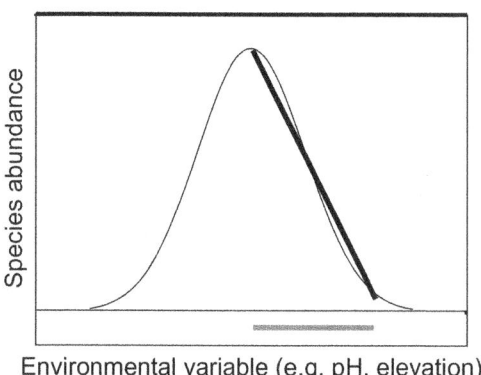 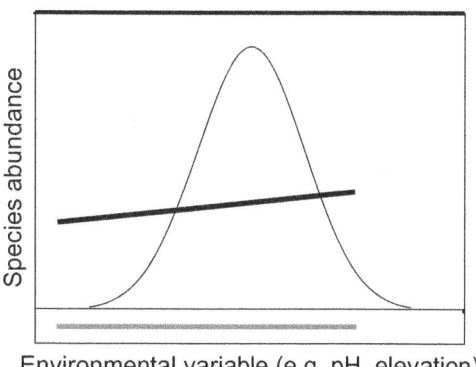

Figure 4–1 Comparison of the suitability of linear approximation of a unimodal response curve over a short part of the gradient (left diagram) and over a larger part of the gradient (right diagram). The grey horizontal bar shows the part of the gradient we are interested in.

The regression can also be a part of the algorithm of multivariate analyses, where many regression models are fitted simultaneously. For this purpose, we need a relatively simple model, which can be easily fitted to the data. Two models of species response to environmental gradient are frequently used: the model of a linear response and that of a unimodal response. The linear response represents the simplest possible approximation, whereas the unimodal response model assumes that the species has an optimum on the environmental gradient. To enable simple estimation of species optimum and tolerance, we usually assume that the response curve is symmetrical around the optimum (see Hutchinson, 1957, for the concept of resource gradients and species optima). When using ordination methods, we must first decide which of the two models should be used. Generally, both models are only approximations, so our decision depends on which of the two approximations is better for our data. Even the unimodal response is a simplification: in reality (Whittaker 1967), the response is seldom symmetrical, and also more complicated response shapes can be found (e.g. bimodal ones). Moreover, the method of fitting the unimodal model imposes further restrictions.

Over a short gradient, a linear approximation of any function (including a unimodal one) works well, but over a long gradient, the approximation by the linear function is poor (Figure 4–1). Even if we have no measured environmental variables, we can expect that, for data with relatively small compositional variation, the underlying gradient is short and the linear approximation appropriate.

4.3 Estimating species optima by weighted averaging

The linear response is usually fitted by the classical method of (least squares) regression. For the unimodal response model, the simplest way to estimate the species optimum is by calculating the weighted average ($WA(Sp)$) of the values of an environmental variable

Basics of gradient analysis

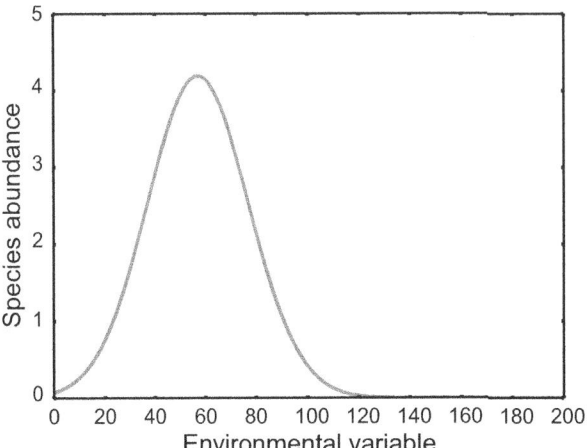

Figure 4–2 Example of a unimodal species response with a range completely covering the response curve.

in the n cases where the species is present. The species importance values (abundances) are used as the weights in calculating the average:

$$WA(Sp) = \frac{\sum_{i=1}^{n} Env_i \times Abund_i}{\sum_{i=1}^{n} Abund_i}$$

where Env_i is the value of environmental variable in the i-th case, and $Abund_i$ is the abundance of the species in the i-th case.[4] If needed, the species tolerance (a measure of the width of the bell-shaped response curve) can be calculated as a square root of the weighted mean of the squared differences between the species optimum and the actual values in each case. The value is analogous to standard deviation (and is a basis for the definition of SD units for measuring the length of ordination axes, see Section 2.3):

$$SD = \sqrt{\frac{\sum_{i=1}^{n} (Env_i - WA(Sp))^2 \times Abund_i}{\sum_{i=1}^{n} Abund_i}}$$

The method of weighted averaging is reasonably good when the whole range of a species distribution is covered by the cases. Consider the dependence displayed in Figure 4–2. If the complete range is covered, then the estimate is not biased (Table 4–2).

On the other hand, if only a part of the range is covered, the estimate is biased. The estimate is shifted in the direction of the tail that is not truncated. An example is presented in Table 4–3.

---

[4] Another possibility is to explicitly state the functional form of the unimodal curve and estimate its parameters by the methods of non-linear regression, but this option is more complicated and not so suitable for the simultaneous calculations that are usually used in ordination methods.

4.3 Estimating species optima by weighted averaging

Table 4–2 Estimation of species optimum from the species response curve displayed in Figure 4–2, using a weighted averaging algorithm when a complete range of species distribution is covered.

Environmental value	Species abundance	Product
0	0.1	0
20	0.5	10
40	2.0	80
60	4.2	252
80	2.0	160
100	0.5	50
120	0.1	12
Total	9.4	564
$WA(Sp) = 564/9.4 = 60$		

Table 4–3 Estimation of species optimum from the response curve displayed in Figure 4–2, using a weighted averaging algorithm when only a part of the range of species distribution is covered.

Environmental value	Species abundance	Product
60	4.2	252
80	2.0	160
100	0.5	50
120	0.1	12
Total	6.8	472
$WA(Sp) = 472/6.8 = 69.4$		

When the covered portion of the gradient is short, most of the species will have their distributions truncated, and the optimum estimates will be biased. The longer the gradient, the more species will have their optima estimated well. We can reasonably expect that the more limited the compositional variation in the data (i.e. the lower their β diversity), the shorter the gradient.

> The techniques based on the linear response model are suitable for data sets with low β diversity. The weighted averaging techniques related to the unimodal response model are suitable for more heterogeneous data.

The decision is usually made on the basis of gradient length in detrended correspondence analysis (DCA), which estimates the heterogeneity in community composition (see Section 2.3).

4.4 Calibration

The goal of calibration is to estimate the values of environmental descriptors on the basis of species composition.[5] This procedure can be used only if we know beforehand the relation of species to the environmental variable to be estimated. Theoretically, the procedure assumes that we have a **training set** with both species and environmental data available, which are used to estimate the relationships of species and environment (e.g. species optima with respect to environmental gradients). Those quantified relationships are then used to estimate the unknown environmental characteristics for cases where only the species composition is known (e.g. in palaeolimnological records, Birks 2012).

The most commonly used calibration method is **weighted averaging**. For this purpose, the estimates of species optima in relation to selected environmental gradients must be available. These estimated optima values are sometimes called **indicator values**. For example in Central Europe, the indicator values of most plant species are available (on relative scales) for light, nitrogen availability, soil moisture, etc. (e.g. Ellenberg 1991).[6] The environmental value for a case ($WA(Case)$) can then be estimated as a weighted average of the indicator values of all the s present species, their abundances being used as weights:

$$WA(Case) = \frac{\sum_{i=1}^{s} IV_i \times Abund_i}{\sum_{i=1}^{s} Abund_i}$$

where IV_i is the indicator value of i-th species (presumably its optimum) and $Abund_i$ is the abundance of i-th species in a case. The procedure is based on the assumption of existence of species optima and, consequently, of the unimodal species responses. Calibration procedures based on the assumption of a linear response also exist, even if they are seldom used in ecology.

An example of the calibration procedure is given in Table 4–4.

The opinion on the use of indicator values and calibration differs among ecologists. It is generally argued that it is more reliable to measure the abiotic environment directly than to use a calibration. However, for historical records (not only relevés (cases) done in the past, but also palynological and paleolimnological evidence, etc.) calibration provides information that might not be completely reliable, but it is often the only available. Also the species composition (and so the calibration) reflects the state of environment over a period of time, which might be more informative than a direct, but snapshot

---

[5] A common-sense example: the presence of polar bears indicates low temperatures, the presence of *Salicornia* species a high soil salinity, and the presence of common stinging nettle a high concentration of soil nitrogen.

[6] However, these species indicator values are **not** based on the analyses of training sets, but rather on personal experience of the experts. As all the values are on a relative scale, the estimates of case environmental values are also on this relative scale.

4.5 Unconstrained ordination

Table 4–4 Estimation of nitrogen availability for two cases, *Case 1* and *Case 2*. The a priori known indicator values for nitrogen are shown in bold style (*Nitrogen IV* column, according to Ellenberg 1991), the collected data and the calculations are shown in standard typeface.

	Nitrogen IV	Case 1	IV * abund.	Case 2	IV * abund.
Drosera rotundifolia	**1**	2	2	0	0
Andromeda polypofila	**1**	3	3	0	0
Vaccinium oxycoccus	**1**	5	5	0	0
Vaccinium uliginosum	**3**	2	6	1	3
Urtica dioica	**8**	0	0	5	40
Phalaris arundinacea	**7**	0	0	5	35
Total		12	16	11	78
Nitrogen (WA)			1.333		7.090
			(=16/12)		(=78/11)

measurement. For example the instant soil moisture depends on the rain in preceding days, but the estimates based on Ellenberg indicator values do not.

Calibration is also one of the steps used in more complicated analyses, such as ordination methods, as we show in the following section.

4.5 Unconstrained ordination

Starting from this section, we return to the use of the terms response variables and explanatory variables, but please keep in mind that community ecologists typically use these to refer to species and the properties of environment, respectively.

The problem of traditional (unconstrained) ordination can be formulated in two ways:

(1) Find a configuration of cases in the ordination space so that the distances between cases in this space correspond best to the dissimilarities of their response data. This is explicitly done by multidimensional scaling (MDS) methods (Legendre & Legendre 2012, Kruskal 1964). The metric MDS (also called Principal Coordinates Analysis, see Section 6.5) considers the cases to be points in a multidimensional space (e.g. where species are the axes and the position of each case on an axis is given by the corresponding species abundance). Then the goal of ordination is to find a projection of this multidimensional space into a space with reduced dimensionality that will result in a minimum distortion of the spatial relationships. Note that the result is dependent on how we define the 'minimum distortion'. In non-metric MDS (NMDS), we do not search for projection rules,[7] but using a numerical algorithm we seek for a configuration of case points that best portrays the rank of inter-case distances (see Section 6.7).

---

[7] NMDS method is usually not considered a gradient analysis and it cannot be formulated in the second way discussed in this section.

(2) Find 'latent' variable(s) (ordination axes) which represent the best predictors for the values of all the response variables (e.g. species). This approach requires the model of response to such latent variables to be explicitly specified: the linear response model is used for linear ordination methods and the unimodal response model for unimodal (weighted averaging) methods. In linear methods, the case score is a linear combination (weighted sum) of the response variable scores. In unimodal methods, the case score is a weighted average of the response variable scores (after some rescaling). This formulation of the ordination task is elaborated below.

The two formulations may lead to the same solution.[8] For example, principal component analysis (PCA) can be formulated either as a projection in Euclidean space, or as a search for latent predictor variables when linear responses are assumed.[9]

Similarly, correspondence analysis (CA) can be viewed as a method projecting data into a space where the chi-square distance metric (see Section 6.2.2) is preserved, but also as a method that computes latent predictor variables for (approximated) models of unimodal response.

In the Canoco program, the approach based on formulation (2) is adopted. The principle of ordination methods can be elucidated by their algorithm. We will use the unimodal methods as an example. We try to construct a 'latent' variable (the first ordination axis) so that the fit of all the response variables using this latent variable as the predictor will be the best possible fit (using unimodal response curves). The result of the ordination will be the values of this latent variable for each case (called the **case scores**) and each response variable – the estimate of the latent variable value, at which its unimodal model achieves its peak – i.e. the optimum of a species (the **response variable scores**). Further, we require that the response variable scores be correctly estimated from the case scores by weighted averaging and the case scores be correctly estimated as weighted averages of the response variable scores. This can be achieved by the following iterative algorithm:

- **Step 1** Start with some (arbitrary) initial case scores $\{x_i\}$
- **Step 2** Calculate new response variable scores $\{u_j\}$ by (weighted averaging) regression from $\{x_i\}$
- **Step 3** Calculate new case scores $\{x_i\}$ by (weighted averaging) calibration from $\{u_j\}$
- **Step 4** Remove the arbitrariness in the scale by standardising case scores (stretch the axis)
- **Step 5** Stop on convergence, else go to Step 2

---

[8] If cases with similar response variable values were distant on an ordination axis, such an axis could not serve as a good predictor of the response data.

[9] The first formulation (maintaining Euclidean distances among cases) also implies the view of PCA as a multidimensional rotation of the original m-dimensional space (of the m response variables, where the value of each variable in a case represents the case coordinate on the corresponding axis). The rotation achieves maximisation of the variance of case positions along the first few (best) axes – principal components.

4.5 Unconstrained ordination

Table 4–5 Calculation of the first ordination axis by the weighted averaging (WA). Further explanation is in the text.

	Case1	Case2	Case3	u.WA1	u.WA2	u.WA3	u.WA4
Cirsium	0	0	3	10.000	10.000	10.000	10.000
Glechoma	5	2	1	2.250	1.355	1.312	1.310
Rubus	6	2	0	1.000	0.105	0.062	0.060
Urtica	8	1	0	0.444	0.047	0.028	0.027
initial score	0	4	10				
x.WA1	1.095	1.389	8.063				
x.WA1resc	0.000	0.422	10.000				
x.WA2	0.410	0.594	7.839				
x.WA2resc	0.000	0.248	10.000				
x.WA3	0.376	0.555	7.828				
x.WA3resc	0.000	0.240	10.000				
x.WA4	0.375	0.553	7.827				
x.WA4resc	0.000	0.239	10.000				

We can illustrate the calculation by an example, presented in Table 4–5. The data table (with three cases and four response variables) is displayed using bold numbers. The initial, arbitrarily chosen case scores are displayed in italics. From those, we calculate the first set of response variable scores by weighted averaging, *u.WA1* (Step 2 above, see also Section 4.3). From these scores, we calculate new case scores (*x.WA1*), again by weighted averaging (Step 3, see also Section 4.4). We can see that the axis is shorter (with the range from 1.095 to 8.063 instead of from 0 to 10). The arbitrariness in the scale has to be removed by linear rescaling and the rescaled case scores *x.WA1resc* are calculated (Step 4 above):

$$x_{rescaled} = \frac{x - x_{min}}{x_{max} - x_{min}} \times length$$

where x_{max} and x_{min} are the maximum and minimum values of x in the non-rescaled data and *length* is the desired length of the ordination axis. The *length* parameter value is arbitrary, 10.0 in our case. This is true for some ordinations but there are methods where the length of the axis reflects the heterogeneity of the data set (see e.g. detrended correspondence analysis, DCA, with Hill's scaling, Section 11.2). Now, we compare the original values with the newly calculated set. Because the values are very different, we continue by calculating new response variable scores, *u.WA2*. We repeat the cycle until the consecutive case scores *x.WA[N]resc* and *x.WA[N+1]resc* have nearly the same value (some criterion of convergence is used, for example we considered in our case values 0.240 and 0.239 to be almost identical).

Table 4–5 was copied from the example file *ordin.xlsx*. By changing the values in the *initial score* row in this file, you can confirm that the final scores are completely independent of the initial ones. In this way, you get the first ordination axis.[10] Depending

---

[10] In the linear methods, the algorithm is similar, except the regression and calibration are not performed using weighted averaging, but in the way corresponding to the linear model – using the least squares algorithm.

on the chosen initial values, the polarity of the computed axis can be reversed, i.e. instead of 0.000, 0.239, and 10.000 we might obtain coordinates 0.000, 9.761, and 10.000. This demonstrates that the polarity of individual ordination axes is arbitrary and consequently its change does not affect the interpretation of ordination results.[11]

The other ordination axes are derived in a similar way, with the additional constraint that they have to be linearly independent of all the previously derived axes.

The whole method can be also formulated in terms of matrix algebra and eigenvalue analysis. For practical needs, we should note that the better the response variables are fitted by the ordination axis (the more variability the axis explains), the less the axis 'shrinks' in the course of the iteration procedure (i.e. the smaller is the difference between range of *x.WA* and *x.WAresc*). Consequently, the value

$$\lambda = \frac{x_{max} - x_{min}}{length}$$

is a measure of the explanatory power of the axis and, according to the matrix algebraic formulation of the problem, it is called the **eigenvalue**. Using the method described above, each axis is constructed so that it explains as much variability as possible, under the constraint of being independent of the previous axes. Consequently, the eigenvalues decrease with the order of the axis.[12]

4.6 Constrained ordination

To explain constrained ordination, it is best to start with the formulation (2) of unconstrained ordination in the preceding section. Whereas in unconstrained ordinations we search for any compositional gradient (ordination axis) that best explains the response data, in constrained ordinations the ordination axes are weighted sums of explanatory variables. Numerically, this is achieved by a slight modification of the above algorithm, in which we add one extra step after Step 3.

> - **Step 3a** Calculate a multiple linear regression of the case scores $\{x_i\}$ on the explanatory variables and take the fitted values of this regression as new case scores $\{x'_i\}$.

Note that the fitted values in a multiple regression are a linear combination (i.e. a weighted sum) of the predictors and, consequently, the new case scores x' are linear combinations of the explanatory variables. The fewer explanatory variables we have,

---

[11] In Canoco 5, you can change the polarity of ordination axes for ordination diagrams (in already performed analysis) using the *Analysis | Plot creation options* menu command and changing the *Flip axes* option in the *General* page of the *Plot Settings* dialog.

[12] A strictly mathematically correct statement is that they do not increase, but for typical data sets they always do decrease.

the stricter is the constraint. If the number of explanatory variables is greater than the number of cases minus 2, then the ordination becomes unconstrained.

The unconstrained ordination axes correspond to the directions of the greatest variability within the data set.[13] The constrained ordination axes correspond to the directions of the greatest data set variability that can be explained by the environmental variables. The number of constrained axes cannot be greater than the number of explanatory variables.[14] When we use just one (numeric) explanatory variable, only the first ordination axis is constrained and the remaining axes are unconstrained.

The existence of Step 3a also explains why the constrained ordination methods produce two types (sets) of case scores: the case scores derived from the response data (x_i, denoted *CaseR* in Canoco 5 output) and calculated in Step 3 from response variable scores, and the case scores derived from the explanatory variable values (x'_i, denoted *CaseE* in Canoco 5 output) and calculated in Step 3a. The *CaseR* and *CaseE* scores shown in ordination results are those calculated in the last loop of the iterative algorithm.[15] Canoco presents in analysis results the correlation between *CaseR* and *CaseE* scores for individual axes, called pseudocanonical correlation. See Canoco 5 manual, sections 5.2.2 and 5.6.4 for additional details.

4.7 Basic ordination techniques

Four basic ordination techniques can be distinguished based on the underlying response model and whether the ordination is constrained or unconstrained (see Table 4-6, based on Ter Braak & Prentice, 1988). The unconstrained ordination is also called *indirect gradient analysis* and the constrained ordination (also called *canonical ordination*) belongs to the group of *direct gradient analysis* methods.

For unimodal (weighted averaging) methods, detrended versions exist, i.e. detrended correspondence analysis (DCA) implemented in the legendary DECORANA program (Hill & Gauch 1980), and detrended canonical correspondence analysis (DCCA); see Section 2.3 for more information about detrending.

For all methods, **partial analyses** exist. In a partial analysis, the effect of covariates is first removed and the analysis is then performed on the remaining variability.

The **hybrid analyses** represent a 'hybrid' between constrained and unconstrained ordination methods. In standard constrained ordinations, there are as many constrained axes as there are independent explanatory variables and only the additional ordination axes are unconstrained. In a hybrid analysis, only a pre-specified number of constrained axes are calculated and any additional ordination axes are unconstrained. In this way, we

---

[13] The 'variability' is defined differently for different ordination methods: it is a standard variance for linear methods and inertia in unimodal methods.
[14] But each factor variable counts here as the number of its levels minus 1.
[15] In fact, even the unconstrained ordination results may contain a second set of case scores: if you have added supplementary variables to such ordination, they are not used when calculating case and response variable scores by the iterative algorithm explained in Section 4.5, but afterwards the case scores derived from response scores (*CaseR* scores) are regressed onto supplementary variables and the fitted values are called *CaseS* scores.

Table 4-6 Basic types of ordination techniques.

	Linear	Unimodal
Unconstrained	Principal Components Analysis (PCA)	Correspondence Analysis (CA)
Constrained	Redundancy Analysis (RDA)	Canonical Correspondence Analysis (CCA)

can specify the dimensionality of the solution of the constrained ordination model. See Section 5.3 for additional comments.

With 'release' from the constraint (after all the constrained axes are calculated), the procedure is able to find a 'latent' variable that may explain more variability than the previous, constrained ones. Consequently, in many cases, the first unconstrained axis explains more than the previous constrained axis and, therefore, the corresponding eigenvalue is higher than the previous one – see for example the snapshot of *Summary* page in Figure 14-4.

4.8 Ordination axes as optimal predictors

In this section, we focus on the interpretation of ordination axes produced by linear ordination methods (PCA and RDA), which closely relate to linear regression. You can find a brief introduction to linear regression in Chapter 8 and more details can be found in any decent statistical textbook. The last few paragraphs of this section explain how the topics presented here apply to unimodal ordination.

The first axis of PCA (first principal component) is found so that the positions of cases (case scores, *CaseR*) represent an optimal predictor variable for separate linear regressions for each of the M variables in the response data table. The optimality of the first PCA axis means that we cannot find a better predictor in terms of the percentage of explained variation (i.e. coefficient of determination, R^2), averaged over all M regression equations.[16] When we denote the case scores on the first PCA axis as x_{1i} (where i refers to individual cases, $i = 1, \ldots, N$), we can describe the implied regression equation for the k-th response variable (e.g. a species in a table describing biotic community composition) as

$$Y_{ik} = b_{0k} + b_{1k}{}^* x_{1i} + e_{ik}$$

But as both the case scores (x) and the response variables (Y) are centred[17] in linear ordination methods, the intercept (b_{0k}) is fixed at zero value and can be ignored.

We can now find the second PCA axis (its case scores) by looking for an additional predictor variable x_2 for the regression equation, which most increases (on average, over all response variables) the R^2 value, with an additional requirement that x_2 has no

---

[16] Strictly speaking, this is correct only if all the response variables have the same variation in the data (i.e. they were standardised) and there are no user weights set for them.
[17] So that their average is equal to 0, see also Section 1.3.2.

correlation with x_1.[18] The resulting equation then looks as follows (with the b_{0k} term already omitted, see above):

$$Y_{ik} = b_{1k}{}^{*}x_{1i} + b_{2k}{}^{*}x_{2i} + e_{ik}$$

The regression coefficients (b_{1k} and b_{2k}) are values specific for each response variable and each ordination axis and represent the response variable scores (*Resp*) that are usually plotted as arrows in a PCA ordination diagram (starting at [0,0] coordinates and ending with an arrowhead at [b_{1k}, b_{2k}] coordinates). It is perhaps interesting to note that the 'multiplication' of case scores by response variable scores represents the perpendicular projection of case points onto a response variable arrow in the **biplot rule** (see Section 11.1).

Ordination axes in constrained methods have the same task as those in unconstrained ordination, but we add a further requirement on their definition: the case scores (x) must represent such predictors that can be interpreted in a straightforward manner using the measured values of explanatory variables. This means that the x predictors cannot have just any values – their values must be defined as follows using the measured explanatory variables (Z):

$$x_{1i} = c_{11}{}^{*}Z_{1i} + c_{12}{}^{*}Z_{2i} + c_{13}{}^{*}Z_{3i}$$

In the above equation, we assume three quantitative explanatory variables, but there can be any number of them and – like in a general linear model (see Section 8.2) – they also include contrast (or 'dummy') variables representing factors.[19] The c coefficients are called the **canonical coefficients** of the explanatory variables (*RegE*), but they are **not** usually plotted in the ordination diagrams.[20] The above equation defines the case scores for the first RDA axis only, but similar definitions apply to higher axes, so that each explanatory variable has a separate canonical coefficient for each constrained axis (multiple coefficients in the case of factor variables).

We may combine the above two equations into a single one, further illustrating the direct relation of constrained ordination to multiple multivariate regression:

$$Y_{ik} = b_{1k}{}^{*}c_{11}{}^{*}Z_{1i} + b_{1k}{}^{*}c_{12}{}^{*}Z_{2i} + \cdots + b_{2k}{}^{*}c_{21}{}^{*}Z_{3i}$$

In this equation (shown incompletely above), there are just the response variables (Y) at the left side and the explanatory variables (Z) at the right side, the b and c are coefficients estimated, respectively, for a particular response or explanatory variable. The expression $b*c$ represents the actual coefficients of a multiple multivariate regression model, describing the dependency of values of a particular response variable (e.g. a species) on the values of explanatory variables chosen for the model. If we fit such models independently for each of the M response variables, we need to estimate $M*P$

---

[18] More than the first two ordination axes can be computed, but we limit our discussion to the first two axes here, reflecting the usual contents of ordination diagrams.

[19] Each factor with q levels is then represented by $q-1$ contrast variables, i.e. by $q-1$ terms in the above equation.

[20] So-called biplot scores (*BipE*) are used, where the case scores are related to each explanatory variable in a separate regression, avoiding possible issues with correlated explanatory variables.

regression coefficients (for P explanatory variables). But in our RDA, the estimated coefficients are constrained by their definition (i.e. they are defined as $b*c$): if we use only the first constrained axis, we need to estimate only $M + P$ parameters (the b_{1k} and c_{1j} coefficients, respectively). To use the first two constrained axes, we need $2 * (M + P)$ parameters.[21] This is, for most data sets, still a large saving compared with $M*P$ parameters.

For unimodal methods, a similar equation predicting response data using case scores and response variable scores can be used:

$$p_{ik} = 1 + u_{1k}{}^{*}x_{1i} + u_{2k}{}^{*}x_{2i} + e_{ik}$$

In this case, however, the response variable scores are labelled u instead of b and they do not represent the slopes of linear dependency of response variable values Y upon the ordination axis, but rather an (approximated) optimum of their unimodal response. This is also reflected in the fact that the response variable in the above equation is not Y_{ik}, but p_{ik} defined as double-standardised Y_{ik} values:

$$p_{ik} = Y_{ik}/(Y_{i+}{}^{*}Y_{+k}/Y_{++})$$

where Y_{i+} and Y_{+k} are, respectively, the row (case) and column (variable) totals of the response data table and Y_{++} is the table's grand total. It is also worth noting that the Y_{i+} and Y_{+k} are the weights used in the weighted-averaging Steps 2 and 3, respectively, in the description of the correspondence analysis algorithm in Section 4.5.

> The weighted averaging algorithm contains an implicit standardisation by both case and response variables. In contrast, we can select in linear ordination the standardised and non-standardised forms.

4.9 Ordination diagrams

The results of an ordination are usually displayed as ordination diagrams. Cases are displayed by points (symbols) in all the methods. Response variables (e.g. species) are shown by arrows in linear methods (the direction in which the variable values increase) and by points (symbols) in weighted averaging methods (estimates of species optima). Quantitative explanatory variables are shown as arrows (representing the direction in which the value of explanatory variable increases). For qualitative explanatory variables (factors), the centroids are shown for individual categories, as the centroids of cases, where the category is present. More information on the interpretation of ordination diagrams can be found in Chapter 11.

Typical examples of ordination diagrams produced from the results of the four basic types of ordination methods are shown in Figure 4–3.

---

[21] This estimate disregards the remaining freedom in b and c values. For example, using 4.b and $c/4$ instead of b and c would yield the same model. A more careful mathematical argument shows that the number of parameters is somewhat different (see Robinson 1973 or Canoco 5 manual, p. 85).

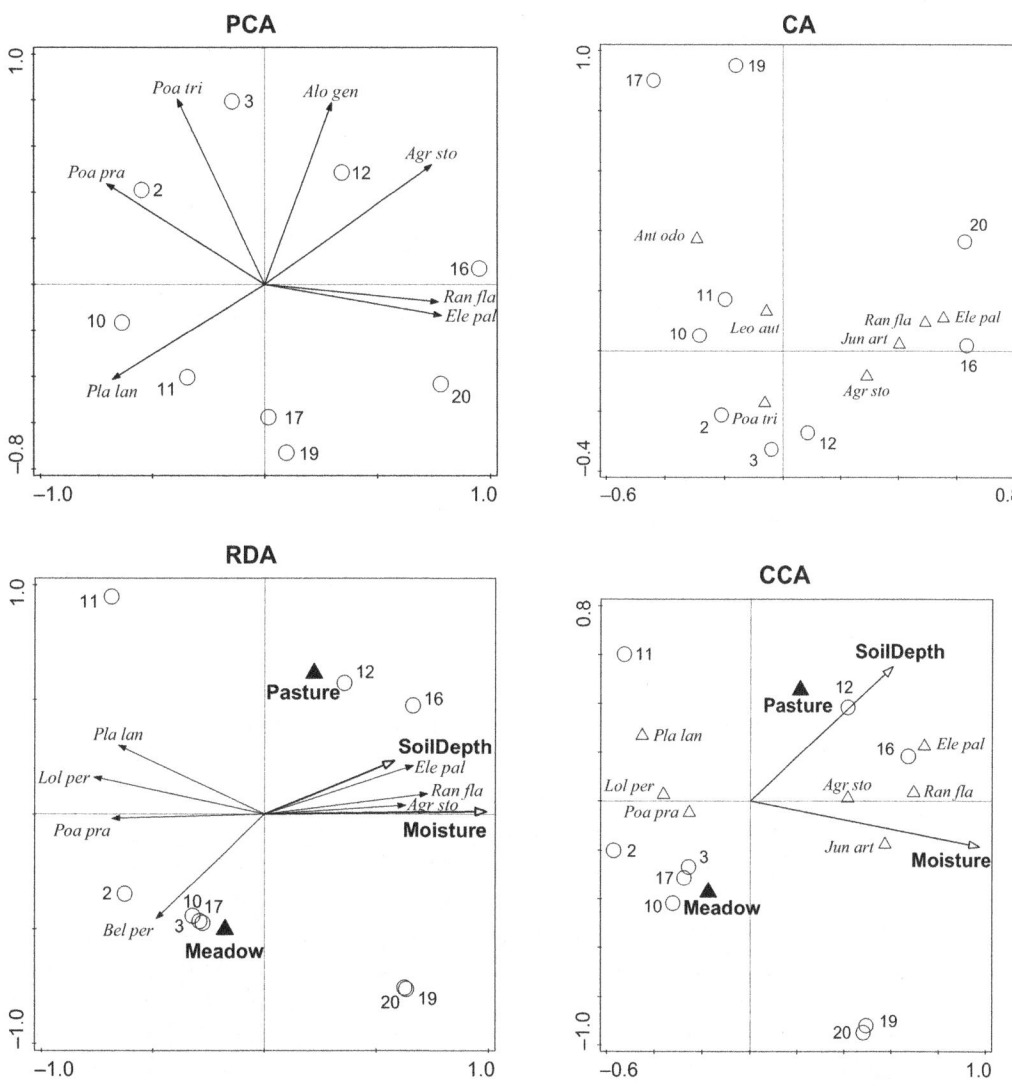

Figure 4–3 Examples of typical ordination diagrams. Analyses of species composition of 20 plots of dune meadow data set (Canoco 5 manual, section 6.2.5, slightly simplified here), in which the vegetation composition was related to the type of management (a factor with two categories – *Meadow* and *Pasture*) and soil properties (depth of A1 horizon and soil moisture). The species are labelled by the first three letters of the generic name and the first three letters of the specific name. Cases (plots) are displayed as circles. For each ordination method (PCA, RDA, CA, and CCA), an ordination diagram (biplot or triplot) is plotted with the first axis shown in horizontal (X) direction and second axis shown in vertical (Y) direction.

4.10 Two approaches

Having both explanatory and response data, we can first calculate an unconstrained ordination with the response data and then calculate a regression of the ordination axes on the measured explanatory variables (i.e. to project the explanatory variables into the ordination diagram[22]) or we can calculate directly a constrained ordination. **The two approaches are complementary and both can be used in the same study**. By calculating the unconstrained ordination first, we do not miss the main part of the variability in the response data table, but we can miss that part of the variability that is related to the measured explanatory variables.[23] By calculating a constrained ordination, we do not miss the main part of the variability explained by the measured explanatory variables, but we can miss the main part of the variability if it is not related to explanatory variables.

When you publish your results, be careful to always specify the method of analysis used. From an ordination diagram, it may be impossible to distinguish between constrained and unconstrained ordinations; because some authors do not follow the convention of using arrows for response variables in linear methods, even the distinction between linear and unimodal methods is not unequivocal.

4.11 Testing significance of the relation with explanatory variables

In an ordinary statistical test, e.g. testing the relation of two variables in a simple linear regression, the value of the test statistic calculated from the data, usually a t–value or F-ratio, is compared with the expected distribution of the statistic under the null hypothesis being tested. Based on this comparison, we estimate the probability of obtaining results as different (or even more different) from those expected under the null hypothesis, as in our data. Or alternatively, we might say that because the null hypothesis is the independence of the two variables, we are estimating the chance that we will get such a 'nicely close' relationship as in our data if the response variable were completely independent of the predictor.

The expected distribution of the test statistic is derived from the assumption about the distribution of the original data (this is why we assume the normal distribution of the response residuals in least squares regression). In constrained ordination, the distribution of the test statistic[24] under the null hypothesis of independence is not known. This distribution depends on the number of response variables, their correlation structure, the distribution of the response data values, etc. But the distribution can be simulated and this is done in a **Monte Carlo permutation test**.

---

[22] When used in this way, Canoco 5 calls the 'explanatory' variables supplementary variables. The explanatory variables term is reserved for their use in constrained ordination.

[23] Particularly if you follow the wrong practice of inspecting the results for just the first two ordination axes.

[24] The pseudo-F ratio, used in the later versions of Canoco, is a multivariate counterpart of the ordinary F-ratio (see Section 5.2 for more details); the eigenvalue was used in earlier versions.

In this test, an estimate of the distribution of the test statistic under the null hypothesis is obtained in the following way. The null hypothesis states that the response data is independent of the explanatory variables. If this hypothesis is true, then it does not matter which observation of response variable values is assigned to which observation of explanatory variables. Consequently, the values of the explanatory variables are randomly assigned to the individual cases of response data, ordination is done with this permuted ('shuffled') data set, and the value of the test statistic is calculated. In this way, both the distribution of the response variables and the correlation structure of the explanatory variables remain the same in the real data and in the null hypothesis simulated data. The significance level (probability of Type I error) of this test is then calculated as

$$P = \frac{n_x + 1}{N + 1}$$

where n_x is the number of permutations where the test statistic was not lower in the random permutation than in the analysis of original data, and N is the total number of permutations. This test is completely distribution-free. This means that it does not depend on any assumption about the distribution of the response data values. More thorough treatment of permutation tests in general can be found in Legendre and Legendre (2012, pp. 25–31).

The permutation scheme can be 'customised' according to the experimental design from which the analysed data set comes. The above description corresponds to the basic version of the Monte Carlo permutation test, but more sophisticated approaches are actually used in the Canoco program – see Chapter 5 and the Canoco 5 manual (section 3.6).

4.12   Monte Carlo permutation tests for the significance of regression

A permutation test can be used to test virtually any relationship. To illustrate its logic, we will show its use for testing the significance of a simple regression model that describes the dependence of plant height on the soil nitrogen concentration.

We know the heights of five plants and the content of nitrogen in the soil in which they were grown (see Table 4–7). We calculate the regression of plant height on nitrogen content. The relationship is characterised by the F-ratio of the analysis of variance of the regression model (10.058 in our example). Under some assumptions (normality of the data), we know the distribution of the F-ratios under the null hypothesis of independence (F–distribution with 1 and 3 degrees of freedom).

Let us assume that we are not able to use this distribution (e.g. normality is violated). We can simulate the distribution by randomly assigning the nitrogen values to the plant heights. We construct many random permutations and for each one we calculate the regression (and corresponding F-ratio) of the plant height on the (randomly assigned values of) nitrogen content. As the nitrogen values were assigned randomly to the plant heights, the distribution of the F-ratios corresponds to the null hypothesis of independence between these two variables.

Table 4–7 Example of permutation test for a simple linear regression.

Plant height	Nitrogen, as measured	1st permutation	2nd permutation	3rd permutation	4th permutation	5th etc
5	3	3	8	5	5	...
7	5	8	5	5	8	...
6	5	4	4	3	4	...
10	8	5	3	8	5	...
3	4	5	5	4	3	...
F-ratio	10.058	0.214	1.428	4.494	0.826	...

The significance of the regression is then estimated as:

$$\frac{1 + no.\ of\ permutations\ where\ (F \geq 10.058)}{1 + total\ number\ of\ permutations}$$

The F-ratio (pseudo-F statistic) in Canoco has a similar meaning[25] as the F-ratio in ANOVA of the regression model and the Monte Carlo permutation test is used in an analogous way.

For our example with five plants, we could even enumerate all their possible permutations (factorial of 5 = 120 possible arrangements) and obtain exact significance estimates as the relative proportion of arrangements that provided a better fit than the original data. But already for 10 plants, there are more than 3.6 million possible permutations. This is why the permutation test must take a random sample of all the possible permutations, earning the label *Monte Carlo* in this way. On the other hand, with 5 plants and only 120 distinct permutations, there is no point doing this test (randomly choosing plant order) for, say, 100 000 times: even when all the 119 possible permutations other than the original arrangement would yield a pseudo–F statistic value smaller than the test statistic for non-permuted data, the non-permuted plant order will appear with the probability of $1/120 = 0.0083$ among them, so 0.0083 is a lower limit[26] for the significance value we might obtain with five randomly permuted units. In fact, there is never any need to use very high numbers of permutations (say > 9999), as the Monte Carlo P-value estimate is unbiased – larger numbers give only slightly more power to the test.

4.13 Relating two biotic communities

If we need to relate two different kinds of biotic community (such as a community of higher plants and a community of soil microorganisms), recorded over identical sets of locations, the method of constrained ordination is usually not appropriate. This is because (a) constrained ordination does not work as expected if the number of explanatory

---

[25] But not necessarily the same distribution.
[26] Apart from random fluctuations. Moreover, the F-ratio does not reflect the sign of the relationship (it gives a two-sided test) and therefore depending on the actual data properties the actual lower limit for P-value might be twice as large.

variables (here the number of taxa in the other community) approaches or exceeds the number of cases[27] and (b) constrained ordination treats the response variables and explanatory variables differently, but when comparing two community kinds, we might prefer a symmetric treatment.[28]

Very often (particularly when sampling over a wider range of ecological conditions, see Section 4.2) the taxa in both compared community data tables are better modelled using a unimodal response to underlying gradients and this provides another reason why the use of constrained ordination is not appropriate, as its ordination axes are defined as linear combinations of explanatory variables. The symmetric co-correspondence analysis (symmetric CoCA, Ter Braak & Schaffers 2004) should be used under such circumstances, as it finds ordination axes (gradients) along which the weighted co-variance among case scores for the two compared communities is maximised. With symmetric CoCA, as implemented in Canoco 5, we can test the hypothesis that the compositional variation is independent between the two communities, using a permutation test.

When we compare two biotic communities with low beta diversity, where linear ordination methods would be appropriate (if each community were to be analysed independently), a linear form of co-inertia analysis (Dray et al. 2003) can be used.[29] The alternative canonical correlation analysis (note this is not the canonical **correspondence** analysis) also maximises linear cross-correlation between two compared data tables, but it cannot be used when the number of variables is larger than the number of cases, which is a common situation in community ecology.

4.14 Community composition as a cause: using reverse analysis

In some studies, we might need to use the species composition of a biotic community to predict (or explain) one or a few other variables. As an example, Křenová (unpublished data) studied the incidence of the endangered butterfly, *Maculinea alcon*, whose caterpillars feed exclusively on the endangered plant species *Gentiana pneumonanthe*. Vegetation composition was recorded on all studied localities to characterise the plant community and the presence of *Maculinea* was recorded. Using the presence of the butterfly as a response and the vegetation composition as predictors would hardly provide satisfactory results: the number of predictors (plant species) exceeds the number of

---

[27] With the number of explanatory variables approaching the number of cases, the constrained ordination turns into an unconstrained one (RDA into PCA and CCA into CA) and the explanatory variables lose their predictive ability.

[28] Sometimes, however, we might prefer the non-symmetric contents, especially when we believe that the organisms in one community affect the other one but not vice versa. Under such circumstances, so-called predictive co-correspondence analysis is recommended (at least when relating two communities with assumed unimodal nature of their response to underlying gradients). This method is not, however, available in Canoco 5.0, so we discuss only the symmetric co-correspondence analysis here.

[29] Co-inertia analysis (CoIA) is a general method for relating two data tables so that their covariance (co-inertia) is maximised. Canoco 5 offers a limited implementation of CoIA, where one data table can be processed in either linear or unimodal context, while the other is always handled linearly. There is also no permutation test available for CoIA in Canoco version 5.0.

observations (localities), abundances of individual plant species will be quite correlated across the sampled localities, and it would be unlikely to find a close relationship with a single plant species (except the host). But the butterfly presence might be well related to the overall community composition.

Probably the best approach we can use here is to reverse the roles of the cause (plant community) and the effect (butterfly occurrence) in their translation into predictor and response variables. After all, as constrained ordination methods (similarly to regression) are mostly about statistical correlation, these models do not necessarily reflect causal relations. So although we can hardly expect that the butterfly would affect the composition of plant community, we can ask whether the plant communities in the localities with *Maculinea* present differ from those where it is missing. To do so, we will use the butterfly presence as a single explanatory variable and the community composition data as the response (in fact, we predict the plant species composition on the basis of the butterfly presence). A similar example from the field of plant ecology is the study of Prentice and Cramer (1990).

A similar reverse analysis was also appropriate for a manipulative field experiment (Blažek, unpublished data) where a hemiparasitic plant was sown into various plant communities. Although the sowing success (quantitative variable representing the percentage of established plants) was dependent on the plant community and not vice versa, the plant species could not be used as predictors for reasons given above and so a reverse analysis with sowing success as explanatory variable was performed instead. It is perhaps interesting to note that in both quoted examples, the swapping of the roles does not necessarily imply reversed direction of causality, because the plant community composition might merely reflect some (non-measured) environmental gradient that affects the target species presence or sowing success.

5 Permutation tests and variation partitioning

In this chapter, we discuss testing hypotheses in a constrained ordination using the Monte Carlo permutation test and two related issues: stepwise selection of a set of explanatory variables and the variation partitioning procedure.

5.1 Permutation tests: the philosophy

Canoco has the ability to test the significance of constrained ordination models described in the preceding section, using **Monte Carlo permutation tests**. These statistical tests relate to a general null hypothesis, stating the independence of the response data (such as community composition) on the values of the explanatory variables (for example environmental factors or human impact). The principles of the permutation test were introduced in Sections 4.11 and 4.12, in an example of testing a regression model and using the simplest possible type of permutation, i.e. completely random permutation.

The null model of the independence between the corresponding rows of the response data table and of the explanatory data table (the rows referring to the same set of cases in both tables) is the basis for the permutation test in Canoco. While the actual algorithm used in Canoco does not employ exactly the approach described here,[1] we use it to better illustrate the meaning of the permutation test.

- We start by randomly re-shuffling (permuting) the cases (rows) in the response data table, while keeping the explanatory data intact. Any combination of the response and explanatory data rows obtained in that way is as probable as the original data set, if the null hypothesis is true.
- For each of the data sets permuted in this manner, we calculate the constrained ordination model and express its quality in a way similar to that used when judging the quality of a regression model. In a regression model, we use an F-statistic, which is the ratio between the response variable variability explained by the regression model (divided by the number of the model parameters) and the residual (unexplained)

---

[1] In particular, there is no need to repeat all the computations needed for a full ordination to determine the pseudo-F statistic for each permutation, and the residuals (not original data table rows) are permuted when covariates are present (see Canoco 5 manual, section 3.6.4).

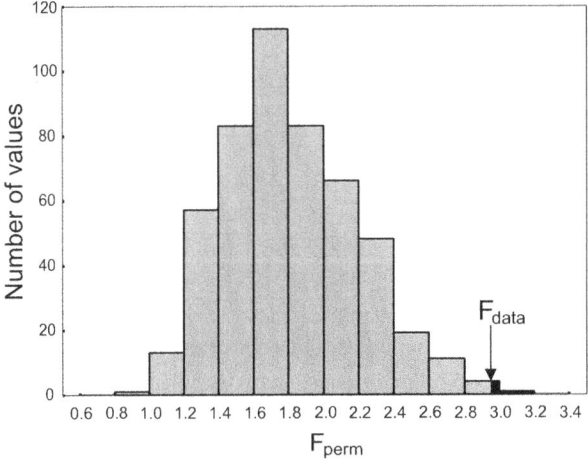

Figure 5–1 The distribution of the pseudo-F statistic values from a Monte Carlo permutation test compared with the statistic value of the original data sets. The black area corresponds to permutation-based values exceeding the pseudo-F statistic value based on the original data.

variability (divided by the number of residual degrees of freedom). In the case of constrained ordination methods, we use a similar statistic, called **pseudo-F statistic** and described in more detail in the following section.

- We record the value of such a statistic for each permuted (randomised) version of the data set. The distribution of these values defines the distribution of the test statistic under the null model (the histogram in Figure 5–1). If it is highly improbable that the test statistic value obtained from the original data (with no permutation of the rows) comes from that distribution (being much larger, i.e. corresponding to an ordination model of higher quality), we reject the null hypothesis. The probability that the 'data-derived' value of the test statistic originates from the calculated null model distribution then represents the probability of a Type I error, i.e. the probability of rejecting a correct null hypothesis.

- When performing a partial test (i.e. permutation test in constrained ordination with at least one covariate), we are testing the effects of explanatory variables on response data adjusted for the effects of the covariates. Instead of permuting the original response data, we permute the residuals of a constrained ordination of the response data with respect to the covariate data. Permuted residuals are then added to the fitted values of that ordination to form the new 'permuted' response data table (see Canoco 5 manual, section 3.6.4).

5.2 Pseudo-F statistics and significance

The previous section described the general principle of permutation tests and here we discuss the possible choices for the test statistics used in such tests. We mentioned that such a statistic resembles the F-statistic used in the parametric significance test of a regression model. But the choice of definition for this statistic in constrained ordination is difficult, because of the multidimensionality of the obtained solution. In general, the variability in response data, described by the explanatory variables, is expressed by more than one constrained (canonical) axis. The relative importance of the constrained axes decreases from the first up to the last constrained axis, but we can rarely ignore all the constrained axes beyond the first one. Therefore, we can either express the accounted variance using all constrained axes or focus on one constrained axis, typically the first one. This corresponds to the two test statistics available in Canoco since its 3.x version and the two corresponding permutation tests:

- Test of the sum of the canonical eigenvalues (test of all constrained axes), in which the overall effect of p explanatory variables, revealed on (up to) p constrained axes is tested. This test uses the following pseudo-F statistic:

$$F_{trace} = \frac{\sum_{i=1}^{p} \lambda_i / p}{RSS/(n - p - q)}$$

The *RSS* term in the formula represents the difference between the total variability in the response data and the sum of eigenvalues (λ_i) of all the canonical ordination axes (adjusted, again, for the variability explained by covariates, if there are any; the number of covariates is q). The number of independent explanatory variables (i.e. the total number of constrained axes) is p. The total number of ordination axes is n.

- Test of the first constrained axis uses a pseudo-F statistic defined in the following way:

$$F_1 = \frac{\lambda_1}{RSS_{-1}/(n - p - q)}$$

The variation explained by the first (constrained) axis is represented by its eigenvalue (λ_1). The residual sum of squares (RSS_{-1}) term corresponds to the difference between the total variation in the response data and the amount of variability explained by the first constrained axis (and also by the covariates, if any are present in the analysis).

As explained in the previous section, the value of one of the two test statistics, calculated from the original data, is compared with the distribution of the same statistic under the null model assumption (with the relation between the response data and the explanatory variables subjected to permutation). This is illustrated by Figure 5–1.

In this figure, we see a histogram approximating the shape of the distribution of the test statistic. The histogram was constructed from the pseudo-F values calculated using the permuted data.[2]

---

[2] The pseudo-F value for the unpermuted data set is also included.

The position of the vertical arrow marks the value calculated from the original (unpermuted) data.[3] The permutations where the corresponding pseudo-F statistic value is above this level represent the evidence in favour of not rejecting the null hypothesis and their relative frequency represents our estimate of the Type I error probability. The actual formula, which was introduced in Section 4.11, is repeated here:

$$P = \frac{n_x + 1}{N + 1}$$

where n_x is the number of the permutations yielding a pseudo-F statistic as large or larger than that from the real data and N is the total number of permutations. The value 1 is added to both numerator and denominator because (under the assumption of the null model) the pseudo-F statistic value calculated from the actually observed data is also considered to come from the null-model distribution and, therefore, to 'vote against' the null hypothesis rejection. The usual choices of the number of permutations (like 499, 999, or 9999) follow from this specific pattern of adding one to both numerator and denominator.

5.3 Testing individual constrained axes

If we use several independent explanatory variables in a constrained ordination, the analysis results in several constrained (canonical) ordination axes. In Canoco, it is easy to test the effect of the first (most important) constrained axis or the effect of the whole set of constrained axes. But we may be interested in reducing the set of constrained axes (the dimensionality of the canonical ordination space) and to do so, we need to find how many constrained axes effectively contribute to the explanation of the response variables (to the explanation of community variation, typically). We must therefore test the effects of individual constrained axes. Their independent effects do not differ from their additional (conditional) effects, of course, because they are mutually linearly independent by their definition.

We will demonstrate the simplest situation, when we have a constrained ordination (RDA or CCA) with no covariates. If we need to test the significance of a second constrained axis, we must use the case scores on the first axis as a covariate[4] and in this modified ordination we must test the significance of the new first constrained axis, which corresponds to the second axis of the original ordination model. Similarly, when testing the third axis, both the first and second axis of the original ordination model must be used as covariates. Legendre et al. (2011) confirmed the validity of this approach through simulations. Canoco 5 provides an analytical template that makes it very easy to test the significance of all individual constrained axes. Without any a priori covariates, we shall choose the *Test-constr-axes* template in the *Standard Analyses* folder in the

---

[3] When you perform a permutation test in Canoco 5, the *Summary* page presents a *Histogram* button that displays a graph, similar to the schematic diagram in Figure 5–1, but with F-values obtained from the performed permutations.

[4] The *CaseE* scores (scores defined as a linear combination of explanatory variables) must be used here.

last page of the New Analysis Wizard, while the *Test-constr-axes-covariates* template (located in the *Advanced Constrained Analyses* folder) tests the individual constrained axes for a model with a priori covariates.

When we find that some of the constrained axes are not significant, it sometimes makes sense to do more than just simply ignore them. This is because the axes following the constrained ones represent the variation not related to the effect of explanatory variables and we might want to inspect them and interpret their meaning (e.g. using the positions of response variables on these axes). But part of this residual variation is diluted within the non-significant constrained axes. If the testing reveals, say, that only the first out of three constrained axes is significant, we can make our constrained ordination model more parsimonious by letting Canoco extract just the first constrained axis and summarise the remaining variation without considering the explanatory variables. This is what the hybrid analyses do and you can set up their use in the *Ordination Options* page of the setup wizard by modifying the *Hybrid analysis* control value, specifying how many constrained axes should be extracted (one to three).

The above-suggested analysis templates for testing higher constrained axes cannot be used in a DCCA (detrended canonical correspondence analysis). If we use detrending by segments, the tests are only approximate and if we select detrending by polynomials the required algorithm is much more complicated. The way the tests of higher constrained axes are performed in DCCA is outlined in section 6.2.4.3 of the Canoco 5 manual.

5.4 Tests with spatial or temporal constraints

The permutation test described in Section 5.1 is justified when the collected set of cases does not have any implied internal structure, namely if the cases are sampled randomly, independently of each other. In this case, we can fully randomly re-shuffle the cases, because under the null model each case's explanatory data values can be matched with any other case's response data with equal probability.

This is no longer true when the 'relatedness' between different cases is not uniform across the whole data set. In such circumstances, we must **restrict** (adjust) the permutations across the set of cases to reflect the sampling/experimental design. This approach is, in fact, just one of the two approaches to permutation testing supported by Canoco and it is called **design-based permutations**. The other, called **model-based permutations**, maintains completely free permutations and the sampling/experimental design is reflected by a choice of covariates in the model. For some types of design, however, only one of these two families is available (or correct). Specifically, with a split-plot design we cannot use the model-based permutations when testing effects at the whole-plot level (unless we average over their split-plots) – see Section 5.5 and Canoco 5 manual (p. 348).

Canoco provides a rich toolbox of specific setups for the design-based permutation tests applied to data sets with a particular spatial, temporal or logical internal structure, which is related to the experimental or sampling design. The three main families of an internal data set structure as recognised by Canoco 5 (and briefly discussed in this and following sections) are represented by the three types of permutation restrictions in the

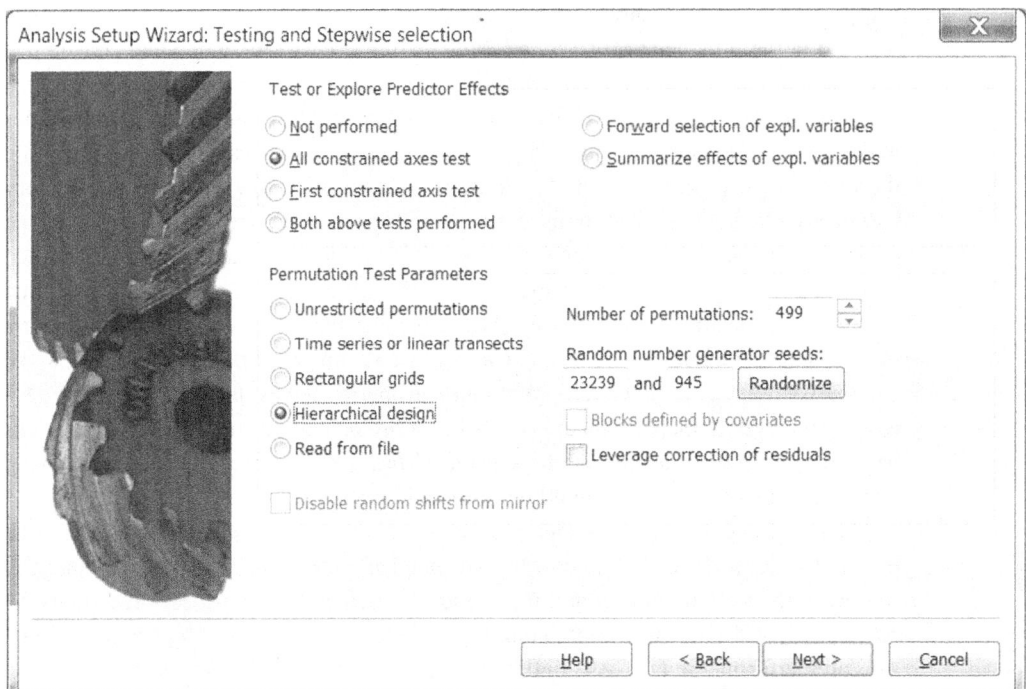

Figure 5–2 Introductory page of the Project Setup Wizard selecting the Monte Carlo permutation type.

lower left part of the Analysis Setup Wizard first page for permutation testing, illustrated in Figure 5–2. The Canoco 5 manual should be consulted for more details of this and other permutation restrictions (see Canoco 5 manual, sections 3.6.3 and 4.4.2.8).

The cases may be arranged along a linear transect through space or along a time axis, with a regular spacing (see Figure 5–3). With such arrangement, the cases cannot be permuted randomly, because we must assume an autocorrelation between the individual observations in both the response and the explanatory data. We should not disturb this correlation pattern during the test because our hypothesis concerns the relations between the response and explanatory data, not the relations within these data sets. To respect this autocorrelation structure, Canoco (formally) bends the sequence of cases in both the response and the explanatory data into a circular form and the permutation is performed by 'rotating' the explanatory data band with respect to the response data band.

The example permutations for a linear transect or time series shown in the lower right part of Figure 5–3 include two permutations in which the direction of the series is mirrored (*perm 4* and *perm 5*). Such permutations are not used when the *Disable random shifts from mirror* option is checked in the analysis setup. But the mirroring of sequences is appropriate for both spatial transects and temporal series (see Canoco 5 manual, p. 74).

When we want to **test for a trend** in community composition (or in the values of explanatory variables) along the transect (or through time), it is recommended to use completely random permutations, because the permutations shown in Figure 5–3 do not

5.4 Tests with spatial or temporal constraints

Plot arrangement in field

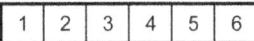

Generated data tables

Community composition (response variables)

Plot	Spc 1	Spc 2	Spc 3	...	Spc M
1	3	1		...	
2		2		...	3
3	1		1	...	
4		1		...	2
5	2		1	...	
6		2	1	...	4

Environmental parameters (explanatory variables)

Plot	EnvV 1	EnvV 2	...	EnvV P
1	3.5	low	...	651
2	2.7	low	...	430
3	1.4	high	...	257
4	1.2	high	...	381
5	4.7	med	...	209
6	0.8	med	...	580

Permutation settings in analysis

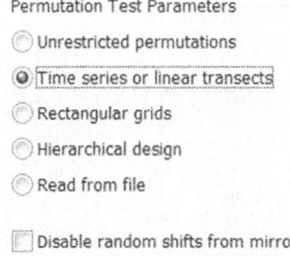

Examples of permuted plot order

in (residualized) response data

perm 1	perm 2	perm 3	perm 4	perm 5	
3	5	2	4	6	...
4	6	3	3	5	
5	1	4	2	4	
6	2	5	1	3	
1	3	6	6	2	
2	4	1	5	1	

Figure 5–3 Data from a linear transect or simple time series and the outline of design-based permutation scheme for testing the effect of explanatory variables. Please note that the number of cases is too small for a real test. The snapshot in the lower left corner shows the relevant choices in the *Testing and stepwise selection page* of the Analysis Setup Wizard.

remove such a trend – it is only broken into two pieces and/or reversed. But when we want to test for the effect of another explanatory variable, it is appropriate to first detrend the data, for example by using a covariate that codes for the position of individual plots (e.g. with values 1–6 in our simple example).

A similar spatial correlation occurs when we generalise the location of cases on a linear transect into the position of cases in space. Canoco supports only the placement of cases on a rectangular grid,[5] as illustrated in Figure 5–4, where a grid with four columns and three rows is illustrated. To maintain correct permutation of whole rows and whole columns, these two entities must be correctly distinguished in terms of their order within the data table, so the choice of '3 rows * 4 columns' is **not** equivalent to the '4 rows * 3 columns' one. A row consists of cases that are consecutive in the data table.

The example permutations of cases coming from a grid, shown in Figure 5–4, include two where a mirrored order of grid columns and/or rows is used (*perm 4* and *perm 5*). Permutation *perm 4* has both columns and rows mirrored, whereas permutation *perm 5*

---

[5] At least in terms of the restricted permutation test. The autocorrelation among irregularly positioned cases can be taken into account as part of the PCNM (principal coordinates of neighbour matrices) method, offered among the Canoco Adviser analysis templates (see Section 10.2).

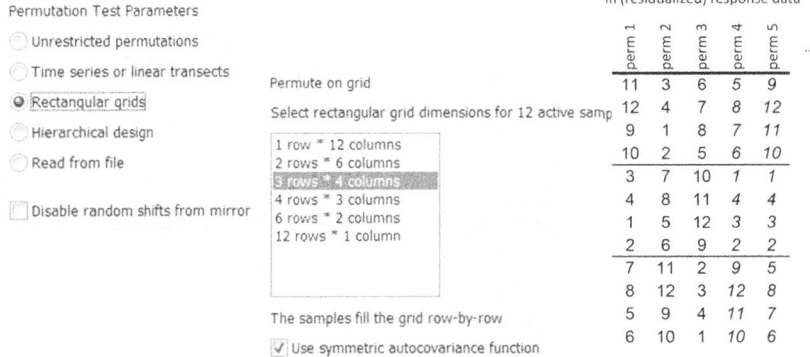

Figure 5–4 Data collected on a rectangular grid and the outline of design-based permutation scheme for testing the effect of explanatory variables. Please note that the number of cases is too small for a real test. The snapshot in the lower left area shows the relevant choices in the *Testing and stepwise selection* page of the Analysis Setup Wizard, followed by a snapshot from the specialised *Permutation Grid Size* page.

has mirrored columns but not rows. Permutations like *perm 5* with mirrored columns but not rows (or vice versa) are not considered when we check the *Use symmetric autocovariance function* option in the *Permutation Grid Size* page. Neither permutation *perm 5* nor permutation *perm 4* is used if the *Disable random shifts from mirror* option is checked (see Canoco 5 manual, p. 74) as it disables all mirroring.

Similarly to the permutation restrictions along a transect or time series, we should remove a suspected trend across the grid using covariates, as this permutation scheme assumes spatial stationarity.[6]

---

[6] Spatial stationarity means (in simple terms) that the statistical properties of the response variable(s) do not change depending on the location: this concerns mean value, but also the variation or cross-correlation among observations.

The most general model of the internal structure for data sets is provided by the **split-plot design**, referred to as *Hierarchical design* in the wizard page illustrated in Figure 5–2. This type of permutation restriction is described in more detail in the following section and examples of its use are shown in Chapters 16 and 17.

All these restricted permutation schemes can be further nested within another level representing blocks. Blocks can be defined in the analysis using factor covariates and they represent groups of cases that are similar to each other more than to the cases from the other blocks. To account in the permutation test for variation due to blocks, check the *Blocks defined by covariates* option in the page illustrated in Figure 5–2. The cases are then permuted only within those blocks, never across the blocks. If we compare a constrained ordination model with a model of the analysis of variance, the blocks can be seen as a random factor with its effect not being interesting for interpretation purposes.

5.5 Tests with hierarchical constraints

The hierarchical (split-plot) design restriction allows us to describe a structure with two levels of variability (with two 'error levels') – see Sections 3.5.2 and 3.6.[7] The higher level of the split-plot design is represented by so-called **whole–plots**. Each of the whole-plots contains **split-plots**, which represent the lower level of the design (see also Figure 3–5). Canoco provides great flexibility in permuting cases at the whole-plot and/or the split-plot levels, ranging from no permutation, through spatially or temporally restricted permutations up to a free exchangeability at both levels. In addition, Canoco allows for the notion of dependence of the split-plot structure across the individual whole-plots. In this case, the particular permutations of the split-plots within the whole-plots are identical across the whole-plots.

We introduce this permutation scheme using a simple split-plot design previously illustrated in Chapter 3 (Figure 3–5). It shows the design of a hypothetical field experiment: effects of bedrock combined with nutrient addition were studied using four large plots (whole-plots), two of them on limestone and two on granite. Within each of the large plots, three smaller subplots (split-plots) were established, differing in the applied nutrient: none (C), nitrogen (N) or phosphorus (P).

In Figure 5–5 we can see that the permutation options in the *Split-plot Arrangement Permutation* page must be set differently when testing the effect of *Nutrient* (the triples of split-plots must be permuted within each whole-plot) and when testing the effect of *Bedrock* (the triples of split-plots are held together and permuted across the four whole-plots). The upper part of the same page is identical for both permutation settings, however, as it specifies the assignment of table rows (split-plots) to individual whole-plots. The values of the two controls (*TAKE* and *SKIP NEXT*) describe the way Canoco should progress through the rows of the data table to find out the cases that form together

---

[7] Another level can be added, in some cases, using permutation within blocks defined by covariates (see the previous section).

Permutation tests and variation partitioning

Generated data tables

Community composition (response variables)

Plot	Spc 1	Spc 2	Spc 3	...	Spc M
1	3		1	...	
2		2		...	3
3	1		1	...	
4		1		...	2
5	2		1	...	
6		2	1	...	4
7	2		1	...	
8	4	1		...	
9	2		2	...	1
10	1			...	
11	4		2	...	5
12	5	1		...	5

Experimental treatments (explanatory variables)

Plot	Nutrient	Bedrock
1	P	limestone
2	N	limestone
3	C	limestone
4	N	granite
5	C	granite
6	P	granite
7	P	limestone
8	C	limestone
9	N	limestone
10	N	granite
11	P	granite
12	C	granite

Examples of permuted plot order in (residualized) response data

Nutrient test

perm 1	perm 2	perm 3
1	3	2
3	1	1
2	2	3
5	6	4
6	5	6
4	4	5
9	8	7
8	9	8
7	7	9
10	10	10
12	11	12
11	12	11

Bedrock test

perm 1	perm 2	perm 3
4	1	4
5	2	5
6	3	6
1	7	10
2	8	11
3	9	12
7	4	7
8	5	8
9	6	9
10	10	1
11	11	2
12	12	3

Permutation settings in analysis

Permutation Test Parameters
- ○ Unrestricted permutations
- ○ Time series or linear transects
- ○ Rectangular grids
- ● Hierarchical design
- ○ Read from file

Split-plot Layout

Number of split-plots in each whole plot: 3

Split-plots forming a whole-plot are collected by this rule:
TAKE 3 plots, then SKIP NEXT 0

Nutrient test

How to permute whole-plots and split-plots

Whole-plot Permutations	Split-plot Permutations
● No permutation	○ No permutation
○ Time series or linear transect	○ Time series or linear transect
○ Spatial grid	○ Spatial grid
○ Freely exchangeable	● Freely exchangeable
☐ Disable shifts from mirror image	☐ Disable shifts from mirror image
	● Independent across whole-plots
	○ Dependent across whole-plots

Bedrock test

How to permute whole-plots and split-plots

Whole-plot Permutations	Split-plot Permutations
○ No permutation	● No permutation
○ Time series or linear transect	○ Time series or linear transect
○ Spatial grid	○ Spatial grid
● Freely exchangeable	○ Freely exchangeable
☐ Disable shifts from mirror image	☐ Disable shifts from mirror image
	● Independent across whole-plots
	○ Dependent across whole-plots

Figure 5–5 Data recorded in a split-plot design experiment (see Figure 3–5) and suggested setup of design-based permutation test. Snapshot in the lower left area shows relevant choices in the *Testing and stepwise selection* page of the setup wizard; snapshots in the lower right area show alternative settings in the *Split-plot Arrangement Permutation* page: upper part is the test for a split-plot level explanatory variable (here Nutrient); lower part has the same settings in the Split-plot Layout area (not shown) and the visible ones correspond to a test of a whole-plot level explanatory variable (Bedrock).

one whole-plot. This recipe (taking n_1 cases as whole-plot members and then skipping another n_2) is repeated until the total number of split-plots given above is reached. Therefore the 'TAKE 3, SKIP NEXT 0' settings specify that the three split-plots of each whole-plot are contiguously placed in the table.

When testing the effect of a whole-plot predictor (such as the Bedrock in Figure 5-5), the above-suggested permutation test cannot be performed for whole-plots with one or more split-plots missing (e.g. when one of the subplots was destroyed or its record lost). In such case, an alternative test must be performed, using averages or sums[8] of split-plots within each whole-plot (i.e. generating, for our example, new data tables with just four 'synthetic' cases) and then permuting them completely at random (as the whole-plots were permuted in the original data set).[9]

There is a price we pay for using restricted permutations to achieve correct Type I error estimates when our observations are not fully independent: we pay by a lower power of the test. For our example in Figure 5-5, we cannot obtain P-value much lower than $2*0.042$,[10] because we do not permute the 12 plots randomly here, but only 4 whole-plots (localities at different bedrock), so there are only 24 possible permutations (see the end of Section 4.12 for additional comments).

The permutation types supporting the split-plot design in Canoco are also important for evaluating data where community composition at sampling plots is repeatedly recorded at multiple occasions (corresponding to so-called longitudinal or repeated measurement data sets). A simple case of such a sampling arrangement is illustrated in Figure 5-6.

As before, the settings are different when we test the whole-plot factor (here the *Mowing* factor) fixed for each permanent plot (whole-plot) or when the split-plot factor (here the sampling year) is tested. Note that our example codes the *Year* as a factor with three levels and such coding supports any type of changes in the response variables (e.g. community composition) with the time. If we assume, however, a directed change of the response data with time, using a quantitative time variable might be more appropriate and leads, under such circumstances, to a more parsimonious model (with a smaller number of model degrees of freedom). See Chapter 16 for additional discussion.

There is one important difference between the split-plot level test as configured in Figure 5-5 and the split-plot level test from Figure 5-6, namely the choice of *Independent across whole-plots* in the former and *Dependent across whole-plots* in the latter. The *Dependent* choice is appropriate for repeated measures because years in different whole-plots may share unobserved effects or noise (e.g. the second year could be wetter than the

---

[8] With averages preferred for linear ordination and sums preferred for unimodal ordination. In the latter, the sum of case values represents the case weight, so it is desirable that an aggregated whole-plot based on a larger number of split-plots should have a higher weight.

[9] Canoco 5 makes the creation of a new project with averaged cases easy with the *Project | Create derived project | Aggregate cases* menu command. You just need to have – in your original project – a factor variable that identifies the membership of individual cases within whole plots and this variable is then used to perform aggregation.

[10] This P-value is, however, achievable only when the tested predictor has a unique value for each whole-plot. In our example with the two levels of the *Bedrock* factor repeated in two whole-plots the minimum P value one can achieve will be much higher, namely 0.333.

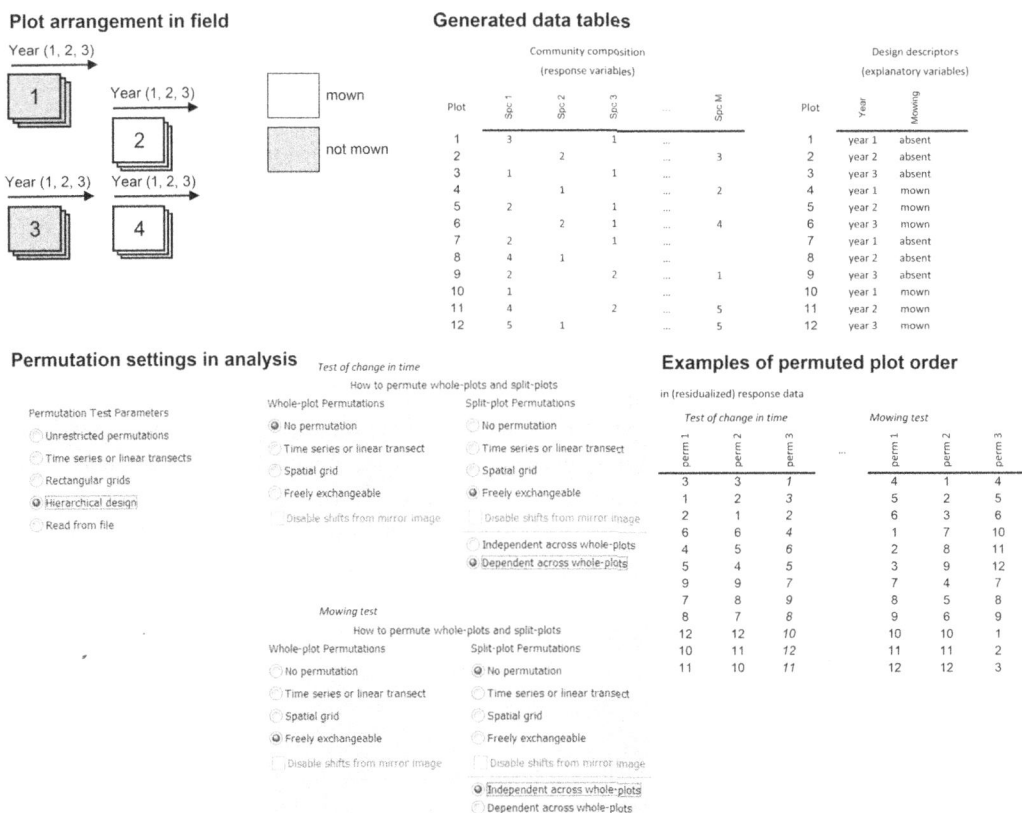

Figure 5–6 Data collected from an experiment with repeated measurements. Each of the four plots (two mown and two unmown) were measured for three years and the three records (cases) coming from the same permanent plot are contiguously placed in the data table. Please note that the number of plots is too small for a real test.

other two) and the dependent permutations preserve these this sharing. In the split-plot design referred to in Figure 5–5, however, the C plot in one area has nothing particular in common with the C plot in another area, except the type of treatment.

Very often, repeated measurements start with the onset of experimental treatment – e.g. in our example, all the plots may have been mown up to the start of experiment and only then were two plots abandoned. In such case, the effects of our experimental treatment develop gradually and consequently we are not so much interested in the main treatment effects, but more in its interaction with time. Under such circumstances, the choice of whether to code the time as a factor or as a quantitative variable is even more important. When coding time as a numeric variable and using it in the interaction term with treatment factor for a data set with baseline measurement (i.e. before the treatments were applied), it is important that the base line year has value 0 (see Section 16.3 for more details).

When time is coded numerically, it will model a linear change in response data, implying a constant pace of change. This might not be appropriate e.g. when we follow a succession of biotic communities, which is known to proceed more quickly at its onset than in the later stages. Such a process might be therefore better modelled with the time variable log-transformed in the analysis.[11]

Chapter 16 provides an extensive case study analysing a repeated-measures field experiment using a constrained ordination with split-plot based permutation restrictions.

It is perhaps interesting to note that the restricted permutations adjust the hypothesis testing with the same task as adopted by mixed-effect models (see Section 8.6), where for example the data set illustrated in Figure 5–6 would be analysed (if it had only one response variable) with fixed-effect factor of Mowing and random-effect factor identifying whole-plot membership. The inclusion of the random effect changes the value of the F statistic for the test of Mowing. With restricted permutations in Canoco, however, the value of the (pseudo-)F statistic does not change, only the reference population of pseudo-F values, generated under the null model (by the permutations), does and consequently the estimated Type I error (model significance) also changes.[12]

5.6 Simple versus conditional effects and stepwise selection

Stepwise selection of explanatory variables represents a procedure by which you find a minimum adequate subset of the available predictor variables. The selection procedure can either start from a model using all predictors and gradually simplify it by removing least 'suitable' predictors (backward elimination), or it can start from an empty ('null') model with no predictors and gradually extend it by adding most 'suitable' predictors (**forward selection**). We find the use of stepwise selection (when performed correctly) useful in the context of exploratory (observational) studies where the search for a subset of important predictors is one of the study tasks. On the other hand, the data from confirmatory (manipulative, experimental) studies are usually not appropriate for stepwise selection.

Forward selection of explanatory variables implemented in the Canoco program uses partial Monte Carlo permutation tests to assess the usefulness of each potential predictor for extending the subset of explanatory variables used in the constrained ordination model.

If we select the *Forward selection of expl. variables* option seen in the upper right corner of the setup wizard page (Figure 5–2), Canoco shows an interactive dialog during the analysis (Figure 5–7).

Figure 5–7 illustrates the state of the forward selection procedure after the two best explanatory variables (Moisture and Manure) were selected (they are displayed in the lower part of the window). Information in the middle part of the window (under the *Term Contribution* heading) shows that the two selected variables account for approximately 61 percent of the total variability explained by all the explanatory variables

---

[11] And when using the interaction of time with experimental treatment, make sure that the log-transformed value of the base-line time is zero, as explained in the preceding paragraph.

[12] Given the fact the *Generalized-linear-model* analysis template in Canoco 5 supports permutation tests for fitted linear (or generalized linear) models, we can see that template as implementing functionality corresponding to (generalized) linear mixed-effect model.

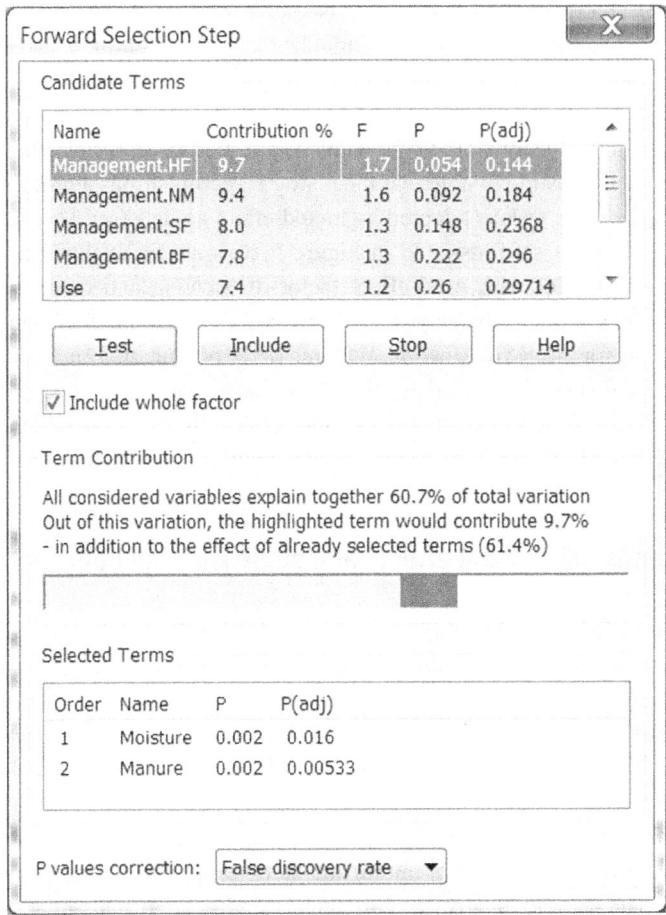

Figure 5–7 Dialog box for the forward selection of explanatory variables.

(i.e. explained when all the variables in the top list are also included in the ordination model) and that the inclusion of the currently selected candidate term (the *HF* level of the *Management* factor) would add another ca 10 percent to the explained variability. The same information is also presented graphically by the three-coloured horizontal bar.[13]

The list of variables in the upper part of the window shows the remaining 'candidate predictors' ordered by the decreasing contribution that the variable would provide when added to the set of variables already selected. We can also see that the forward selection treats individual dummy variables (automatically generated for factor variables by Canoco, see near the end of Section 2.2) representing factor levels as separate entities. The forward selection window offers, however, an option to include all factor levels automatically, after one level was selected (*Include whole factor*).

---

[13] Using cyan colour for the variation explained by already selected terms and red colour for the variation explained by the currently selected candidate term.

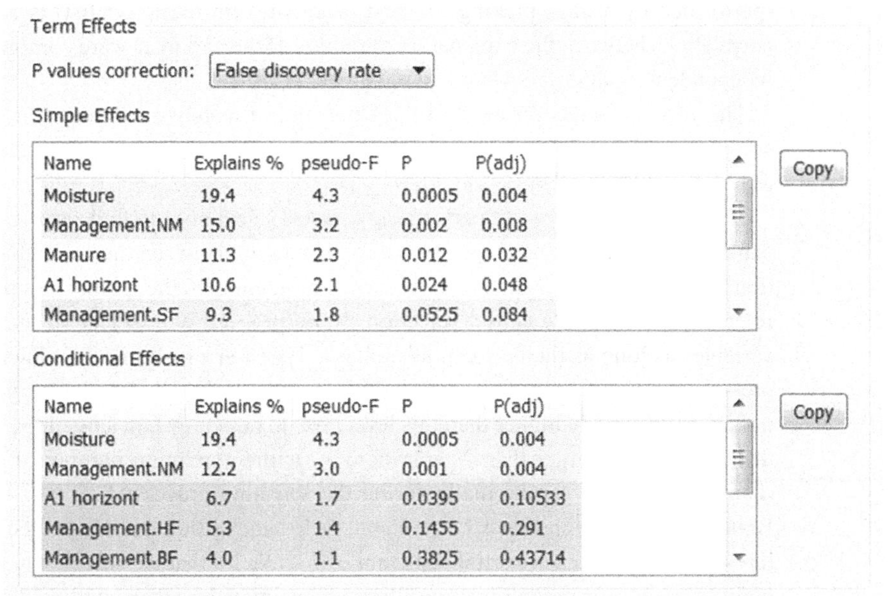

5.6 Simple and conditional effects and stepwise selection

Term Effects

P values correction: False discovery rate ▼

Simple Effects

Name	Explains %	pseudo-F	P	P(adj)
Moisture	19.4	4.3	0.0005	0.004
Management.NM	15.0	3.2	0.002	0.008
Manure	11.3	2.3	0.012	0.032
A1 horizont	10.6	2.1	0.024	0.048
Management.SF	9.3	1.8	0.0525	0.084

Conditional Effects

Name	Explains %	pseudo-F	P	P(adj)
Moisture	19.4	4.3	0.0005	0.004
Management.NM	12.2	3.0	0.001	0.004
A1 horizont	6.7	1.7	0.0395	0.10533
Management.HF	5.3	1.4	0.1455	0.291
Management.BF	4.0	1.1	0.3825	0.43714

Figure 5–8 Term Effects of predictors in a constrained ordination: output of the *Summarize effects of expl. variables* option.

To judge whether the increase in explained variation due to selecting a candidate predictor is larger than a random contribution, we can use a **partial Monte Carlo permutation test**. In this test, the candidate variable is used as the only explanatory variable (resulting in an ordination model with just one constrained axis) and variables already selected are used as covariates. If the null hypothesis is rejected for this partial test, we can include that variable in the subset. Canoco 5 precomputes the tests for all (visible) candidate explanatory variables and shows the result in the *F* and *P* columns in the upper part of the dialog.

The effect of the variable tested in such a context is called its **conditional** (or partial) **effect** and its size and significance depend on the variables already selected. But at the start of the forward selection process, when no explanatory variable has yet entered the selected subset, we can test each variable separately, to estimate its independent, **simple** (marginal) **effect**. This is the amount of variability in the response data that would be explained by a constrained ordination model using that variable as the only explanatory variable.

We can easily compare the simple and conditional effects of explanatory variables by choosing, in the Analysis Setup Wizard page illustrated in Figure 5–2, the *Summarize effects of expl. variables* option instead of the *Forward selection of expl. variables* one. As a result, new information is shown in the *Summary* page of the analysis notebook (see Figure 5–8).

The discrepancy between the order of variables sorted by their simple effects (in the upper list in Figure 5–8) and the order corresponding to a 'blind' forward selection (performed by always picking the best candidate) (in the lower list) is caused by the correlations between the explanatory variables. If the variables were completely linearly independent, both orders would be identical.

The forward selection method does not, unfortunately, guarantee that the selected subset of predictors is the best one among all possible subsets of the same size (Legendre & Legendre 2012, p. 562).

If the primary purpose of performing forward selection is to find a sufficient subset of explanatory variables that represents e.g. the relation between the community composition and environmental data, then we have a problem with the 'global' significance level referring to the whole subset selection. If we proceed with selecting the explanatory variables as long as the best candidate has a Type I error estimate (P) lower than some pre-selected significance level α (say 0.05), then the 'global' Type I error probability is, in fact, considerably higher than this level. We do not know how large it is, but we know that it cannot be larger than $N_c \cdot \alpha$, where N_c is the maximum number of distinct tests (comparisons) that can be made during the selection process. This is the basis of the Bonferroni correction (Rice 1989; Cabin & Mitchell 2000). It adjusts the significance threshold levels in each partial test from α to α/N_c, so that we select only the variables with a Type I error probability (significance) estimate less than α/N_c.

Use of the Bonferroni correction is a controversial issue: some statisticians do not accept the idea of 'pooling' the risk of multiple statistical tests and in addition, the Bonferroni correction is 'overacting': the resulting uncertainty about the selected subset of explanatory variables will be never larger than α, hence it will be typically smaller. So we advise the reader to use the other methods of correcting the significance levels in partial tests; some of which are reviewed in Wright (1992). They are slightly more complicated than Bonferroni correction, but result in more powerful tests. Some of them can be applied during the selection procedure in Canoco 5, with the help of the option available at the bottom of the forward selection dialog. Our example snapshots (Figures 5-7 and 5-8), for example, use the transformation of the Type I errors into false discovery rate (FDR) estimates. Significance adjustment methods available in Canoco 5 are reviewed in Section 5.8 of this chapter.

We now provide a simple explanation of the problem of selecting a subset of 'good' explanatory variables from a larger pool: the probability that a variable will be a significant predictor (e.g. of the species composition) purely due to chance equals our pre-specified significance level. If we have, in a constrained ordination, 20 explanatory variables available and select them with the α threshold equal to 0.05, we can expect that on average one of them (20·0.05) will be judged significant even if the response data are not related to any of them. If we select a subset of the best predictors, drop the other variables, and then test the resulting model, we will very probably get a significant relationship, regardless of whether the explanatory variables explain the species composition or not.

Let us illustrate the situation with a simple example. For real data on species composition of plots on an elevation gradient (Lepš et al. 1985, file *tatry.xlsm*, the *tatry* sheet), we have generated 50 random variables with a uniform distribution of values between zero and a random number between 0 and 100 (sheet *tatrand*).[14] You can also check an example Canoco project, *TatryRandom.c5p*. We suggest that you use those 50 variables as the explanatory data. Use CCA (canonical correspondence analysis) with the default settings and the forward selection. During the forward selection, use the 'standard' procedure, i.e. always select the best variable, and stop the forward selection if the best variable has the significance level estimate P > 0.05 (not using any significance adjustment). In this way, three variables are selected (*var20*, *var31*, and *var47*) – see example analysis *Constrained-forward-selection*. If you now test the resulting model (with the three explanatory variables) in a separate constrained ordination (analysis *Constrained-global-test*), the global permutation test estimates the Type I error probability P=0.001 (with 999 permutations). Clearly, the forward selection (or any stepwise procedure) is a powerful tool for model building, but when you need to test the significance of the built model, use of an independent data set is necessary. If your data set is large enough, then the best solution is to split the data into a part used for model building, and a part used for model testing (Hallgren et al. 1999).

Additional discussion of the above-mentioned intricacies of forward selection can be found in Blanchet et al. (2008). The authors also suggest two solutions protecting us (partially) against the selection of non-significant predictors: (a) before we perform forward selection, constrained ordination using all candidate variables should be performed and its constrained axes tested for significance; and (b) we should restrict the subset of selected predictors to the size that does not lead to an adjusted coefficient of determination (R^2_{adj}) greater than the same statistic for the constrained ordination of the point (a), i.e. with all candidate predictors included. Both suggestions[15] are implemented in two specialised analysis templates in Canoco 5: *Interactive-forward-selection* and (for constrained analyses with a priori covariates) *Interactive-forward-selection-covariates*. The false discovery rate approach (discussed above, see also Section 5.8) is an alternative way of guarding against the selection of too many explanatory variables.

Another difficulty that we might encounter during forward selection occurs when we use one or more factors (coded internally as dummy variables by Canoco, see above) as explanatory variables. The forward selection procedure treats each dummy variable as an independent predictor, so we cannot evaluate the contribution of the whole factor at once. This is primarily because the whole factor contributes more than one degree of freedom to the constrained ordination model. In a regression model, a factor with K levels contributes $K - 1$ degrees of freedom. In a constrained ordination, $K - 1$ constrained axes are needed to represent the contribution of such a factor.

---

[14] The macro used to generate the values of explanatory variables is included in the same Excel file so that you can check how the variables were generated.

[15] These suggestions are not always usable, namely when you start with a pool of explanatory variables larger than the number of cases in your data, such as the example in the preceding paragraph: try to summarise your predictors with a PCA (with both centring and standardisation of variables) and use its case scores (on first few axes) instead.

While this independent treatment of individual factor levels can make interpretation difficult, it also provides an opportunity to evaluate the extent of differences between the individual case classes defined by such a factor. The first selected class is apparently most different from all others, the second selected class is most different from the first one and the others, etc. The outcome is partly analogous to the multiple comparison procedure in analysis of variance.[16] Alternatively, we can check an option in the selection dialog box so that when one of the factor levels is significantly different from the others (and so we select it), the other factor levels are selected as well.

5.7 Variation partitioning

In the previous section, we explained the difference between the conditional and simple effects of individual explanatory variables upon the response data. We also stated that the discrepancy in the importance of explanatory variables, as judged by their simple effects and their conditional effects, is caused by the correlations between those explanatory variables. Any two explanatory variables that are correlated share part of their effect exercised[17] upon the response data. The amount of the explanatory power shared by a pair of explanatory variables (A and B, say) is equal to the difference between variable A's simple effect and its conditional effect evaluated in addition to the effect of variable B.[18]

This concept forms the basis of the **variation partitioning**[19] procedure (Borcard, Legendre & Drapeau 1992; Legendre 2007). But in this procedure, we usually do not analyse the effects of just two explanatory variables: rather, we attempt to quantify the effects (and their overlap) of two or more **groups** of explanatory variables representing some distinct, interpretable phenomena. A separation of the variability of community composition due to environmental conditions and due to human impact might serve as a good example, because the overlap of these two effects is expected as the human impact modifies the environmental conditions as well.

We will describe the variation partitioning procedure using the simplest example with two groups of explanatory variables (X_1 and X_2). Each group contains one or multiple explanatory variables. The diagram in Figure 5–9 presents the breakdown of the total variability in the response data according to those two groups of variables.

The whole rectangle in Figure 5–9 represents the total variation in the response data, which is a classical variance in the case of linear ordination methods and so-called **inertia** in unimodal ordination. The area marked as D corresponds to the residual variability,

---

[16] But here we compare the partial effect of a particular factor level with a pooled effect of the not-yet-selected factor levels.
[17] 'exercised' in a statistical, not necessarily causal, sense.
[18] The relation is symmetrical – we can label any of the two considered variables as A and the other as B.
[19] It was called **variance** partitioning in the original paper, but we prefer, together with Legendre and Legendre (2012), the more appropriate name referring to *variation*, as we include also the unimodal ordination methods in our considerations.

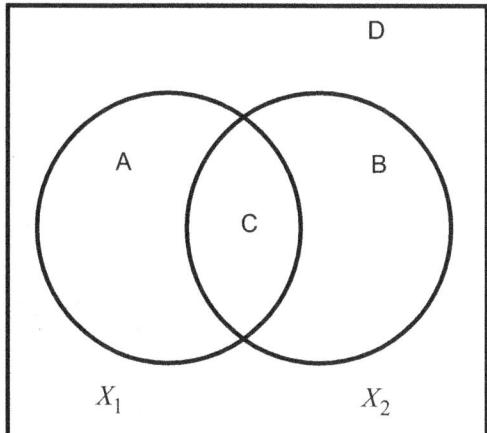

Figure 5-9 Partitioning of the total variation in the response data into the contributions of two subsets of explanatory variables (*A*, *B*, and shared portion *C*) and the residual variation (*D*).

i.e. the variability not explained by the ordination model which includes the explanatory variables from both the X_1 and the X_2 groups. Fraction *A* represents the partial effect of the variables from group X_1, similarly fraction *B* represents the partial effect of the variables in group X_2. The effect shared by both groups is the fraction *C*.[20] The amount of variability explained by group X_1, when ignoring variables from group X_2, is equal to *A+C*. We estimate the individual fractions *A*, *B*, and *C* using constrained ordinations, which might be partial or without covariates. We illustrate the partitioning procedure using both approaches in the following paragraph.

We estimate *A* from the analysis where variables from X_1 are used as explanatory variables and the variables from X_2 as covariates. Similarly, we estimate *B* as the sum of the eigenvalues of constrained axes from the analysis where X_2 act as explanatory variables and X_1 as covariates. We then calculate the size of *C* by subtracting the sum of *A* and *B* from the amount of variability explained by an ordination model with the variables from both X_1 and X_2 acting as explanatory variables. Instead of calculating the fractions *A* to *C* using partial analyses, as suggested above, we can alternatively compare the variation explained in analysis using both X_1 and X_2 groups (and so representing *A+B+C* fractions) with analyses using only the X_1 variables or only the X_2 variables (representing, respectively, *A+C* and *B+C* fractions).

When we define a group of explanatory variables (such as X_1 and X_2 in our example above), we can either choose them a priori or select them from a larger subset using forward selection. The pre-selection of each group member can be also used to make the size (number of variables) of each group as similar as possible. This is important

---

[20] Please note that the *C* does **not** represent an 'interaction' between X_1 and X_2. In fact, when each of the two compared groups represents one factor and the data come from a balanced factorial experiment, the size of *C* will be zero, because the two factors will be uncorrelated, while their interaction might be quite large and significant.

Table 5–1 Overview of Canoco 5 analysis templates for a general variation partitioning procedure with two or three variable groups

Number of groups	Variation partitioning computed with		
	Partial analyses	Non-partial analyses	Non-partial analyses with forward selection
Two	Var-part-2groups-Conditional-effects-tested	Var-part-2groups-Simple-effects-tested	Var-part-2groups-Simple-effects-tested-FS
Three	Var-part-3groups-Conditional-effects-tested	Var-part-3groups-Simple-effects-tested	Var-part-3groups-Simple-effects-tested-FS

because the amount of variation explained by a group of variables (even if none of them has a real effect) can be expected to grow linearly with the group size.[21]

Occasionally, a shared variation fraction (like *C* in our example) might have a negative value, indicating that the joint explanatory effect of the two groups of variables is stronger than the sum of their marginal effects. See Legendre and Legendre (2012, p. 573) for a more detailed discussion of this issue.

In Canoco 5, you do not need to run separate analyses to calculate variation partitioning for your data. The program offers instead a multitude of analysis templates that perform the routine work for you, including the calculation of variation partitioning fractions and set up of appropriate ordination diagrams to illustrate the various fractions of explained variation. The names and functionality of these templates are differentiated based on the following criteria (see Table 5–1): number of predictor groups (two or three), use of partial or non-partial analyses, and (for non-partial variant only) the preselection of predictors using forward selection.

Additionally, Canoco offers three specialised analysis templates (with their names starting with *Var-part-PCNM*) for the variation partitioning with two variable groups, where one represents PCNM (recently called db-MEM) spatial eigenfunctions. Spatial eigenfunctions allow us to model spatial patterns in the data (see Section 10.2 and the case study in Chapter 19). They not only represent the spatial variation, but their use as covariates also takes out the spatial correlation in the response data and makes the significance tests valid. In this way, the approach adopted in the PCNM method is similar to the method of phylogenetic correction, proposed by Desdevises et al. (2003) and illustrated in section 6.3.7 of the Canoco 5 manual.

In each of the analysis templates available for variation partitioning, Canoco 5 setup wizard also allows, for calculating the variation fractions, the use of **adjusted explained variation** percentages,[22] corresponding to adjusted R^2 computed in linear regression and recommended for variation partitioning by Legendre (2007) and Peres-Neto et al. (2006).

---

[21] One possible strategy would be to select a subset of variables independently for each group and then unify the number of variables across all groups using the size of smallest group after selection, say *K*, retaining always the *K* variables with highest conditional effect during stepwise selection.

[22] This option is offered on a separate *Variation Partitioning* page of the setup wizard, as a single *Adjust explained variation* option.

The reason for its preference lies in the fact that the standardly computed percentage of explained variation does not represent an unbiased estimate of the sampled statistical population value. It is typically overestimated – the more so the closer the number of model parameters (number of explanatory variables and covariates in a constrained ordination) is to the number of independent observations (cases).

5.8 Significance adjustment for multiple tests

When we select a subset of explanatory variables to be used in a model (of multiple regression or constrained ordination – see Section 5.6) or when we compare effects of individual predictors, the commonly adopted approach is that the null hypothesis is rejected (i.e. the predictor is claimed to have an effect) when $P \leq 0.05$, or another a priori selected threshold (α). But this is valid only for individual tests. The Type I error estimate (P) is inflated when we interpret the results at the level of a family of tests, e.g. for enumerating significant predictors or interpreting the selected explanatory variables in stepwise selection as a group of significant predictors (see Section 5.6).

This problem of Type I error inflation has been recognised as an important issue in statistical modelling for rather a long time and blind application of variable selection approaches is either discouraged entirely or family-wise corrections of P-values estimated for individual tests are recommended. Canoco 5 offers in relevant situations several widely used corrections. Among them, the Bonferroni correction is provided (in our perspective) just for educational purposes; it likely overestimates the real Type I error probabilities and leads to a large loss of power and a corresponding increase in Type II error (Nakagawa 2004).

For real-world tasks, however, Canoco 5 offers two other correction procedures. The first one is a traditional, powerful Type I error adjustment for a family of dependent tests, known as Holm's correction (Holm 1979, see also Legendre & Legendre 2012, section 1.2). Canoco optionally presents the corrected values alongside the ones originally estimated during permutation test for a particular predictor. The other approach focuses on limiting the frequency of false (i.e. incorrect) discoveries among all discoveries (i.e. the rejections of null hypothesis), known as the *false discovery rate* (FDR, Verhoeven et al. 2005). This approach adjusts the P-values in a way that limits the false discovery rate for a whole family of tests to a specified threshold (such as 0.05). When calculating FDR estimates, Canoco 5 offers the approach of Benjamini & Hochberg (1995).

But some statisticians believe that the adjustment of Type I error estimates is not necessary in protected tests, for example when stepwise selection of explanatory variables is only carried out if a global permutation test, with all candidate explanatory variables, is significant. Such a protected test (see also Section 5–6 above) is performed in Canoco 5 when we use one of the *Interactive-forward-selection* analysis templates.

6 Similarity measures and distance-based methods

In many multivariate methods, one of the first steps is to calculate a matrix of similarities (resemblance measures) either between the cases or between the variables. Although this step is not explicitly done in all the methods, in fact each of the multivariate methods works (even if implicitly) with some similarity measure. The linear ordination methods can be related to several variants of Euclidean distance, while the unimodal (weighted averaging) ordination methods can be related to chi-square distances. The resemblance functions are reviewed in many texts (e.g. Orloci 1978; Gower & Legendre 1986; Ludwig & Reynolds 1988; Legendre & Legendre 2012), so here we will introduce only the most important ones.

In this chapter, we will use the following notation: we have n cases (e.g. relevés), containing m response variables (e.g. species). Y_{ik} represents the value (abundance) of the k-th variable ($k = 1, 2, \ldots, m$) in the i-th case ($i = 1, 2, \ldots, n$). In this chapter, we will also refer to (response) variables of the analysed data table as species, to reflect the common nature of data and increase the readability of the text.

It is technically possible to transpose the data matrix (or exchange i and k in the formulae), thus any of the resemblance functions can be calculated equally well for rows or columns (i.e. for cases or species). Nevertheless, there are generally different kinds of resemblance functions suitable for expressing the (dis)similarity among cases, and those suitable for describing similarity among the species.[1] This is because the species set is usually a fixed entity: if we study e.g. vascular plants, then all the vascular plants in any plot are recorded – or at least we believe we have recorded them. But the set of cases is usually just a (random) selection, something which is not fixed. Accordingly, the similarity of (the vascular flora of) two cases is a meaningful characteristic, regardless of the set of cases with which we are working.[2]

On the other hand, the similarity of two species has a meaning only within a particular data set. As an example, imagine that we study the relationship between two plant species – *Oxycoccus quadripetalus* and *Carex magellanica* – using 1 m² quadrats. Very

---

[1] By **similarity of cases**, we mean the similarity in their values (e.g. species composition of plots). By **similarity of species** we mean the similarity of their distribution among the cases, which presumably reflects the similarity of the ecological behaviour of species (those are often called species association measures).

[2] When the set of cases is extended, new species might appear in the whole data set, but all of them are absent from the cases already present in the original set.

likely, these two species will exhibit a high distributional similarity within a landscape, because they are found together on peat bogs and their common occurrence will be more frequent than expected by chance. But they will have a low similarity **within** a peat bog, because *Oxycoccus* is found on hummocks and *Carex* in hollows and, consequently, they are found in common less frequently than expected by chance.

The set of analysed cases is considered a random selection of all possible cases in the studied area, i.e. a sample of a population of sampling units in classical statistical terminology. The population of sampling units is defined by the purpose of a study: the result will be different if we ask whether two species are positively or negatively *associated* (i.e. found more or less frequently than expected by chance) within a large landscape area, and if we ask the same question in the context of particular peat bog. The similarity among species in the analysed data set can be considered an estimate of the species similarity in the studied area. Therefore, standard errors of this estimate can be (and in some cases have been) derived for the species similarities (see, e.g. Pielou 1977, p. 210).

Also, there is a reasonable null model for species similarity: the species occur independently of each other. This corresponds to the null hypothesis for a 2 × 2 contingency table or for a correlation analysis. Consequently, many measures of species similarity are centred around zero, with the zero value corresponding to species independence (all the coefficients mentioned in the following text). For similarities of cases, there is no similar null model.

Finally, when comparing two cases, the fact that a species is missing from both does not provide reliable information about their similarity, particularly when we have a heterogeneous data set. For example, in a data set describing terrestrial community composition along a gradient of soil wetness, particular species might be missing from two cases because the soil is too dry in one of them and too wet in the other one. This is so-called **double-zero problem** and it is the reason why the measures ignoring the double zeros when comparing two cases are preferred when measuring similarity among cases (particularly when studying community composition, not e.g. in taxonomic studies), but not among species (see the next section).

6.1 Similarity measures for presence–absence data

There are two classical similarity indices ignoring double zeros that are widely used in community studies for evaluating the **similarity of cases**: the **Jaccard coefficient** (Jaccard 1901) and the **Sørensen coefficient** (Sørensen 1948), both available for distance-based ordination methods in Canoco 5.[3] To be able to present their formulae, we first define values of a, b, c and d using a 2 × 2 frequency table:

---

[3] Canoco 5 converts these two similarity coefficients into distance values by taking the square root of their one-complement (i.e. $D = \sqrt{(1 - S)}$ or $\sqrt{(1 - J)}$): in this way, the resulting distances are not only metric, but also Euclidean (in the meaning used by Legendre & Legendre 2012, p. 296).

		Species in case 2	
		present	absent
Species in case 1	present	a	b
	absent	c	d

Therefore, a is the number of species present in both cases being compared, b is the number of species present in case 1 only, c is the number of species present in case 2 only, and d is the number of species absent from both cases. The **Sørensen coefficient** is then defined as

$$S_{1,2} = \frac{2a}{2a+b+c}$$

and the Jaccard coefficient as

$$J_{1,2} = \frac{a}{a+b+c}$$

Although there are some differences between these two coefficients, the results based on them are roughly comparable. For both coefficients, the similarity of identical cases equals 1 and the similarity of cases sharing no species is 0. Note that these coefficients do not use the d value. The number of species absent from both cases does not convey any ecologically interesting information, except in some specific cases.

The **simple matching coefficient** described by Sokal and Michener (1958) is one of the most widely used coefficients for binary data, not ignoring the double zeros (i.e. the value of d). It is defined as

$$SM_{1,2} = \frac{a+d}{a+b+c+d}$$

and expresses, therefore, the relative fraction in 0/1 values agreeing in both compared cases. It is widely used when evaluating binary data produced by molecular biology techniques, where shared presence or absence of a feature has equivalent importance.[4]

For calculating the **similarity of species** (more generally of response variables) a, b, c and d are defined as follows:

		Species B	
		present	absent
Species A	present	a	b
	absent	c	d

---

[4] Canoco 5 offers this coefficient for distance-based methods and allows its transformation into distance either as $(1 - SM)$ or as $\sqrt{(1 - SM)}$, with only the latter one having full metric properties.

6.1 Similarity measures for presence–absence data

so that a is the number of cases where both species are present, b is the number of cases only supporting species A, c is the number of cases with species B only, and d is the number of cases where neither of the two species is present. As a measure of species similarity, we usually employ the measures of 'association' in a 2×2 table.[5] The most useful measures are probably the V coefficient (based on the classical chi-square statistic for 2×2 tables):

$$V_{A,B} = \frac{ad - bc}{\sqrt{(a+b)(c+d)(a+c)(b+d)}}$$

and the Q coefficient:

$$Q_{A,B} = \frac{ad - bc}{ad + bc}$$

Values of both V and Q range from -1 to $+1$, with 0 corresponding to the situation where the two species are found in common exactly at the frequency corresponding to the model of their independence (i.e. where the probability of common occurrence P_{ij} is a product of the independent probabilities of occurrences of the species: $P_{ij} = P_i \cdot P_j$). Positive values mean a positive association, negative values mean a negative association.

The difference between the V and Q measures is in their distinguishing between complete and absolute association (see Pielou 1977). **Complete association** is the maximum (minimum) possible association under given species frequencies. This means that for a complete positive association, it is sufficient that either b or c is zero (i.e. the rarer of the species is always found in the cases where the more common species is present). Similarly, for complete negative association, a or d must be zero. **Absolute positive association** occurs when both species have the same frequency and are always found together (i.e., $b=c=0$). Absolute negative association means that each of the cases contains just one of the two species (i.e. $a=d=0$). V is also called the **point correlation coefficient** because its value is equal to the value of the Pearson correlation coefficient of the two species, when we use the value of 1 for the presence of species and 0 for its absence. This is a very useful property, because qualitative data are usually coded this way and the classical Pearson correlation coefficient is readily available in virtually all computer programs.[6]

Note that, unlike the similarity of cases, the d value is absolutely necessary for the similarity of species (contrary to recommendations of, for example, Ludwig & Reynolds, 1988, and others). We illustrate this by an example:

---

[5] Canoco 5 does not implement any specific coefficient of association between species, as the implemented methods are based on an (implied) matrix of distances among cases. When direct comparison of species would be needed, based on community composition data, the simple matching coefficient (using the double-absence frequency as well) for binary data or chi-square distance for quantitative data would be an appropriate choice.

[6] However, we cannot trust here the related statistical properties of the Pearson correlation coefficient, such as confidence intervals, statistical significance, etc., because their derivation is based on the assumption of normality. For statistical properties of V and Q coefficients, see Pielou (1977, p. 210).

	Table A				Table B	
		Species B				Species B	
		present	absent			present	absent
Species A	present	50	50	Species A	present	50	50
	absent	50	1000		absent	50	5

In Table A, the two species, both with a low frequency, are found together much more often than expected under the hypothesis of independence (this might be the example of *Carex* and *Oxycoccus* at the landscape scale). Accordingly, both V and Q are positive. On the other side, in Table B, both species have relatively high frequencies, and are found much less in common than expected under the independence hypothesis (e.g. *Carex* and *Oxycoccus* within a peat bog). In this case, both V and Q are negative. Clearly, the species similarity has a meaning only within the data set we are analysing. Therefore, the similarity indices that ignore the d value are completely meaningless. Our experience shows that their value is determined mainly by the frequencies of the two species being compared. For further reading about species association, Pielou (1977) is recommended.

6.2 Similarity measures for quantitative data

6.2.1 Measuring and transforming the quantity

When we use the (dis)similarities based on species presence–absence, the information on species quantity in cases is lost. From an ecological point of view, it may be very important to know whether the species is a subordinate species found in a single specimen or whether it is a dominant species in the community. There are various quantitative measures of species representation (sometimes generally called importance values): number of individuals, biomass, cover, or frequency in subunits. Often the quantity is estimated on some semi-quantitative scale (e.g. the Braun-Blanquet scale for data about vegetation). The quantitative data are sometimes transformed and/or standardised – see Section 1.3.

In vegetation ecology, the two extremes are presence–absence and non-transformed species cover (similar results would be obtained using non-transformed numbers of individuals, biomass or basal area values). In between, there are measures such as an ordinal transformation of the Braun-Blanquet scale (i.e. the replacement of the original scale values r, +, 1 ... with the numbers 1, 2, 3, ...), which roughly corresponds to a logarithmic transformation of the cover data. In many vegetation studies, this is a reasonable and ecologically interpretable compromise. Similarly in animal ecology, the two extremes are presence–absence, and numbers of individuals or biomass. The log transformation is again often a reasonable compromise.

Van der Maarel (1979) showed that there is a continuum of transformations that can be approximated by the relation $Y_{\text{transf}} = Y^z$ (see also McCune & Mefford 1999). When $z = 0$ the data values are reduced to presence and absence. As the value of z increases, the final result is increasingly affected by the dominant species. For similar purpose, the Box-Cox transformation (Sokal & Rohlf 1995) can also be used.

Lepš and Hadincová (1992) have shown that when the analysis is based on plant cover values, ignoring all species in a record that comprise less than 5 per cent of the total cover does not significantly affect the results of multivariate analyses (based implicitly or explicitly on a measure of similarity). The results in this case are completely dependent on the dominant species. Because of the typical distribution of species importance values in most ecological communities (i.e. few dominants and many subordinate species, whose counts and/or biomass are much lower than those of the dominant ones), the same will be true for non-transformed counts of individuals or biomass data. A $\log(Y + 1)$ transformation (or $\log(AY + B)$) might be used to increase the influence of subordinate species.

6.2.2 Similarity or distance between cases

A very common measure of dissimilarity between cases is the **Euclidean distance** (ED). For a data matrix with m species, with the value of k-th species in the i-th case written as Y_{ik}, the ED is the Euclidean distance between two considered case points:

$$ED_{1,2} = \sqrt{\sum_{k=1}^{m} (Y_{1k} - Y_{2k})^2}$$

If we consider the cases to be points in multidimensional space, with each dimension corresponding to a species, then ED is the Euclidean distance between the two points in this multidimensional space. ED is a measure of dissimilarity, its value is 0 for completely identical cases and the upper limit (when the cases have no species in common) is determined by the quantity of the species in the cases. Consequently, the ED might be relatively low, even for cases that do not share any species, if the abundances of all present species are very low. This need not be considered a drawback, but it must be taken into account in our interpretations. For example, in research on seedling recruitment, the plots with low recruitment can be considered similar to each other in one of the analyses, regardless of their species composition, and this is then correctly reflected by the values of ED without standardisation (see Chapter 15).

If we are interested mainly in species composition, then standardisation by case norm is recommended. With standardisation by case norm, we obtain the so-called **chord distance** or **standardised Euclidean distance** (Orloci 1967 and Legendre & Legendre 2012, p. 301). Its upper limit is $\sqrt{2}$ (i.e. the distance of two perpendicular vectors of unit length).

Legendre and Gallagher (2001) discuss various data transformations and their effect on the Euclidean distance. Sometimes a special name is coined for the ED after a particular

Table 6–1 Hypothetical table with cases 1 and 2 containing one species each and cases 3 and 4, containing three equally abundant species each (for standardised data the actual quantities are not important). Case 1 has no species in common with case 2 and case 3 has no species in common with case 4. The cases with *t* in their labels contain values standardised by the total; those with *n* are cases standardised by case norm. For cases standardised by total, $ED_{1,2} = 1.41$ ($=\sqrt{2}$) and $ED_{3,4} = 0.82$, whereas for cases standardised by case norm, $ED_{1,2} = ED_{3,4} = 1.41$. Note that when we apply Hellinger standardisation (i.e. divide by sum, but then take a square root), it leads in this example to the same results as the standardisation by case norm.

Species	Cases											
	1	2	3	4	1t	2t	3t	4t	1n	2n	3n	4n
1	10		5		1		0.33		1		0.58	
2		10	5			1	0.33			1	0.58	
3			5				0.33				0.58	
4				5				0.33				0.58
5				5				0.33				0.58
6				5				0.33				0.58

data transformation is used. This is the case of **Hellinger distance** (HD) originally described by Rao (1995):

$$HD_{1,2} = \sqrt{\sum_{k=1}^{m}\left[\sqrt{\frac{Y_{1k}}{Y_{1+}}} - \sqrt{\frac{Y_{2k}}{Y_{2+}}}\right]^2}$$

where Y_{1+} and Y_{2+} represent the sums of values for case 1 and case 2, respectively. Unlike the standard Euclidean distance, but similar e.g. to chord distance, HD is not sensitive to the double-zero problem, while maintaining the metric properties.[7]

The use of ED is not recommended with standardisation by case totals.[8] With this standardisation, the length of the case vector decreases with the species richness and, consequently, the cases become (according to ED) more and more similar to each other. We illustrate this situation by an example in Table 6–1. The decrease in similarity with species richness of the case is quite pronounced. For example, for two cases sharing no common species, with each case having 10 equally abundant species, the ED will be only 0.45.

Among the ordination methods, the linear methods (PCA, RDA) reflect the Euclidean distances between the cases (with the corresponding standardisation used in the particular analysis).

---

[7] Canoco 5 implements HD (as well as the chord distance) for distance-based methods (PCoA, NMDS, db-RDA), but it also offers so-called Hellinger standardisation (suggested by Legendre & Gallagher, 2001) of the data when the linear ordination methods (PCA and RDA) are used (see Canoco 5 manual, p. 123). In PCA or RDA, chord distance is implied by the standardisation by case norm.

[8] This does not concern HD or Hellinger standardisation, however: see Table 6–1 caption.

6.2 Similarity measures for quantitative data

Another very popular measure of case similarity is **percentage** (or **proportional**) **similarity** (PS). Percentage similarity of cases 1 and 2 is defined as

$$PS_{1,2} = \frac{2\sum_{k=1}^{m} \min(Y_{1k}, Y_{2k})}{\sum_{k=1}^{m}(Y_{1k}+Y_{2k})}$$

Its value is 0 if the two cases have no common species, and it is 1.0 for two identical cases (sometimes it is multiplied by 100 to produce values on the percentage scale). If needed, it can be used as the **percentage dissimilarity** PD = 1 − PS (or PD = 100 − PS, when multiplied by 100), also called the **Bray–Curtis distance** (Bray & Curtis 1957).[9] When used with presence–absence data (0 = absence, 1 = presence), it is identical to the Sørensen coefficient. If it is used with data standardised by the case total, then its value is

$$PS_{1,2} = \sum_{k=1}^{m} \min(Y'_{1k}, Y'_{2k})$$

with

$$Y'_{ik} = \frac{Y_{ik}}{\sum_{k=1}^{m} Y_{ik}}$$

These resemblance measures are used under a plethora of names (see, for example, Chapter 7 in Legendre & Legendre 2012, or McCune & Mefford 1999).

Chi-square distance is rarely used explicitly in ecological studies. But as the linear ordination methods reflect Euclidean distance among cases, unimodal (weighted-averaging) ordination methods (CA, DCA, CCA) reflect chi-square distance. In its basic version, the distance between cases 1 and 2 is defined as:

$$\chi^2_{1,2} = \sqrt{\sum_{k=1}^{m} \frac{S_{++}}{S_{+k}} \left[\frac{Y_{1k}}{S_{1+}} - \frac{Y_{2k}}{S_{2+}} \right]^2}$$

where S_{+k} is the total of the k-th species values over all cases:

$$S_{+k} = \sum_{i=1}^{n} Y_{ik}$$

S_{i+} is the total of all the species values in the i-th case:

$$S_{i+} = \sum_{k=1}^{m} Y_{ik}$$

---

[9] Canoco 5 implements both the original Bray–Curtis distance and its variant where the PD values are further square-root transformed, to achieve metric properties (Legendre & Legendre 2012, p. 312).

and S++ is the grand total:

$$S_{++} = \sum_{i=1}^{n} \sum_{k=1}^{m} Y_{ik}$$

The chi-square distance is similar to Euclidean distance, but it is weighted by the inverse of the species totals. As a consequence, the species with low frequencies are over-emphasised.[10] The chi-square distance value for two identical cases is 0 and its upper limit depends on the distribution of the species values. Legendre and Gallagher (2001) have shown that the chi-square distance is actually used when the Euclidean distance is calculated after a particular type of data standardisation, which we show using the p_{ik} symbol at the end of Section 4.8.

Gower distance (GD) is a dissimilarity measure particularly suitable for comparing dissimilarity of cases in data where some of the variables are factors or where variables can take negative values. It is defined as

$$GD_{1,2} = \frac{1}{m} \sum_{k=1}^{m} d_{12k}$$

where d_{12k} is defined for numeric variables as $|Y_{1k} - Y_{2k}|/R_k$, with R_k being the range of k-th variable values (i.e. $\max(Y_k) - \min(Y_k)$). For factor variables d_{12k} is set to 0 for the identical value (factor level) of the variable in both compared cases and to 1 for value difference. What we describe here is the symmetrical type of Gower coefficient, described in Legendre and Legendre (2012, pp. 278–280). This makes it unsuitable for data where the double-zero problem (see the introductory part of this chapter) is expected, but very appropriate e.g. for morphometric (taxonomical) studies or when comparing e.g. the functional trait values among species (see also Section 6.4 and Section 16.8.1 for an example).

6.2.3 Similarity between species

For species similarity based on quantitative data, the classical **Pearson correlation coefficient** is often a reasonable choice. When it is used for qualitative (presence–absence) data, it is identical to the V coefficient. This shows that it is a good measure of covariation for a wide range of species abundance distributions. An alternative choice is provided by one of the rank correlation coefficients, Spearman or Kendall coefficients. Values of all the correlation coefficients range from -1 for a deterministic negative dependence to $+1$ for a deterministic positive dependence, with the value of 0 corresponding to independence. Both rank correlation coefficients are independent of the standardisation by cases.

Alternatively, we can use the chi-square distance, which is quite appropriate to compare the distribution of relative frequencies (profiles) between species.

---

[10] This is why the optional 'downweighting of rare species' is available in unimodal methods.

See also Section 6.4 for a discussion of issues concerning similarity between species, based on their (functional) traits.

6.3 Similarity of cases versus similarity of communities

The above-described case similarity measures describe just the similarities between two selected samples. It is often expected that the cases (samples) are representative of their communities (defined here operationally as all the organisms in a specified area or space). In plant ecology, this is usually correct: either we believe in the minimal area concept (see Moravec 1973),[11] or we are interested in the variation in species composition at the spatial scale corresponding to the size of our cases (quadrats).

The situation is very different in many insect community studies. There, the individuals in a case (sample) represent a selection (usually not a completely random one) from some larger set of individuals forming the community. This is especially true in the tropics, where species richness is high and not only a fraction of individuals but also a small fraction of taxa is represented in the sample. Even in large samples, many species are represented by single individuals and consequently the probability that the species will be found again in a sample from a similar community is not high. For example, in samples (greater than 2000 individuals) of insects feeding on tree species in Papua New Guinea over 40 per cent of species were represented by singletons (Novotný & Basset 2000). Consequently, the similarity of the two random samples taken from the community might be low and depends on the sample size.

There are methods that attempt to measure the similarity of the sampled communities, and so to adjust for the size of the compared samples.[12] One such measure is the **normalised expected species shared** index (NESS, Grassle & Smith 1976). It is calculated as the expected number of species in common between two random subsamples of a certain size drawn from the two compared larger samples without replacement, normalised (divided) by the average of the expected number of shared species in two subsamples taken randomly from the first sample and in two subsamples taken randomly from the second sample. These measures, however, are not computationally simple and are not available in common statistical packages. But new indices based on samples and corrected for unseen species were recently suggested (Chao et al. 2005) and are available in the EstimateS software (Colwell 2011).[13]

The Morisita index (Morisita 1959), popular in zoological studies, is a special case of the NESS index. It measures the probability that two individuals selected randomly from different subsamples will be the same species. This probability is standardised by

---

[11] And consequently the relevés are taken accordingly and we believe that they are good representatives of the respective communities and so there is no need to worry.

[12] Note that in some research areas, there is often a standard to count a fixed number of individuals, cells, etc. (e.g. in paleolimnological plankton research), whereas in other disciplines, this is not possible, as the number of individuals is not controlled by the researcher, for example the number of insect individuals caught in a light trap.

[13] Available, when this book went to print, at http://viceroy.eeb.uconn.edu/estimates/.

the probability that two individuals selected randomly from the **same** subsample will be the same species. It is calculated as:

$$Morisita_{1,2} = \frac{2 \sum_{k=1}^{m} Y_{1k} Y_{2k}}{(\lambda_1 + \lambda_2) S_{1+} S_{2+}}$$

where

$$\lambda_i = \frac{\sum_{k=1}^{m} Y_{ik}(Y_{ik} - 1)}{S_{i+}(S_{i+} - 1)}$$

The index can be used for numbers of individuals only, so:

$$S_{i+} = \sum_{k=1}^{m} Y_{ik}$$

is the total number of individuals in the i-th sample.

6.4 Similarity between species in trait values

Dissimilarity among biotic species can be expressed in terms of single and multiple traits. Such trait dissimilarity can be used as input data for a distance-based method (such as the principal coordinates analysis) or when calculating functional diversity (i.e. the extent of trait differences among species coexisting in a community). To compute trait dissimilarity we need to consider several specific issues that we discuss here.

Data on species traits usually include both quantitative (numeric) and categorical (factor) variables and the quantitative variables are measured using different scales. To allow the combination of such a variety of trait types, we can use the Gower distance (discussed in Section 6.2.2), which copes with a combination of factor and numeric variables and standardises the effect of individual variables to a value between 0 and 1. With numeric variables we can also perform the standardisation separately and then use other distance measures, such as Euclidean distance (implied in linear ordination methods such as PCA). This allows us to use alternative standardisation methods, such as the standardisation by maximum value in the data set or by a potential maximum for the trait found in the literature.

Both standardisation by maximum and by range have, however, some undesirable consequences particularly for the calculation of functional diversity. The dissimilarity among two species becomes context-dependent and so do the estimates of functional diversity. For example, if we extend a study of grassland vegetation to include also wooded meadows (with a regular presence of adult trees), the range (and maximum) of plant height values will be substantially extended, with additional, but smaller effects on the mean and variation of that trait. Such an extension will then lead to a considerable decrease of the importance of plant height in the calculation of functional diversity among the grasslands with no trees.

Mouillot et al. (2005) and Lepš et al. (2006) offered a solution for this problem, via measuring species (functional) dissimilarity by a complement of their trait overlap. To do so, we must first characterise the distribution of trait values for each species–trait combination, and then we can calculate the overlap of these distributions for each pair of compared species. The overlap is measured on identical scales (in the range between 0, for no overlap, and 1, for identical distributions) and the species dissimilarity for a particular trait is then the *1-overlap* value. The same approach can be also used for factor variables, including those with fuzzy coding. De Bello et al. (2013) demonstrated that this approach is useful for calculating functional diversity. But it has a major practical disadvantage, as the knowledge of the distribution of trait values (i.e. the intraspecific variability for each trait and species) is required, which is often not an option for today's online databases of traits. Even when collecting trait data as part of a study, we need a sufficient number of measurements to reliably estimate the trait mean and variability as the simplest parameters of their distribution.

6.5 Principal coordinates analysis

Principal coordinates analysis (PCoA or PCO, also called metric multidimensional scaling, and described by Gower 1966) is a multivariate method that attempts to represent in the Cartesian coordinate system a configuration of n objects (cases), defined by a $n \times n$ matrix of dissimilarities (distances) or similarities between the objects. Unlike the principal component analysis (PCA), which can be also defined in similar way, PCoA is able to represent a wide range of similarity (or distance) measures in the space of principal coordinates. Similarly to PCA, the first axis (principal coordinate) of the PCoA solution represents the true distances in the matrix in the best possible way that can be achieved with one dimension. As in PCA, each axis has an eigenvalue that indicates its importance. Therefore, starting from the first axis, the importance of principal coordinates decreases in the order of their decreasing eigenvalues.

There is, however, one additional 'problem', which often occurs in the PCoA solution. Not all the (dis)similarity measures can be fully represented in the Euclidean space. If this happens, PCoA produces one or more axes with negative eigenvalues. Because the eigenvalues represent the variance explained on the corresponding principal coordinates and this variance must be non-negative for standard real numbers, the scores of cases (objects) on those axes are complex numbers (with an imaginary component). These cannot be plotted, due to their non-Euclidean properties. Consequently, the sum of absolute values of such negative eigenvalues divided by the sum of absolute values of all the eigenvalues represents the distortion of the original distance matrix properties caused by projection into Euclidean space. Anyway, in the standard applications of PCoA, we do not typically plot even all the principal coordinates corresponding to the positive eigenvalues. Rather, only the first few axes are interpreted, as in the other ordination methods.

To eliminate the presence of negative eigenvalues, at least two different correction methods were developed. These methods adjust the original distance matrix to enable its full embedding in the Euclidean space of PCoA. Instead of using such non-intuitive adjustments, we prefer to choose a different (dis)similarity measure or to perform its simple, monotonic transformation. For example, when the distances are based on a complement-to-1 of a non-metric similarity coefficient (such as the Sørensen coefficient, see Section 6.1), then the square-root transformation of such distance often has the required metric properties.[14]

To calculate PCoA in the Canoco 5 program, you can use a special analysis template for this method offered in New Analysis Wizard. When specifying the analysis options, you can use either a pre-computed matrix of distances (or similarities) or you can specify a data table and select one of the distances offered by the Analysis Setup Wizard.

We will illustrate the application of PCoA using a data set of 14 vegetation records (relevés) from an altitudinal transect in Nízké Tatry Mts, Slovakia. Relevé 1 was recorded at an altitude of 1200 m a. s. l., relevé 14 at 1830 m a.s.l. Relevés were recorded using the Braun-Blanquet scale (r, +, 1–5, see Mueller-Dombois & Ellenberg 1974). For calculations, the scale was converted into numbers 1 to 7 (ordinal transformation, Van der Maarel 1979, see also Section 1.3.1). Data were entered as a classical vegetation data table (file *tatry.xlsx*), from which you can start if you want to create your own project. This data set is also used in Chapter 7 to demonstrate various classification methods.

The corresponding analysis (together with two others demonstrating NMDS and db-RDA methods, described in the following two sections) is stored in the example *Dist.c5p* project (analysis *Principal-coordinates*). To calculate PCoA, click the *New* button below the list of analyses, keep the default choices in the first two wizard pages, and select the *Principal-coordinates* analysis template in the *Specialized Analyses* folder of the last page of the New Analysis Wizard.

In the first page of the Analysis Setup Wizard[15] (illustrated in Figure 6–1), you should specify that the distances must be calculated and select the required type of dissimilarity measure. We recommend the Bray–Curtis (i.e. percentage dissimilarity) distance here, but with the additional square-root transformation which makes it fully metric. The checkbox at the bottom of the dialog requests the response variables (plant species) to be projected into PCoA ordination space, in the role of supplementary variables. After you finish the setup wizard and the analysis is executed, close the Graph Wizard with the *Cancel* button. When you check the *Summary* page of the analysis notebook, you can see that the analysis has actually two steps: the first one calculates the PCoA, while the second step performs rotation of its axes by analysing the case scores on principal coordinates with a principal component analysis. This step also allows you to project plant species into the ordination space.

You will now create a diagram with the first two principal coordinates and the positions of individual cases, as well as the scores of individual plant species. Note that the variables representing the species play here the role of supplementary, rather than

---

[14] In the specialised use of PCoA to calculate spatial eigenfunction predictors, the axes with negative eigenvalues are treated with more reverence, as they represent the predictors for negative spatial correlation (see Section 10.2).
[15] We assume here that you have the QuickWizard option set in Canoco 5.

6.5 Principal coordinates analysis

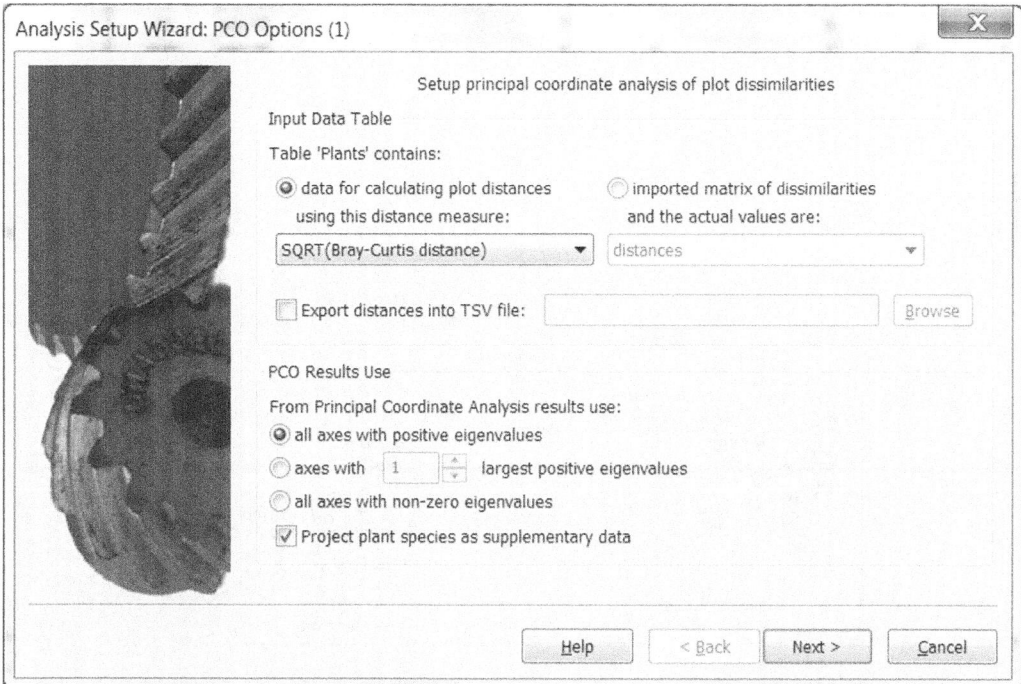

Figure 6–1 *PCO Options* of the Analysis Setup Wizard shown with *Principal-coordinates* template.

response variables. They would be therefore shown as arrows, implying linear change of their abundances across the ordination plane. But given the substantial beta-diversity in this data set, plotting their centroids might be preferable.[16] To achieve this, you must select the *Analysis | Plot creation options* menu command and check the *Customize plant species* option in the lower part of the *General* page of the *Plot settings* dialog box. Additionally, before closing this dialog, you might like to switch to its *Predictor Selection* page and limit (using the two control fields at the top) the set of displayed species to those with correlation with the ordination axes smaller than −0.5 or larger than 0.5. This selects 26 plant species out of the total of 48 species. Then you can create the plot using the *Graph | Biplots | Plots and plant species* menu command. Given our earlier choice, Canoco 5 will display a dialog named *Customize predictor presentation*, where you must select all the shown plant species in the left-hand list and move them to the right-hand list using the ≫ button, so that the selected plant species are represented by centroids. The resulting diagram (with further adjustment of symbols size and label font and positions) is shown in Figure 6–2.

---

[16] The centroids of supplementary variables are also calculated by averaging the case positions, so they bear similarity with the response variable positions in unimodal (weighted-averaging) ordination.

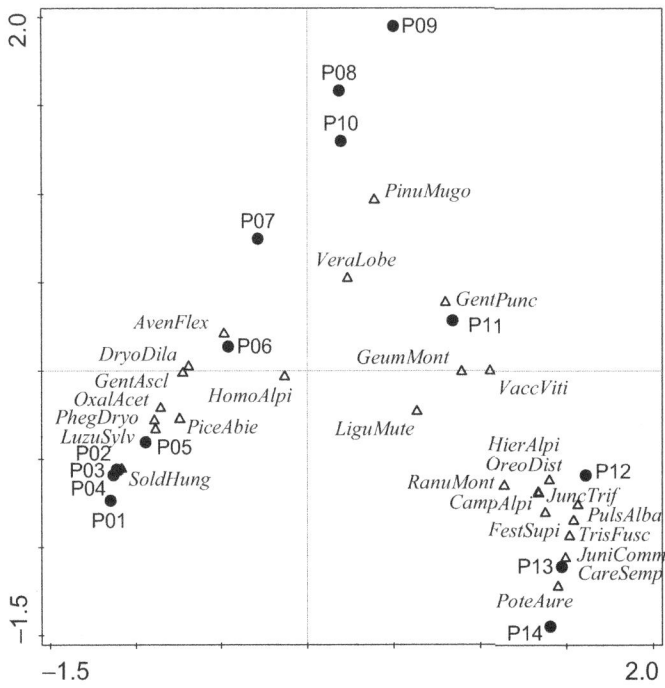

Figure 6–2 Ordination diagram with the first two axes of principal coordinates analysis (PCoA).

This data set seems to have just one interpretable, altitudinal gradient, represented by the first axis, as documented by the artificial character of the second axis, displaying the arch effect. See Section 7.1 for additional interpretation of the revealed pattern.

6.6 Constrained principal coordinates analysis (db–RDA)

A constrained PCoA, representing a canonical form of the analysis of a matrix of (dis)similarities, was proposed by Legendre and Anderson (1999) under the name **distance-based RDA** (db-RDA). It is also known as Canonical Analysis of Principal Coordinates (CAP), as suggested by Anderson and Willis (2003). To perform this type of analysis, the analysis of principal coordinates must be performed on the matrix of case (dis)similarities and the resulting case coordinates are used as response data in a redundancy analysis (RDA).

This method enables us to test hypotheses about the effects of explanatory variables on community composition, while (almost) freely selecting the measure of (dis)similarity among the cases. Distance-based RDA can use covariates and also reflect non-random sampling or experimental design during the Monte Carlo permutation tests.

But we pay for the flexibility in choice of distance measure by losing the model of response variable (species) relation with the (constrained) ordination axes, which is otherwise fruitfully used when interpreting the standard constrained ordinations (RDA or CCA) – see Chapter 11. While we can project selected variables a posteriori into a db–RDA diagram, there is no implied model of the response. Therefore, our decision to show variables as arrows (implying linear change of variable values across the db-RDA space) or as centroids (implying unimodal model) is an arbitrary choice, with no guaranteed coherence with the selected (dis)similarity coefficient.

In Canoco 5, you can perform db-RDA with the help of the Canoco Adviser, offering the *Distance-based-RDA* analysis template in the *Specialized Analyses* folder at the last page of the New Analysis Wizard (example analysis is in the *Dist.c5p* project, with the name *Distance-based-RDA*). As in the case of PCoA and NMDS analysis templates, this one also allows you to either work with a pre-computed matrix of (dis-)similarities or to calculate it during the analysis using one of the 11 offered distance measures. If you want to perform partial db-RDA (i.e. to include covariates), you must set up the new analysis with the QuickWizard mode switched off.

6.7 Non-metric multidimensional scaling

The method of ordination can be formulated in several ways (see Section 4.5). A very intuitive one is that it finds a configuration of cases in the ordination space so that the distances among cases in this space best correspond to dissimilarities of their composition. Generally, n objects could be ordinated in $n - 1$ dimensional space without any distortion of dissimilarities. But the need for clear visual presentations dictates the necessity of reducing the dimensionality (two-dimensional ordination diagrams are most often presented in journals).

This reduction could be done in a metric way, by explicitly looking for projection rules (using principal coordinates analysis, see previous section), or in a non-metric way, by non-metric multidimensional scaling (NMDS, see Shepard 1962; Kruskal 1964; Cox & Cox 1994). This method analyses the matrix of dissimilarities between n objects (i.e. cases) and aims to find a configuration of these objects in k-dimensional ordination space (k is a priori determined), so that those distances in ordination space correspond to dissimilarities. A statistic termed 'stress' is designed to measure the 'lack of fit' between distances in ordination space and dissimilarities.

$$\text{stress} = \sum [d_{ij} - f(\delta_{ij})]^2$$

where d_{ij} is the distance between case points in the ordination diagram, δ_{ij} is the dissimilarity in the original matrix of distances, calculated from the data, and $f()$ is a non-parametric monotonous transformation. In this way, the 'correspondence' is defined in a non-metric way, so that the method reproduces the general rank-ordering of dissimilarities, but is not required to reproduce exactly the dissimilarity values. The algorithm attempts to place the objects in the ordination space to minimise the stress criterion value. The algorithm is necessarily an iterative one, and its convergence depends on the initial configuration; also, the global minimum is not always achieved and, consequently, it is worthwhile trying various initial configurations. The NMDS method is available in most general statistical packages, as well as in specialised multivariate packages. The implementation details differ among these programs – for example, it is often possible to start from various initial configurations and automatically select the result with lowest stress value.

In contrast to other ordination methods (except PCoA and db-RDA illustrated in previous sections), the input data is not the original data table, but a matrix of dissimilarities among objects.[17] Consequently, virtually any measure of (dis)similarity could be used. Unlike other methods, in NMDS the number of ordination axes (i.e. the dimensionality of the ordination space) is given a priori. In real applications, various dimensionalities (number of axes, k) can be tried and the 'proper' number of dimensions is then decided according to the 'quality' of resulting models. The quality is measured by the stress of the resulting configuration. Because NMDS is normally used as an indirect ordination, the interpretability of the results is usually one of the criteria. As with the other indirect methods, the axes could be a posteriori interpreted using measured explanatory variables, which could be projected to the ordination diagram. But NMDS axes do not have any special optimum properties as do the axes of other ordination methods. Many implementations of NMDS therefore rotate the axes to, e.g., maximise the variance of case scores along the axes (PCA rotation, as done also by Canoco 5).

Similarly to PCoA, Canoco 5 offers a specialised analysis template for NMDS. To create such analysis (you can also check the *NMDS* analysis in the example project file *Dist.c5p*), click the *New* button below the list of analyses in the project, keep the default choices at the first two wizard pages and then select the *NMDS* template in the *Specialized Analyses* folder at the last page of the New Analysis Wizard. After you click the *Finish* button, the Analysis Setup Wizard (when run in QuickWizard mode) shows as its first page a specialised *NMDS Options* page, similar to the page illustrated in the previous section in Figure 6–1. But there are some differences in the offered options and you are advised to check the online help (by clicking the *Help* button at the wizard page bottom) to see their explanation. We suggest that you keep the Bray-Curtis distance measure selected here[18] and change the *NMDS solution based on* option from 3 to 2 calculated NMDS axes. You can keep the other options with their default values. After you close the setup wizard with the *Finish* button, the NMDS solution is calculated and you can this time accept the diagram creation offered by Graph Wizard. The resulting graph is shown in Figure 6–3.

You can see that the method reasonably reconstructed the elevation gradient: the first axis is strongly correlated with the elevation. The second axis is difficult to interpret, and is clearly a quadratic function of the first axis – even the NMDS does not escape the arch effect to some extent.

6.8 Mantel test

In some studies, we are interested in the relationship of two similarity or distance matrices. For example, we have n individuals in a plant population, each individual

---

[17] Recall that in the response-model-based ordinations this matrix is implicitly determined by the selection of linear/unimodal model.
[18] This is not meant as general advice, the choice of distance measure depends on data properties and the questions we ask.

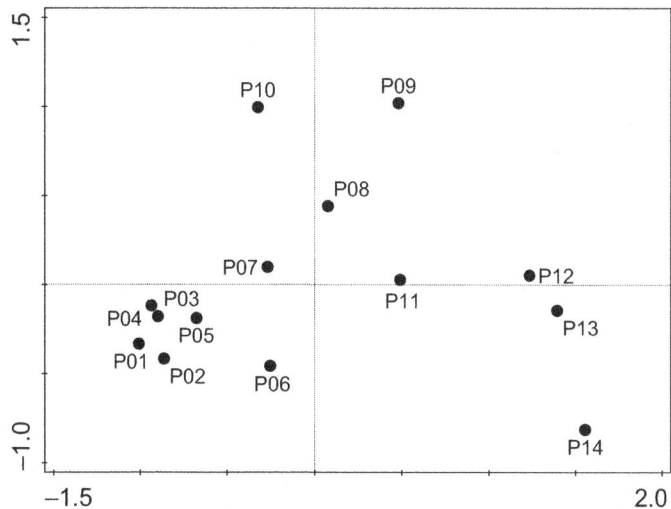

Figure 6–3 Ordination diagram from the NMDS method, displaying its two computed axes.

being characterised by its genetic constitution (e.g. allelic composition, determined by methods of molecular biology such as PCR or RFLP), and also by its position at a locality (as X and Y coordinates). We want to test whether the distance between plants reflects genetic similarity. We can use the X and Y variables as predictors and the allelic composition as a response matrix and apply some constrained ordination. But this would test for a spatial trend in the genetic composition, and would show no relationship if the population were composed of spatially restricted groups of genetically similar individuals scattered over the locality.

An alternative approach is to calculate two matrices: the matrix of genetic similarities and the matrix of physical distances between the individuals. We can then calculate the classical (Pearson) correlation coefficient (r) between the corresponding values of physical distance and genetic similarity (or we can use a regression of genetic similarity on the physical distance). Such analyses provide reasonable measures of how closely the two matrices are related (r) or how much of the genetic similarity is explained by the physical proximity (coefficient of determination R^2). However, the parametric test of significance for r (or the estimate of standard error of r) will not be correct, because both are based on the assumption that the observations are independent, which is not true: the number of degrees of freedom is greatly inflated, as we have $n(n-1)/2$ pairs (distances) that enter the analysis, but only n independent observations. If there is an individual which is genetically different from the others, all its $n-1$ similarities to other individuals will be very low, all based on a single independent observation.

The best solution is to estimate the distribution of the test statistics (e.g. the Pearson linear correlation r) under the null hypothesis using the Monte Carlo permutations. If there is no relationship between the spatial distance and genetic similarity, then the locations of individuals can be randomly permuted, and all the spatial arrangements are equally likely. Consequently, in the permutation, not the individual similarities (distances), but the identity of individuals is permuted in one of the matrices (it does not matter which one). In this way, the internal dependencies within each of the matrices are preserved in all permutations, and only the relationship between the spatial coordinates and genetic properties is randomised.

One-tailed or two-tailed tests could be used, depending on the nature of the null hypothesis. In the above example, we would very probably use the one-tailed test, i.e. the null hypothesis would be that the genetic similarity is either independent of, or increases with, physical distance, and the alternative hypothesis would be that the genetic similarity decreases with the distance (we can hardly imagine mechanisms that cause the genetic similarity to increase with physical distance between individuals). Because the alternative hypothesis suggests negative r values, the estimate of Type I error probability is

$$P = \frac{n_x + 1}{N + 1}$$

where n_x is number of simulations where $r < r_{\text{data}}$; for a two-tailed test, n_x would be the number of simulations where $|r| > |r_{\text{data}}|$.

In practical calculations, the sum of products (which is quicker to calculate) is used instead of r (which is, on the other hand, more intuitive), but it can be shown that the results are identical. Instead of permutations, an asymptotic approximation can be used. Also, any measure of relationship between the two sets of distances can be used (e.g. rank correlation) or the distances can be transformed, as needed. For example, the compositional similarity is expected to be negatively related to the log of physical distance among the units (so-called distance decay) and so we can log-transform the physical distances prior to their use in the test.

One of the two compared distance (or similarity) matrices can also be constructed using arbitrary values, reflecting the alternative hypothesis in the test. For example, when studying differences between two or more groups of cases, we can use a distance matrix, where the within-group distances are set to 0 and the between-group distances are set to 1. See Section 10.5 of Legendre and Legendre (2012) for further discussion.

The other examples of the Mantel test use in ecology are as follows:

1. It can be expected that species with similar phenology compete more than species with different phenology. Species that compete more can be expected to have negative 'inter-specific association', i.e. they are found together less often than expected, when recorded on a very small scale in a homogeneous habitat (because of competitive exclusion). Two matrices were calculated, the matrix of phenological similarity of species and the matrix of species associations, and their relationship was evaluated by the Mantel test (Lepš & Buriánek 1990). No relation was found in this study.
2. In a tropical forest, a herbivore community of various woody plant species was characterised by all the insect individuals (determined to the species) collected on various individuals of the woody species in the course of one year. Then, abundance of all the investigated woody species in about 100 quadrats was recorded. The two matrices – matrix of similarities of woody species' herbivore communities, and matrix of 'inter-specific associations' of woody species, based on the quadrat study – were compared using the Mantel test. It was shown that woody species with similar

distribution also had similar herbivore communities, even though the relationship was weak (Lepš et al. 2001).

The Mantel test is available in specialised statistical packages, such as the PC-ORD or PRIMER programs (see Appendix C), as well as in multiple R packages. For example, the *vegan* package has functions *mantel* for a standard Mantel test and *mantel.partial* implementing partial Mantel test (with a third matrix of dissimilarities serving as covariates).

Beside the Mantel test, there also exist more specialised methods based on dissimilarity matrices and testing differences among groups of cases: multiple response permutation procedure (MRPP, Mielke & Berry 2007), analysis of similarities (ANOSIM, Clarke 1993) or more general PERMANOVA (Anderson 2001). As an example, we can compare the spatial segregation among plant species (measured, say, using the V-coefficient, see Section 6.1) with their functional type. The limiting similarity hypothesis suggests that we should expect the species belonging to the same functional type to be more segregated than the species from different functional groups and this can be tested by the Mantel test described above, but alternative, perhaps more plausible hypotheses cannot. We can e.g. hypothesise that two tussock perennial species will be highly segregated (with low V values) due to inter-species competition, but two annual species will appear together in the gaps between perennial tussocks, and consequently will be aggregated (higher V values), and the pairs of species from different functional groups will be somewhere between these two extremes. ANOSIM is then the method to be used.

The Mantel and related tests are sometimes applied also to problems where the two matrices are calculated from a species composition data table and a table with explanatory variables. Therefore, the various constrained ordination methods (CCA or RDA) can also be applied to the same problem, with the extra advantage of visualising the relation between the community and the environment in terms of individual species. In addition, the use of constrained ordination was claimed (Legendre & Fortin 2010) as a more appropriate approach when the focus is on relating the original variables in the two data tables that become turned into dissimilarity matrices for the Mantel test. These authors recommend that the 'domain of application of the methods of comparison based on distance matrices (Mantel test, partial Mantel test, ANOSIM, ...) is the set of questions that are originally formulated in terms of distances. Testing the distance predictions of a hypothesis of isolation by distance in genetics is one of these questions.'

7 Classification methods

The aim of classification is to obtain groups of objects (cases, variables) that are internally homogeneous and distinct from the other groups. When the variables (such as biological species) are classified, the **homogeneity** can be interpreted as their positive correlation, implying for species similar ecological behaviour, as reflected by the similarity of their distributions. The classification methods are usually categorised as in Figure 7–1.

Historically, numerical classifications were considered an objective alternative to subjective classifications, such as the classification of vegetation types by the Zürich–Montpellier phytosociological system (Mueller-Dombois & Ellenberg 1974; van der Maarel & Franklin 2013). It should be noted, however, that the results of numerical classifications are objective just in the sense that the same method gives the same results. Nevertheless, the results of all numerical classifications depend on the methodological choices, as we discuss in Section 7.3.1.

7.1 Example data set properties

The various possibilities of data classification will be demonstrated using vegetation data of 14 cases ('relevés') from Nízké Tatry Mts, already introduced in Section 6.5. Data were imported from the Excel file into a Canoco 5 project (*TatryDCA.c5p*). The primary data table was then exported into the condensed Cornell format (file *tatry.dta*) used by earlier versions of Canoco, to enable use of the TWINSPAN for Windows program. The data table present in the Excel file was also imported into the R software as a data frame called *tatry*, using the *read.delim* function.

First, we look at the similarity structure displayed by the DCA ordination (see Section 4.7), present in the *TatryDCA.c5p* example project as the *DCA* analysis. The resulting diagram with species and cases is shown in Figure 7–2.

The graph (Figure 7–2) demonstrates that there is a strong unidirectional pattern of variation in the data, corresponding to the altitudinal gradient, similar to patterns revealed by PCoA and NMDS methods, as illustrated, respectively, in Figure 6–2 and Figure 6–3. Cases 1 to 5 are from the spruce forest (characterised by plant species *Picea abies, Dryopteris dilatata, Avenella flexuosa*, etc.), cases 12 to 14 are from typical alpine grassland (e.g. *Oreochloa disticha*), and there is also a dwarf-pine (*Pinus mugo*) zone in between (see Figure 7–3, containing a plot with species response curves). Now we

7.2 Non-hierarchical classification (K-means clustering)

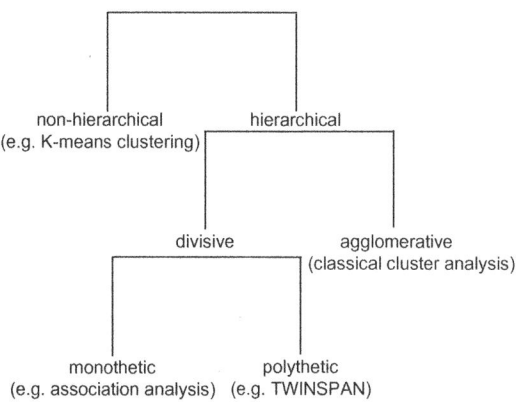

Figure 7–1 Types of classification methods.

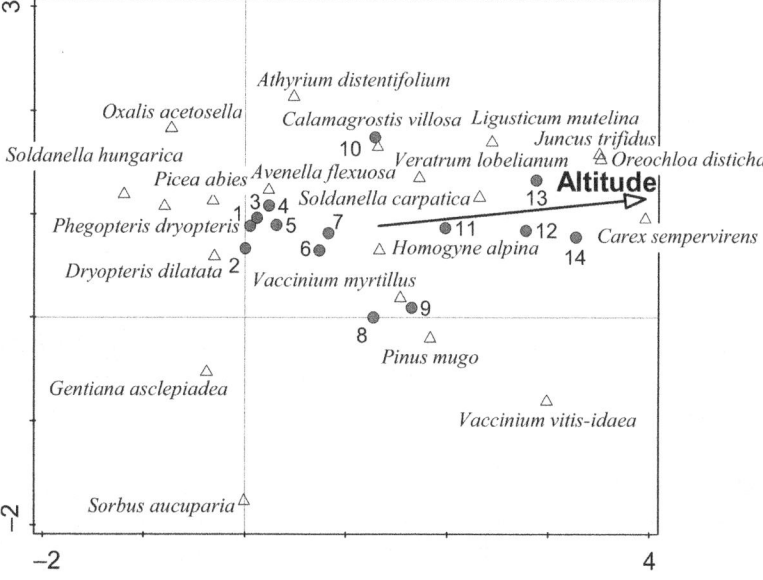

Figure 7–2 Species-cases ordination diagram with results of DCA of data from an altitudinal transect in Tatry Mts; the first two axes are plotted, altitude is passively projected into the diagram. Only the species with highest weight (most frequent ones) were selected for display.

will explore how the particular classification methods divide this gradient into vegetation types and also demonstrate how to perform the individual analyses.

7.2 Non-hierarchical classification (K-means clustering)

The goal of this approach is to form a predetermined number of groups (clusters). The groups should be internally homogeneous and different from each other. All the

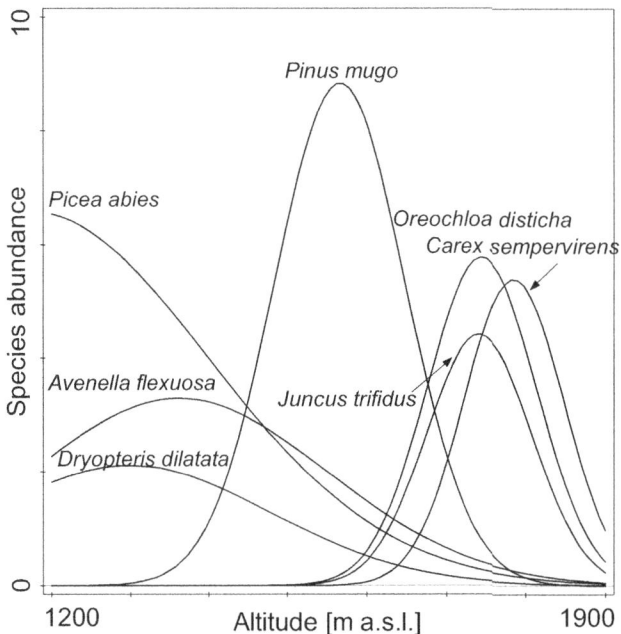

Figure 7–3 Response curves of important species on the elevation gradient, fitted by a second-order polynomial predictor (GLM, see Section 8.3). *m a.s.l.* means metres above sea level.

groups are on the same level, there is no hierarchy. The space of observations (with its dimensions defined by the variables of the analysed data table) is simply divided into multiple sub-volumes – this is why these techniques are also called partitioning methods. Here, you will use K-means clustering as the most widely used method of non-hierarchical classification.

For the computation, an iterative relocation procedure is applied. The procedure starts with k (desired number of) clusters, and then moves the cases among clusters to minimise the within-cluster variability and maximise the between-cluster variability. When the clusters are different, then ANOVA for (at least some) variables (here species) shows significant results, so the procedure can be thought of as 'ANOVA in reverse', i.e. forming groups of cases to achieve the most significant differences in ANOVA for the maximum number of variables (species).

The relocation procedure stops when no move of a case improves the criterion. You should be aware that you might get a local optimum with this algorithm, and you can never be sure this is the global optimum. It is therefore advisable to start with different initial clusters and check if the results are the same for all of them.

You will use the R function *kmeans*, requesting creation of three clusters and starting from ten different, randomly chosen initial conditions (locations of cluster centres). Simply printing the object returned by *kmeans* summarises the results.

7.2 Non-hierarchical classification (K-means clustering)

```
tatry.km <- kmeans( tatry, 3, nstart=10)
tatry.km
K-means clustering with 3 clusters of sizes 6, 5, 3

Cluster means:
  PiceAbie SorbAucu PinuMugo SaliSIle JuniComm VaccMyrt OxalAcet
1 5.416667     1.75 0.3333333      0.0      0.0 4.166667 2.666667
2 1.600000     1.40 6.4000000      0.8      0.0 4.400000 1.000000
3 0.000000     0.00 1.3333333      0.0      1.5 4.000000 0.000000
  HomoAlpi SoldHung AvenFlex CalaVill GentAscl DryoDila PhegDryo
1 3.166667 2.333333 2.833333 2.583333 1.833333      2.0 1.833333
2 2.800000 0.000000 2.400000 3.000000 0.800000      0.8 0.400000
3 2.333333 0.000000 0.000000 2.333333 0.000000      0.0 0.000000
...
Clustering vector:
 [1] 1 1 1 1 1 1 2 2 2 2 2 3 3 3
Within cluster sum of squares by cluster:
[1] 117.79167 206.30000  90.33333
 (between_SS / total_SS =  53.4 %)
```

The summary shows the average values of individual species within each of the three clusters. Clearly, *Picea abies* is common in cluster 1 (spruce forest) and missing in cluster 3 (alpine meadows). The dwarf pine *Pinus mugo*, on the other hand, is rare outside the middle ('krumholz') zone. Further, the membership of individual cases in the three clusters is shown (and further used below). The ratio of between-cluster sum of squares to the total sum of squares (53.4%) provides an indication of the extent of cluster separation.

The membership of cases can be shown in more readable form using the following command:

```
tatry.kclus <- as.factor( tatry.km$cluster)
split(1:14,tatry.kclus)
$'1'
[1] 1 2 3 4 5 6
$'2'
[1]  7  8  9 10 11
$'3'
[1] 12 13 14
```

Cases 1 to 6 are in the first cluster, cases 7 to 11 in the second cluster, and cases 12 to 14 in the third one. The fact that the members of each cluster are contiguous in numbering is caused by the linear character of variation in our data, as already illustrated by the arrangement of cases in Figure 7–2.

You can further supplement this general impression about the difference of individual clusters in the abundance of individual species by an ANOVA-based test as shown below for two species:

```
with(tatry,summary(aov(PinuMugo~tatry.kclus)))
              Df Sum Sq Mean Sq F value   Pr(>F)
tatry.kclus    2  107.7   53.83   38.95 1.02e-05 ***
Residuals     11   15.2    1.38
...
with(tatry,summary(aov(VaccMyrt~tatry.kclus)))
              Df Sum Sq Mean Sq F value Pr(>F)
tatry.kclus    2  0.324  0.1619   0.089  0.916
Residuals     11 20.033  1.8212
```

You should not interpret the $Pr(>F)$ values in these tests as ordinary Type I error probabilities, because the clusters were defined to maximise the differences between them. Nevertheless, these values provide a useful indication as to which species differ considerably among clusters in their abundance (or, alternatively, on which species the classification is based). *Pinus mugo* abundance differs considerably among the clusters, while *Vaccinium myrtillus* average abundance does not (this species is relatively common along the whole transect).

It might be a reasonable decision to standardise the data by the case norm first (see Section 1.3). Such standardisation can be achieved in R e.g. using the following command (and then using *tatry.std* instead of *tatry* in the above commands).

```
tatry.std <- t(apply( tatry,1,function(x)x/sqrt(sum(x^2))))
```

7.3 Hierarchical classification

In hierarchical classifications, groups are formed that contain subgroups, so there is a hierarchy of levels. When the groups are formed from the bottom (i.e. the method starts with joining the two most similar objects), then the classification is called **agglomerative**. When the classification starts with division of the whole data set into two groups, which are further split, the classification is called **divisive**. The term **cluster analysis** is often used for agglomerative methods only.

7.3.1 Agglomerative classification (cluster analysis)

The aim of these methods is to form a hierarchical classification (i.e. groups containing subgroups) which is usually displayed as a **dendrogram**. The groups are formed 'from the bottom', i.e. the most similar objects are first joined to form the first cluster, which is then considered as a new object, and the joining continues until all the objects are joined in a final cluster, containing all the objects. The procedure has two basic steps: in the

7.3 Hierarchical classification

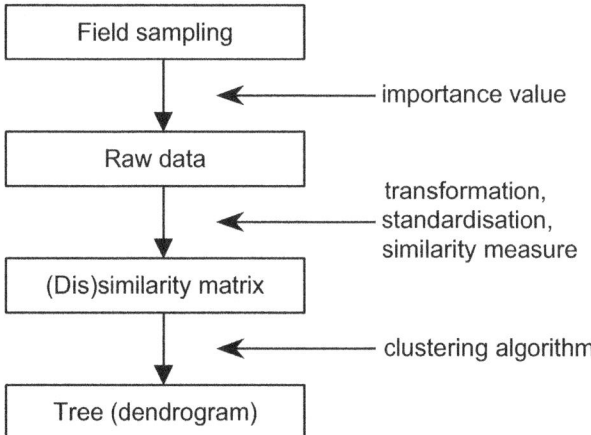

Figure 7-4 The methodological decisions (listed at the right side) for individual analytical steps, affecting the results of an agglomerative hierarchical classification.

first step, the similarity matrix is calculated for all the pairs of objects.[1] In the second step, the objects are clustered (amalgamated) so that, after each amalgamation, the newly formed group is considered to be an object and the similarities of the remaining objects to the newly formed one are recalculated. The individual methods (algorithms) differ in the way they recalculate the similarities.

One should be aware that there are several methodological decisions affecting the final result of a classification (see Figure 7-4).

All classification methods are prone to certain instability caused by the selection of classified objects. Even in the best methods, a single intermediate observation may change completely the topology of the tree. Thus the classification methods are typically more sensitive to the issues of (unavoidable) unevenness in sampling than are the ordination methods.

Agglomerative hierarchical classifications are readily available in most statistical packages. We will demonstrate their use in the R software, but the analytical steps are similar in other software. R, similar to other programs, allows for a direct input of the similarity (or dissimilarity – distance) matrix, but the matrices using distance types widely used in ecology can be easily computed using the *vegdist* function in the *vegan* package. For calculating agglomerative hierarchical clustering, we recommend the use of function *agnes* in the *cluster* package (installed with the R program as a default).

In its basic form, the procedure used in the R program is quite simple:

1. Calculate distances using an appropriate measure (in our examples, you will use the Bray–Curtis distance, which is a complement of percentage similarity).
2. Specify the clustering (amalgamation) method in the call to function *agnes*. You will select the *Complete Linkage* method. There are other possibilities and the decision

---

[1] The matrix is symmetrical, and on the diagonal there are either zeroes – for dissimilarity – or the maximum possible similarity values.

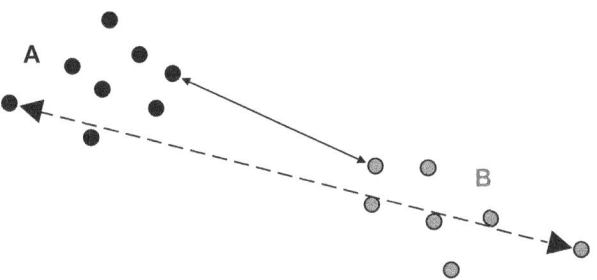

Figure 7–5 Distance of two clusters, A and B, as defined in the single linkage (solid line) and in the complete linkage (dashed line) algorithms.

affects the resulting dendrogram. For example, in the so-called 'short hand' methods (e.g. *Single Linkage*) the distance between the clusters is defined as the distance between the closest points in the two compared clusters. These methods produce dendrograms characterised by a high level of *chaining*. To perform clustering with the Single Linkage algorithm, set the *method* argument to 'single'. The 'long-hand' methods (e.g. *Complete Linkage*), in which the distance of two clusters is defined as the distance of the furthest points, tend to produce compact clusters of roughly equal size (Figure 7–5).

3. After completing the analysis, visualise the resulting classification using a dendrogram.

```
library(cluster)
library(vegan)
d <- vegdist( tatry, method="bray")
tatry.agnes <- agnes(d,method="complete")
plot(tatry.agnes,which=2,sub="",main="")
```

There are many other methods whose underlying approach usually falls somewhere between the two above-described methods. Among them, the **average linkage** method used to be very popular, but this term was used inconsistently. The term **unweighted-pair groups** method (UPGMA) describes its most common variant (set the *method* parameter in the call to *agnes* function to 'average'). In ecology, the 'short hand' (Single Linkage) methods are usually of little use. Compare the resulting dendrograms on the left and right side of Figure 7–6. The results of the complete linkage algorithm (part B) are more ecologically interpretable: when these groups are compared to our external knowledge of the nature of the data, they can be judged as better reflecting the elevation gradient. See Sneath (1966) for a comparison of various clustering methods.

Also, one of the most important decisions greatly affecting the similarity values is data transformation: according to our experience (Kovář & Lepš 1986), the decision whether to use the original measurements (abundance, biomass or cover), the log-transformed values, or the presence–absence data influences the resulting classification more than the selection of clustering method. The standardisation by cases has a major effect when the case totals are very different.

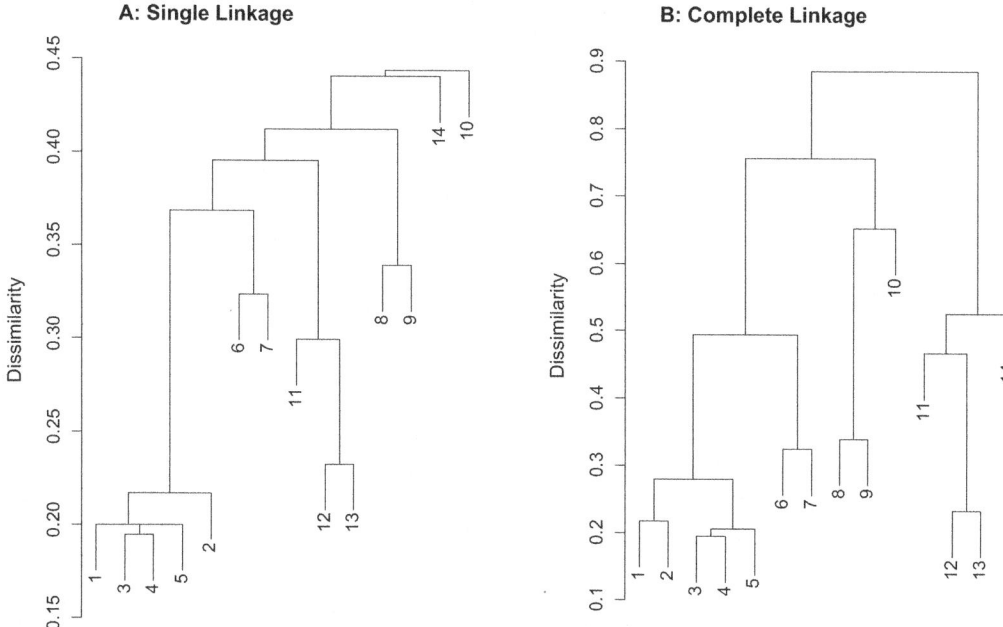

Figure 7–6 Comparison of single linkage (A) and complete linkage (B) clustering of the cases. Note the higher degree of chaining in the dendrogram from single linkage (cases 14 and 10 do not belong to any cluster, but are chained to the large cluster containing all the other cases). The complete linkage results can be interpreted more easily.

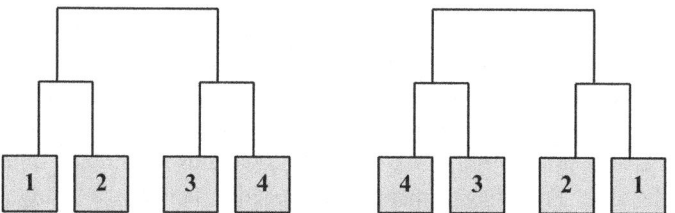

Figure 7–7 Two different dendrograms that represent the same result of cluster analysis. The order of subgroups within a group is arbitrary.

For most ordinary agglomerative classifications the same result can be presented by many different dendrograms. What matters in a dendrogram is which entities (e.g. cases or case joining groups) are joined. It does not matter at all, which of the two groups is on the left and which on the right side. Figure 7–7 shows two dendrograms that represent exactly the same result of cluster analysis. We cannot say, for example, whether cases 1 and 4 are more similar to each other than cases 2 and 3. Unlike in the TWINSPAN method (described below), the orientation of the subgroups in the agglomerative classification dendrogram is arbitrary (usually depends on the order in which the data are entered) and, therefore, should not be interpreted as being a result of the analysis.

It is interesting to compare the results of the classifications with DCA diagram. The results of the detrended correspondence analysis (Figure 7–2) suggest that there is a fairly

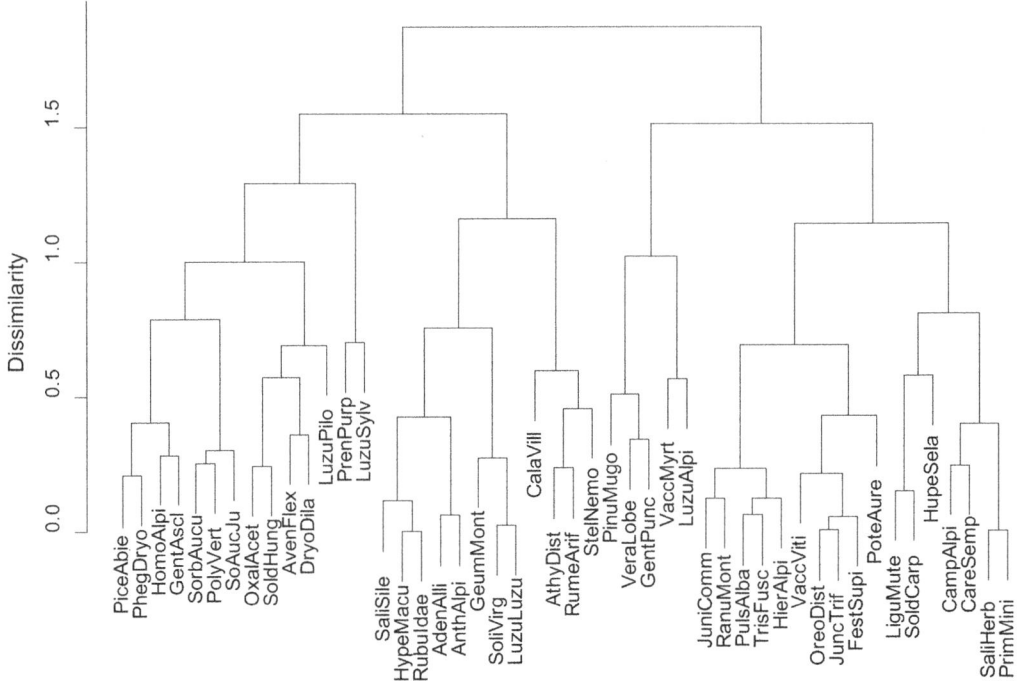

Figure 7–8 The classification of species. The procedure distinguished a group of alpine grassland species reasonably well (*Primula minima, Salix herbacea, Carex sempervirens, Campanula alpina*) on the right side of the diagram. Similarly, species of the spruce forest are on the left side.

homogeneous group of cases 1 to 5 (recorded in spruce forest). All the classification methods also recognised this group as being distinct from the remaining cases. DCA further suggests that there is rather continuous variation with increasing elevation. The individual classifications differ in the way this continuous variation is split into groups. For example, case 11 is in an intermediate position between the sub-alpine shrubs and alpine meadows in the ordination diagram and the classifications differ in their 'decision': the case is allocated to the sub-alpine group in K-means clustering (using implicitly the Euclidean distance) and to the alpine group in hierarchical clustering using the Bray–Curtis distance.

You can also perform the cluster analysis of variables (i.e. of the species in our example). In this case, the correlation coefficient will probably be a reasonable measure of species distributional similarity (to convert it to dissimilarity, $1 - r$ is used). Appropriate similarity measures differ according to whether you cluster cases or species (see also Chapter 6).

```
d2 <- as.dist(1 - cor( tatry))
tatry.spc.agnes <- agnes( d2, method="complete")
plot( tatry.spc.agnes, which=2, sub="", main="")
```

The results of the clustering of species are displayed in Figure 7–8.

7.3.2 Divisive classification

In **divisive classification**, the whole set of objects is divided 'from the top': first, the whole data set is divided into two parts, each part is then considered separately and is divided further. When the division is based on a single attribute (variable, e.g. using a single species), the classification is called **monothetic**, and when based on multiple variables, the classification is **polythetic**. The significance of monothetic methods is mostly a historical one.[2] The classical 'association analysis' was a monothetic method (Williams & Lambert 1959). The advantage of the divisive methods is that each division is accompanied by a rule used for the division, e.g. by a set of species typical for either part of the dichotomy.

The most popular among the divisive classification methods is the TWINSPAN method, described thoroughly in the following section.

7.4 TWINSPAN

The TWINSPAN method (from Two Way INdicator SPecies ANalysis, see Hill 1979, Hill & Šmilauer 2005) is a very popular method (and the program of the same name) among community ecologists and it was partially inspired by the classificatory methods of classical phytosociology (use of indicators for the definition of vegetation types). We treat it separately, based on its frequent use. The idea of an **indicator species** is basically a qualitative one. Consequently, the method works with qualitative data only. In order not to lose the information about the species abundances, the concepts of **pseudo-species** and pseudo-species **cut levels** were introduced. Each species can be represented by several pseudo-species, depending on its quantity in the case. A pseudo-species is present, if the species quantity exceeds the corresponding cut level. Imagine that we selected the following pseudo-species cut levels: 0, 1, 5 and 20. Then the original table is translated to the table used by TWINSPAN as follows:

	Species	Case 1	Case 2
Original Table	*Cirsium oleraceum*	0	1
	Glechoma hederacea	6	0
	Juncus tenuis	15	25
Table with pseudo-species	Cirsoler1	0	1
used in TWINSPAN	Glechede1	1	0
	Glechede2	1	0
	Junctenu1	1	1
	Junctenu2	1	1
	Junctenu3	1	1
	Junctenu4	0	1

In this way, quantitative data are translated into qualitative (presence–absence) data.

---

[2] At least in this context: note that the classification and regression trees (CART) method (Section 8.7) uses recursive division of the data set based on a single variable at each division point; but this is not a classification method in the sense used here, as its aim is to predict response variable values.

Classification methods

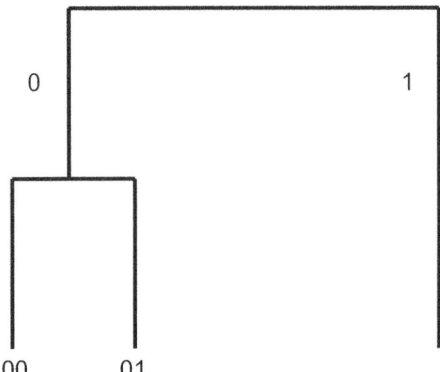

Figure 7-9 Case splits in a TWINSPAN classification.

In TWINSPAN, the dichotomy (division) is constructed on the basis of a correspondence analysis (CA) ordination (see Section 4.7). The ordination is calculated and the cases are divided into the left (negative) side and the right (positive) side of the dichotomy according to their score on the first CA axis. The axis is divided at the centre of gravity (centroid). However, there are usually many cases near the centre of gravity. Consequently, many cases are close to the border, and their classification would depend on many factors. Therefore, a new ordination is constructed, which gives a higher weight to indicator species, i.e. the species preferring one or the other side of the dichotomy. The algorithm is rather complicated, but the aim is to get a polarised ordination, i.e. an ordination where most of the cases are not positioned close to the centre of gravity. Therefore, the classification of the cases is not based so much on the species common to both parts of the dichotomy, but mostly on the species typical of one part of the dichotomy and consequently (in concordance with phytosociological tradition) these species can be considered to be good indicators of particular ecological conditions.

In the first division, the polarity (i.e. which part of the dichotomy will be negative and which positive) is arbitrarily set. In the subsequent divisions, polarity is determined according to the similarity of dichotomy parts to the 'sister' group in higher-level division. For example, in the dendrogram in Figure 7-9, group 01 is more similar to group 1 than group 00 is to group 1. A result of this process is that when the cases are ordered in the final table, we get a table that is similar to an ordered phytosociological table.

Also, the determined 'splitting rule' (i.e. the set of species that were important for a particular division) is printed by the program at each step. This greatly increases the interpretability of the final results. The classification of cases is complemented by a classification of species and the final table is based on this two-way classification.

7.4.1 TWINSPAN analysis of the Tatry data

In the following tutorial, you will use the TWINSPAN method for the analysis of 14 cases (relevés) from the altitudinal transect that you have already used in previous sections. TWINSPAN is useful for large data sets, but the small data set is used here for simplicity.

You will use the file *Tatry.dta*, i.e. the file exported from Canoco 5 project *tatry.c5p* using the Cornell condensed format.[3]

You can use the TWINSPAN for Windows, version 2.3 software (Hill & Šmilauer 2005), freely available from the CEH site.[4] In the sample analysis, you will ask for six cut-levels (in the second, *Cut Levels* wizard page) with the values 0, 1, 2, 3, 4, and 5, keeping all with the weight 1.0 and allowing them to be considered as indicator species.

Because the data were converted into numeric values using ordinal transformation, there is no reason to further 'downweight' the higher values. When the data are in the form of estimated percentage cover, it is reasonable to use the default cut levels, i. e. 0, 2, 5, ... The cut levels 0, 1, 10, 100, 1000 give results corresponding to a logarithmic transformation and are useful when the data are numbers of individuals, differing in order of magnitude.

The following wizard pages allow you to omit from the analysis individual cases or individual species or to change their relative weights, as well as their use as indicator species. While it might be useful to e.g. exclude species that are difficult to identify correctly, we warn you against omitting (or decreasing weight of) species that you expect to be less important for distinguishing community types, as this would be a nice example of circular reasoning.

Other options worth mentioning were chosen in the *TWINSPAN Options* wizard page, as illustrated in Figure 7–10.

The *Minimum group size for division* option implies that a group containing less than five cases is terminal, i.e. it cannot be further divided. For small data sets, it might be reasonable to decrease this value to 4. The *Maximum number of division levels* represents an alternative way to control the divisions, but controlling it by the previous option is a better solution. The value 6 is sufficient even for large data sets.

The *Maximum number of indicators per division* option represents the number of indicator species characterising the polarity of each division. In most cases the default value is reasonable. If you have more than 100 species, the value in the *Number of species in final tabulation* limits the set of species shown in the final sorted table to the most common ones.

After you perform the TWINSPAN project, its execution is logged in the TWINSPAN window, including information about the data and the options chosen. The description of results starts with the information about the first division:

```
DIVISION    1   (N=  14)            I.E. GROUP *
 Eigenvalue  .565  at iteration   1
 INDICATORS, together with their SIGN
  Oreo Dist1(+)
```

The indicator for the first division is *Oreochloa disticha* (the 1 at the end means that the first pseudo-species cut level was used, i.e. presence of the species is enough for the indicator to be considered present).

---

[3] Use the *Data | Export table* menu command (with the table to be exported active in the Canoco 5 workspace) and in the export dialog box select the 'Canoco condensed format'.
[4] Present URL is www.ceh.ac.uk/products/software/wintwins.html

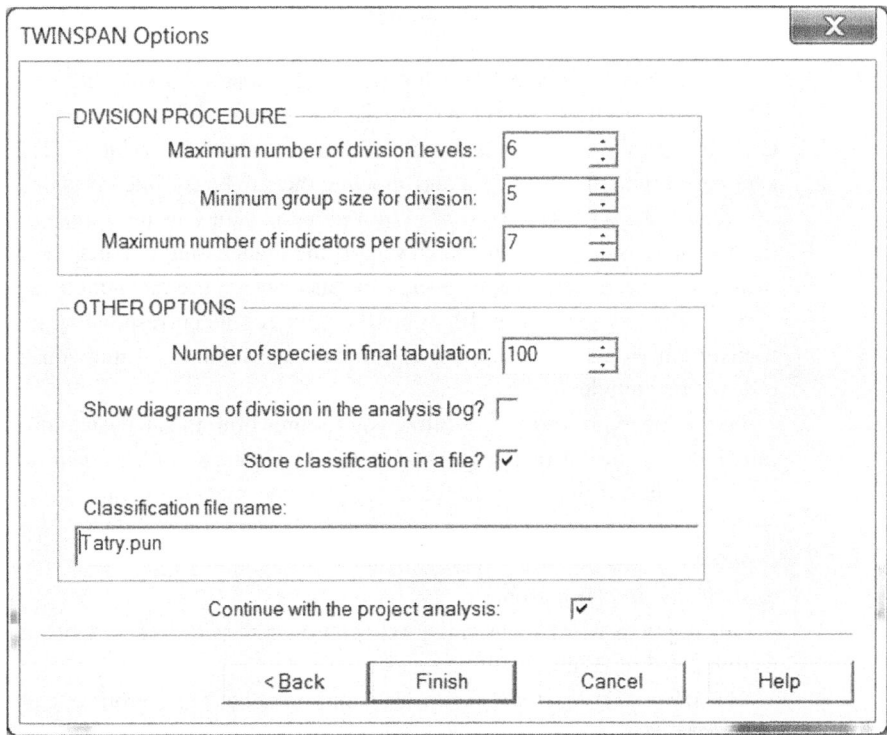

Figure 7-10 TWINSPAN Options chosen for the sample analysis.

```
Maximum indicator score for negative group   0      Minimum indica-
tor score for positive group    1

  Items in NEGATIVE group    2   (N=   10)        i.e. group *0
  P01    P02     P03     P04    P05    P06    P07    P08    P09    P10

  BORDERLINE negatives     (N=   1)
  P09

  Items in POSITIVE group    3   (N=    4)        i.e. group *1
  P11      P12       P13      P14
```

The output displayed above represents the division of cases. Note the warning for case 9 that it was on the border between the two groups.

Next, the species preferring one side of the dichotomy (preferentials) are listed, with the number of occurrences in each of the groups (e.g. *Picea abies* was present in seven cases from the negative group and in one case from the positive group). Note that the preferentials are determined with respect to the number of cases in each group and also for each pseudo-species cut level separately.

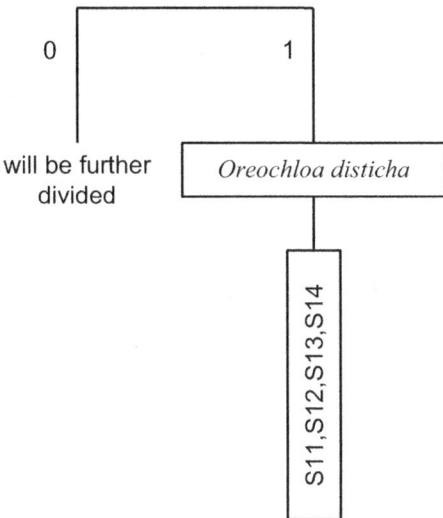

Figure 7–11 The first division in the TWINSPAN example.

```
NEGATIVE PREFERENTIALS
  Pice Abie1(  7,  1) Sorb Aucu1(  7,  0) Oxal Acet1(  7,  0) Sold Hung1(  5,  0)
  Aven Flex1( 10,  1) Gent Ascl1(  8,  0) Dryo Dila1(  8,  0) Pheg Dryo1(  6,  0)
  Pren Purp1(  2,  0) Poly Vert1(  3,  0) SoAu cJu 1(  3,  0) Luzu Pilo1(  2,  0)
etc ...
POSITIVE PREFERENTIALS
  Juni Comm1(  0,  2) Ligu Mute1(  4,  4) Sold Carp1(  2,  4) Ranu Mont1(  1,  2)
  Hupe Sela1(  3,  3) Geun Mont1(  2,  2) Vacc Viti1(  2,  4) Puls Alba1(  0,  2)
  Gent Punc1(  2,  3) Soli Virg1(  1,  1) Luzu Luzu1(  1,  1) Oreo Dist1(  0,  4)
etc ...
NON-PREFERENTIALS
  Pinu Mugo1(  5,  2) Vacc Myrt1( 10,  4) Homo Alpi1( 10,  4) Cala Vill1(  8,  3)
  Rume Arif1(  4,  1) Vera Lobe1(  5,  3) Pinu Mugo2(  5,  2) Vacc Myrt2( 10,  4)
  Homo Alpi2( 10,  4) Cala Vill2(  8,  3) Rume Arif2(  4,  1) Vera Lobe2(  5,  3)
etc ...
     End of level   1
```

You can now start drawing the dendrogram.[5] The part that is now clear is displayed in Figure 7–11.

The positive group is terminal, because it has less than five cases, which is the selected minimum size for division. The next level follows (without the preferentials shown here):

```
DIVISION    2   (N=  10)         I.E. GROUP *0
   Eigenvalue  .344  at iteration   1
```

---

[5] TWINSPAN for Windows does not, unfortunately, provide the possibility of drawing the dendrogram. Instead, you must do it in some general drawing software, such as Adobe Illustrator™.

Classification methods

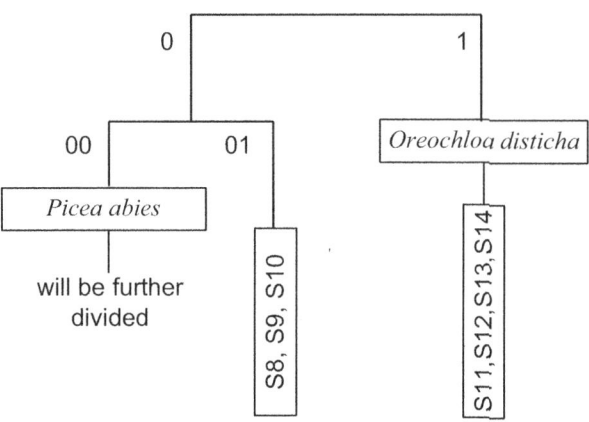

Figure 7–12 The second-level division in the TWINSPAN example.

```
INDICATORS, together with their SIGN
Pice Abie1(-)
Maximum indicator score for negative group   -1     Minimum indica-
tor score for positive group    0

Items in NEGATIVE group    4   (N=    7)         i.e. group *00
P01          P02         P03         P04         P05         P06         P07
Items in POSITIVE group    5   (N=    3)         i.e. group *01
P08          P09         P10
DIVISION    3   (N=    4)           I.E. GROUP *1
DIVISION FAILS - There are too few items
    End of level    2
```

In a similar way, you can continue to construct the dendrogram (Figure 7–12). The next division level is displayed here:

```
DIVISION    4   (N=    7)           I.E. GROUP *00
Eigenvalue   .279   at iteration    1
INDICATORS, together with their SIGN
Pinu Mugo1(+)
Maximum indicator score for negative group    0    Minimum indica-
tor   score for positive group    1

Items in NEGATIVE group    8   (N=    5)         i.e. group *000
P01          P02         P03         P04         P05
Items in POSITIVE group    9   (N=    2)         i.e. group *001
P06          P07
```

Note the meaning of indicator species *Pinus mugo* here. It is an indicator within group 00 (containing seven cases), where it is present just in two of the cases, 6 and 7 (forming

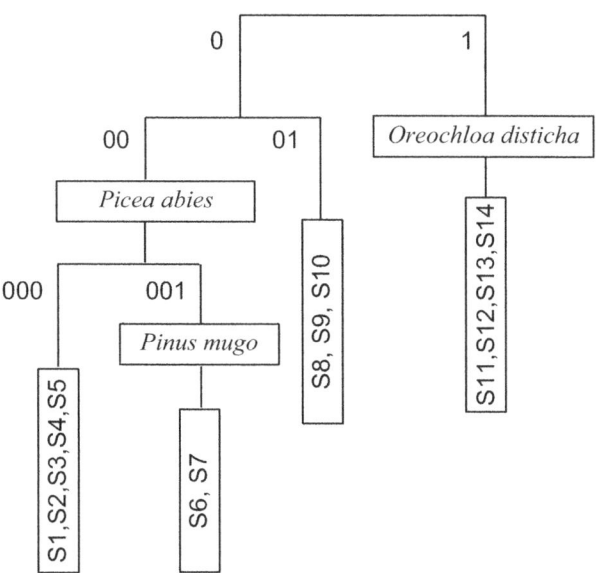

Figure 7–13 Final state of the TWINSPAN classification for the sample data.

the positive group 001). But it is also common in cases 8, 9, 10, 11 and 12. This illustrates that the indicators are determined and should be interpreted just in a particular division context, not for the whole data set. Group 001 is characterised by the presence of *Pinus mugo* in contrast to group 000, and not within the whole data set. The advantage of TWINSPAN becomes clear here, as the orientation of groups (i.e. which group will be negative and which positive) depends on which of the groups is more similar to group 01 (and this group, among others, has *Pinus mugo* in all of its cases).

We will not show further divisions, (the group 000 contains five cases and still can be divided further), and you can finish the dendrogram at this level (Figure 7–13). The fact that each division is accompanied by only one indicator is rather rare (it is a consequence of the small data size). Usually, each division is accompanied by several indicators for each part of the dichotomy. The indicators are the species used in the divisions. They might be very few and might be available only for one side of the dichotomy. This does not mean that there are no more species that characterise one of the dichotomy sides. If you want to characterise the division by more species, you should use the 'preferentials' from the TWINSPAN output.

In a similar way, TWINSPAN also calculates hierarchical classification for species.

Finally, the TWINSPAN output contains the data table, sorted based on classification results. It resembles a classical ordered phytosociological table and is accompanied by the classification of both cases and species (Table 7–1).

You can read the three levels of division and memberships of all the cases in corresponding groups from the bottom three lines. For example, cases 11–14 are in group 1 (which is not further divided).

Table 7–1 Sorted data table at the end of TWINSPAN output.

```
                                PPPPPPPPPPPPPP
                                00000000011111
                                21345678901234
 4      Sali    Sile            ---------5----      0000
29      Anth    Alpi            -----2---3----      0000
30      Hype    Macu            ------2--3----      0000
31      Rubu    Idae            -----2---3----      0000
28      Aden    Alli            -----2---2----      0001
 1      Pice    Abie            6666665---5---      001000
 7      Oxal    Acet            55344-4--3----      001001
 9      Sold    Hung            43444---------      001001
18      Luzu    Pilo            2-2-----------      001001
20      Luzu    Sylv            --3243--------      001001
12      Gent    Ascl            23333333------      001010
14      Pheg    Dryo            4333-33-------      001010
15      Pren    Purp            2----3--------      001011
16      Poly    Vert            3----33-------      001011
22      Stel    Nemo            ---2--3-------      001011
 2      Sorb    Aucu            42-23444------      00110
10      Aven    Flex            34543443343---      00110
13      Dryo    Dila            333333-3-3----      00110
17      SoAu    cJu             3-----32------      00110
19      Athy    Dist            3-23335--53---      00111
 6      Vacc    Myrt            54646666636653      01
 8      Homo    Alpi            44454454334334      01
11      Cala    Vill            33365-54-6445-      01
21      Rume    Arif            --23--4--33---      01
 3      Pinu    Mugo            -----3666665--      10
23      Vera    Lobe            ----2333-4322-      10
27      Hupe    Sela            -----223--22-3      10
36      Soli    Virg            ---------2--2-      10
33      Vacc    Viti            -------33-3343      1100
35      Gent    Punc            -------3-4333-      1100
37      Luzu    Luzu            ---------3--4-      1100
24      Ligu    Mute            ----233--23333      1101
25      Sold    Carp            -----54---3333      1101
26      Ranu    Mont            -----2-----33-      1101
32      Geun    Mont            ------2--3-33-      1101
 5      Juni    Comm            -----------24-      111
34      Puls    Alba            -----------32-      111
38      Oreo    Dist            ----------5564      111
39      Fest    Supi            ----------3444      111
40      Camp    Alpi            ----------34-4      111
41      Junc    Trif            ----------4453      111
42      Luzu    Alpi            ----------33--      111
43      Hier    Alpi            ----------233-      111
44      Care    Semp            -----------545      111
45      Tris    Fusc            -----------33-      111
46      Pote    Aure            ------------32      111
47      Sali    Herb            -------------5      111
48      Prim    Mini            -------------4      111
                                00000000001111
                                0000000111
                                0000011
```

8 Regression methods

Regression models allow us to describe the dependence of (usually) one response variable (which might be quantitative or qualitative) on one or more predictor variables. These can be quantitative variables and/or factors. In this broad view, regression models also include statistical methods such as the analysis of variance (ANOVA) and the analysis of contingency tables. During the 1980s, many new types of regression models were suggested, usually extending in some way the well-established ones, and some of them are also briefly summarised in this chapter.

In the concluding, tutorial part of this chapter, we focus on the use of some of the introduced methods in a posteriori analysis of the patterns of individual response variables (such as biological species) or predictors (e.g. environmental descriptors) in ordination space, or generally in exploration of ecological data, using the Canoco 5 program.

8.1 Regression models in general

All regression models share some assumptions about the response variables and the predictors. To introduce these concepts, we will restrict our attention to the most frequently used kind of regression model, where exactly one response variable is modelled using one or more predictors.

The simplest way to describe any such type of regression model is the following:

$$Y = EY + e$$

where Y refers to the values of response variable, EY is the value of the response variable expected for particular values of the predictor(s) and e is the variability of the true values around those expected values EY. The expected value of the response can be formally described as a function of the predictor values:

$$EY = f(X_1, \ldots, X_p)$$

The EY part is often called the **systematic component** of the regression model, while the e part is called the **stochastic component** (or error component) of the model. The properties and roles they have, when we apply regression models to our data, are compared in Table 8–1.

Table 8-1 Summary of the differences between systematic and stochastic components of a statistical model

Systematic component (EY)	Stochastic component (e)
Is determined by our research hypothesis	Mirrors a priori assumptions of the model
Its parameters (e.g. regression coefficients) are estimated by fitting the model	Its parameter(s) are estimated during or after fitting (variance of response variable)
We interpret it and perform (partial) tests on it (its individual parameters)	We use it to judge the model quality (using the methods of **regression diagnostics**)

When we fit a regression model to our data, our assumptions about its stochastic component are fixed,[1] but we can manipulate the content and complexity of the systematic component. In the simplest example of the classical linear regression model with one response variable Y and one predictor X, the systematic component can be specified as:

$$EY = f(X) = \beta_0 + \beta_1 \cdot X$$

But we can, in fact, achieve a larger complexity in modelling the dependence of Y on X by a polynomial model. We can shift the model complexity along a range starting with a **null model** $EY = \beta_0$, through the linear relation given above, to the quadratic polynomial relation $EY = \beta_0 + \beta_1 \cdot X + \beta_2 \cdot X^2$, up to a polynomial of the n-th degree, where n is the number of data observations we have available, minus one. This most complex polynomial model (representing the so-called **full model**) goes exactly through all the data points, but provides no simplification of the reality (which is one of the basic tasks of a statistical model): we have simply replaced the n data points with n regression parameters ($\beta_0, \ldots, \beta_{n-1}$). The null model, on the other hand, simplifies the reality so much that we do not learn anything new from such a model,[2] and the advance of our knowledge is another essential service provided by statistical models.

From our discussion of the two extremes of regression model complexity, we can clearly see that the selection of model complexity spans a gradient from simple, not so precise models to (overly) complex models. In fact, this issue is more general than the gradient between a linear model and n-th degree polynomial of a single predictor X. A similar gradient exists whenever you choose between a model with just one or a few predictor variables and models with many predictors. Also in this case, using $n - 1$ (or more) different predictors in a model fitted from a data set with n cases guarantees that the model explains all the variation in the response variable, but it is useless at the same time.[3]

---

[1] Usually assumptions about the distributional properties, the independence of individual observations, or a particular type of cross-dependence between the individual observations.
[2] Not a completely correct statement: from a null model we learn about the average value of the response variable.
[3] Similarly for a constrained ordination, using $n - 1$ or more explanatory variables for a data set with n cases means that the ordination (RDA or CCA) ceases to be constrained and the resulting ordination space (as far as the size of variation explained by individual axes and the scores of cases and response variables are concerned) becomes identical with an unconstrained ordination (PCA or CA).

The more complicated models have another undesired property: they are too well fitted to our data sample, but they provide unreliable (too variable) predictions for the whole statistical population we have sampled.[4] Our task is, in general terms, to find a compromise, a model as simple as possible for it to be useful, but neither more simple nor more complex than that. Such a model is often referred to as a **parsimonious** model or the minimal adequate model.

8.2 General linear model: terms

The first important stop on our tour over the families of modern regression methods is the general linear model. Note the word **general** – another type of regression uses the word **generalized** in the same position, but meaning something different. In fact, the generalized linear model (GLM), discussed in the next section, is based on the general linear model (discussed here) and represents its generalisation.[5]

What makes the general linear model different from the traditional linear regression model is, from the user's point of view, mainly that both the quantitative variables and qualitative variables (factors) can be used as predictors.[6] Therefore, the methods of analysis of variance (ANOVA) belong to the family of general linear models. For simplicity, we can imagine that any such factor is replaced (encoded) by $k - 1$ 'dummy', 0/1 variables, if the factor has k different levels.

In this way, we can represent the general linear model by the following equation:

$$Y_i = \beta_0 + \sum_{j=1}^{p} \beta_j \cdot X_{ji} + \varepsilon$$

but we must realise that a factor is usually represented by more than one predictor X_j and, therefore, by more than one regression coefficient.[7] The symbol ε refers to a random variable representing the stochastic component of the regression model. In the context of general linear models, this variable is most often assumed to have a zero mean and a constant variance and independence in its values (but see footnote 6); the parametric tests often assume even its Gaussian (normal) distribution.

The equation of the general linear model immediately shows one very important property of the general linear model – it is **additive**. The effects of individual predictors are mutually independent.[8] If we increase, for example, the value of one of the predictors

---

[4] This is the bias–variance trade-off along the gradient of model complexity (see Hastie et al. 2002).
[5] We mean this sentence seriously, really ☺.
[6] For many statisticians, a general linear model also offers the possibility of specifying covariances (correlations) among individual cases, i.e. to drop the assumption of independence among individual observations.
[7] Additionally, effects of individual variables can be described in more complex ways than as a single linear term, e.g. using polynomials introduced in Section 8.1. Each polynomial term is then treated as a separate predictor term in the model, at the same level as the terms referring to other variables.
[8] This does not mean that the predictors cannot be correlated, only the model formulation does not account for the correlation. Additionally, we ignore here the use of interaction terms that describe mutual dependency of predictor variables.

by one unit, this has a constant effect (expressed by the value of the regression coefficient corresponding to that variable), independent of the values the other variables have and even independent of the original value of the variable we are incrementing.[9]

The above-given equation refers to the (theoretical) population of all possible observations, which we sample when collecting our actual data set. Based on such a finite sample, we estimate the true values of the regression coefficients β_j, and these estimates are usually labelled as b_j. Estimation of the values of regression coefficients is what we usually mean when we refer to **fitting** a model. When we take the observed values of the predictors, we can calculate the **fitted** (predicted) values of the response variable as

$$\hat{Y} = b_0 + \sum_{j=1}^{p} b_j \cdot X_j$$

The fitted values allow us to estimate the realisations of the random variable (ε_i, representing the stochastic component); such a realisation for the i-th case is called the **regression residual** and labelled as e_i:

$$e_i = Y_i - \hat{Y}_i$$

Therefore, the residual is the difference between the observed value of the response variable and the corresponding value predicted by the fitted regression model.

The variability of the response variable can be expressed by the **total sum of squares**, defined as

$$TSS = \sum_{i=1}^{n} (Y_i - \bar{Y})^2$$

where $\bar{Y}$ is the mean of Y.

From the point of view of a fitted regression model, this amount of variability can be further divided into two parts – the variability of the response variable explained by the fitted model – the **model sum of squares**, defined as

$$MSS = \sum_{i=1}^{n} (\hat{Y}_i - \bar{Y})^2$$

and the **residual sum of squares** defined as

$$RSS = \sum_{i=1}^{n} (Y_i - \hat{Y}_i)^2 = \sum_{i=1}^{n} e_i^2$$

Obviously $TSS = MSS + RSS$. We can use these statistics to test the significance of the model. Under the global null hypothesis ('the response variable is independent of

---

[9] The content of this paragraph seems obvious, but we find people are often applying regression models in a context which is at odds with these assumptions. If we for example model the effects of nutrient availability on species abundance, we can often expect a multiplicative effect of the predictor variable: its unit change is reflected in a constant proportional (relative) change in the species abundance. Such a relation would agree with the linear model nature only when we log-transform the species abundance values (see also Section 1.3.1).

the predictors') MSS is not different from RSS if both are divided by their respective number of degrees of freedom.[10]

Additionally, we can use the proportion of explained variation out of the total one (i.e. MSS/TSS) as a measure of model quality – the proportion of variation explained by the model's systematic part. This measure is called **coefficient of determination** and usually labelled as R^2.[11] Unfortunately, the coefficient of determination is biased (over-estimated) when the number of observations used to fit the regression model is not substantially larger than the number of model parameters. In fact, when the number of model parameters approaches the number of observations, R^2 approaches value 1, irrespectively of the actual quality of predictors. It is therefore advisable to replace it, when reporting model quality, with the so-called **adjusted coefficient of determination** (R^2_{adj}), first proposed by Ezekiel (1930).

8.3 Generalized linear models (GLM)

Generalized linear models (McCullagh & Nelder, 1989) extend the general linear model in two important ways.

First, the **expected** values of the response variable (EY) are not always assumed to be directly equal to the linear combination of the predictor variables. Rather, the scale of the response depends on the scale of the predictors through some simple parametric function called the **link function**:

$$g(EY) = \eta$$

where η is the **linear predictor** and is defined in the same way as the whole systematic component of the **general** linear model, namely as:

$$\eta = \beta_0 + \Sigma \beta_j X_j$$

The use of the link function g has the advantage that it allows us to map values from the whole real-valued scale of the linear predictor (reaching, generally, from $-\infty$ to $+\infty$) into a specific interval making more sense for the response variable (such as non-negative values for counts or values between 0 and 1 for probabilities).

Second, generalized linear models have more flexible assumptions about the stochastic component compared to general linear models. The variance needs not be constant but can depend on the expected value of the response variable, EY. This mean–variance relation is usually specified through the statistical distribution assumed for the stochastic part (and, therefore, for the response variable). But note that the mean–variance

---

[10] This division leads to so-called mean squares; see any statistical textbook for more details, e.g. Sokal & Rohlf (1995).
[11] This notation reminds you that the coefficient of determination can be alternatively calculated as the second power of the linear correlation between observed (Y) and fitted ($\hat{Y}$) values of the response variable.

Table 8–2 Summary of useful combinations of link functions and types of response variable distributions. The assumed relation between the variance (V) of the stochastic part and the expected values of response variable (EY) is also indicated.

Type of variables	'Typical' link function	Reference distribution	Variance function (mean-variance relation)
counts (frequency)	log	Poisson	$V \propto EY$
probability (relative frequency)	logit or probit	binomial	$V \propto EY^*(1 - EY)$
dimensions, ratios	inverse or log	gamma	$V \propto EY^2$
quite rare type of measurements[12]	identity	Gaussian ('normal')	$V = $ const

relation (described by the **variance function**, see Table 8–2), not the specific statistical distribution, is the essential property of the model specification.[13]

The options we have for the link functions and for the assumed type of distribution of the response variable cannot be combined independently, however. For example, the **logit** link function maps the real scale onto a range from 0 to +1, so it is not a good link function for, say, an assumed Poisson distribution (see Table 8–2), and it is useful mostly for modelling probability as a parameter of the binomial distribution. Table 8–2 lists some typical combinations of the assumed link functions and the expected distribution of the response variable, together with a short characteristic of the response variables matching these assumptions. As with the classical linear regression models and ANOVA, it is not assumed that the response variable has the particular types of distribution, but that it can be **reasonably approximated** by such a distribution.

With this knowledge, we can summarise what kinds of regression models are embraced by the GLMs:

- 'classical' general linear models (including simple types of analysis of variance): they correspond to the identity link function and Gaussian distribution in Table 8–2
- extension of those classical linear models to variables with non-constant variance (counts, relative frequencies)
- analysis of contingency tables (using log-linear models)
- models of survival probabilities used in toxicology (probit analysis)

The generalized linear models extend the concept of a residual sum of squares. The extent of discrepancy between the true values of the response variable and those predicted by the

---

[12] With this tongue-in-cheek suggestion, we do not claim that classical linear regression or ANOVA are useless, but rather that for many real-world response variables, there are better choices, as suggested by other entries in Table 8–2.

[13] The ability to further parameterise or specify the variance function beyond the one implied by the assumed distribution is known as the quasi-likelihood approach to generalized linear modelling. This finds use e.g. for count data where the variance is not equal to mean (as one would expect for Poisson distribution), being instead proportional to it. When the coefficient of proportionality is substantially larger than 1, we say that the response variable is *overdispersed*. This often happens, for biological data, as a manifestation of gregarious spatial distribution of the individuals.

model is expressed by the model's **deviance**. Therefore, to assess the quality of a model, we use statistical tests based on the **analysis of deviance**, quite similar in concept to an analysis of variance of a regression model, despite not necessarily employing the same statistic for testing hypotheses.[14]

An important property of the general linear model, namely its **linearity**, is retained in the generalized linear models on the scale of the linear predictor. The effect of a particular predictor variable is expressed by a single parameter – the regression coefficient, which describes a linear transformation. Similarly, the model additivity is kept on the linear predictor scale. On the scale of the response variable, things might look differently, however. For example with a logarithmic link function, the additivity on the scale of the linear predictor corresponds to a multiplicative effect on the response variable scale.

Canoco 5 offers generalized linear models (and being part of them, the general linear models as well) as a part of its attribute plots, where the fitted models are presented as contour plots or as curves, depending on the plot type (see Figure 8–6 or Figure 11–18(d)). GLM is also offered as a specialised analysis template. The range of available tools for further work with fitted GLMs (namely those related to regression diagnostics) in Canoco is limited, when compared with general statistical packages such as the R software.

8.4 Loess smoother

The term **smoother** is used for a method of deriving a regression function (often less parametric) from observations. The fitted values produced by smoothing (i.e. by application of a smoother) are less variable than the observed ones (hence the name 'smoother').[15]

There are several types of smoothers, some of them not very good, but simple to understand. An example of such a smoother is the moving average smoother. An example of a better smoother is the **loess smoother** (earlier also named *lowess*). This smoother is based on a locally weighted linear regression (Cleveland & Devlin 1988; Hastie & Tibshirani 1990). The size of the area (band for a single predictor) around the estimation point, which is used to select data for the local regression model fit, is called the **bandwidth** and it is specified as a fraction of the total size of the available data set. Therefore, bandwidth value $\alpha = 0.5$ specifies that at each estimation point half of the observations (those closest to the considered combination of the predictors' values) are used in the regression.

The complexity of the local linear regression model is specified by the second parameter of the loess smoother, called **degree** (λ). Typically, only two degree values are

---

[14] When testing models with a response variable with assumed binomial or Poisson distribution, a χ^2 statistic is used for testing, except for the quasi-likelihood context with notable over-dispersion or under-dispersion, when the F statistic must be used.

[15] We can therefore see even the linear models discussed in previous sections as smoothers, but there the resulting smoother model is very rigid, as it can be described by very few numeric parameters (regression coefficients).

considered: 1 for a linear regression model and 2 for a second-order polynomial model. The choice of the degree parameter may be based on the complexity of the expected response curve (or surface): for a simple response, interpretable in a context of a general ecological theory (e.g. monotonous increase of species abundance or its increase to a certain limit, or unimodal response – even non-symmetric), the choice of degree = 1 is usually appropriate; for complex and/or sudden changes of response values with the values of predictor variables, degree = 2 would work better.[16]

Furthermore, the data points used to fit the local regression model do not enter it with the same weight. Their weights depend on their distance from the considered estimation point in the predictors' space. A data point, which has exactly the required values of the predictors, has its weight set to 1.0, and the weight smoothly decreases to 0.0 at the edge of the smoother bandwidth. The exact form of the weight function might vary, but most implementations use the so-called tricubic weight function.

An important feature of the loess regression model is that we can express its complexity using the same units as in traditional linear regression models – the number of degrees of freedom (DF) taken from the data by the fitted model. It is, alternatively, called the **equivalent number of parameters**. Further, because the loess model produces fitted values of the response variable (like other models), we can work out the variability in the values of the response accounted for by the fitted model and compare it with the residual sum of squares. As we have the number of DFs of the model estimated, we can calculate the residual DFs and calculate the sum of squares per one degree of freedom (corresponding to the mean square in an analysis of variance of a classical regression model). Consequently, we can compare loess models using an analysis of variance in the same way we do so for the general linear models.

Instead of the loess model, we can use various types of smoothing splines, see Eubank (1988) for more details.

Canoco 5 offers fitting of the loess smoother in its visualising tools, namely as a non-parametric response surface in ordination space or in XYZ diagrams or as a curve in XY diagrams, including species-response curve plots (see Figure 8–3 or Figure 11–18c). The *locfit* package in R, however, offers a greater range of functionality, including its ability to select optimum smoothing parameters (bandwidth and/or degree) based on a cross-validation procedure.

8.5 Generalized additive models (GAM)

The generalized additive models (GAMs, Hastie & Tibshirani 1990) provide a useful extension to generalized linear models (GLMs). The strength of GAMs lies in their less rigid description of the shape of the effects that the individual predictors have on the modelled response variable. In fact, GAMs cannot be easily summarised numerically, in

---

[16] The choice of degree parameter interacts with the model bandwidth choice: for degree value 2, the bandwidth must be larger (i.e. the model more global) to achieve the same model complexity as for degree value 1. We recommend the reader starts with degree choice and then adjusts the bandwidth value.

contrast to generalized linear models where their primary parameters (regression coefficients) summarise the shape of the regression model. A fitted GAM is best summarised by plotting the estimated smooth terms representing the relation between the values of a predictor and its effect on the modelled response variable.

The so-called **additive predictor** (η_A below) replaces here the linear predictor of a GLM. It is also defined as a sum of independent contributions of the individual predictors, but the effect of a particular predictor variable is not summarised using a simple regression coefficient. Instead, it is expressed – for the j-th predictor variable – by a smooth function s_j, describing the transformation from the predictor values to the (additive) effect of that predictor upon the expected values of the response variable:

$$\eta_A = \beta_0 + \Sigma s_j(X_j)$$

The additive predictor scale is again related to the scale of the response variable via a link function g.

GAMs include generalized linear models (GLMs) as a special case, where for each of the predictors the transformation function is defined as:

$$s_j(X_j) = \beta_j \cdot X_j$$

In the more general case, smooth transformation functions (usually called 'smooth terms') are fitted using a loess smoother (discussed in the preceding section) or a cubic spline smoother[17]. When fitting a GAM, we do not prescribe the shape of the smooth functions of the individual predictors, but we often specify the complexity of the individual curves, in terms of their degrees of freedom. We also need to select the type of smoother used to find the shape of the smooth transformation functions for individual predictors.

With GAMs, we can do a stepwise model selection including not only a selection of the predictors used in the systematic part of the model but also a selection of complexity in their smooth terms.

As for the loess smoother, Canoco 5 offers fitting GAMs as a tool to visualise response surfaces or response curves in its various attribute plots (see Figure 8–11 or Figure 16–14). We recommend their use particularly when visualising the relation of individual response variables (or various community characteristics, e.g. diversity) to ordination axes or to measured explanatory variables. Full implementation of GAMs, including automatic selection of smooth term complexity using a cross-validation approach, is available in the *mgcv* package of the R software.

8.6 Mixed-effect models (LMM, GLMM and GAMM)

The regression models discussed in the preceding sections (LM, GLM, loess, GAM) all assume (in their standard setup and always in their implementation in Canoco 5) that

---

[17] Cubic splines are related in a more straightforward way to the linear models than the loess smoother, and this is one of the reasons why they are nowadays preferred when fitting GAMs.

the individual observations (cases) are mutually independent. But such an assumption is often not appropriate for real data sets in the field of ecology: plots recorded in the same block of a completely randomised block design, cases measured at the same location at different times, or nearby plots on a linear transect or rectangular grid will tend to have their values (in both response and predictor variables) correlated, when compared with the plots of different blocks, cases measured in different locations, or distant plots on a transect or a grid – see Chapters 3 and 5. When we decide to ignore such correlations existing in our data set, we pay a heavy price in the bias of significance estimates that are so central in classical inferential statistics: assuming independence among cases which are in fact non-independent leads to the true probabilities of Type I errors being higher than computed, and so our conclusions about tested hypotheses are not correct.

In the general linear models using factors as predictors (i.e. in various ANOVA models), the above-mentioned situations were accounted for by various specialised forms of ANOVA, with varying names assigned by various software packages (split-plot ANOVA, repeated measures ANOVA, advanced linear model, etc.). In all these specialised ANOVA models, the internal structure of the set of available cases (presence of blocks, identity of repeatedly measured plots, identity of lines and rows in Latin square grid-like design) is described with additional factors, usually referred to as (factors with) **random effects**. This reflects the fact that the individual levels of such factors (plot, block, row, or column identity) are not interesting per se, as they usually represent a random selection from a potentially infinite number of instances. Using these random effects in linear models is an alternative path to specify correlations among individual data set cases.[18]

For regression models using quantitative (numeric) predictors or a mixture of numeric predictor and factors, the realisation of the need to account for dependence among groups of cases came later than for ANOVA models. Because regression models (similarly to ANOVAs) do usually have also some fixed effect factors (experimentally manipulated or measured numeric predictors, management types, sampling time, etc.), the inclusion of random effects, usually (but not always) reflecting only the design details, changed the name of original regression models into so-called **mixed-effect regression models**: mixed-effect linear models (LMM or LME) instead of the original (general) linear model, generalized mixed-effect models (GLMM) instead of GLM, or generalized additive mixed-effect models (GAMM). An interesting use of GLMM in the context of multivariate analysis of ecological data was suggested by Ives and Helmus (2011) and Jamil et al. (2013).

It is perhaps interesting to note that the concern about correctly estimating Type I error probabilities when testing hypotheses using data sets with non-trivial sampling design is exactly the same that led to the development of the advanced (restricted)

---

[18] In the simplest case of a single random effect factor defining blocks of observations, the membership of cases in the same group defines a constant value of cross-correlation among them, with its size depending on the relation between the variance within the blocks versus the variance among the blocks.

permutation types in CANOCO software (see Sections 5.2 and 5.3). The adopted path is of course different in its implementation details,[19] but the achieved results are essentially equivalent. Canoco 5 implements restricted permutations also for generalized linear models,[20] so that one can say that it offers an overall significance test of the systematic part (with fixed-effect predictors) of the model, equivalent to tests coming from LMM and GLMM. The restriction to balanced complete data in Canoco is lifted in LMM and GLMM.

For a full functionality (including tests of random effect factors and of individual fixed effect predictors) and range of diagnostic tools, we recommend one of the packages available in the R software: the *lme4* package is a good start for (generalized) mixed-effect models and the *nlme* package is a good starting choice for non-linear mixed-effect models or for explicit modelling of error structure, although other existing packages might be preferred depending on specific circumstances or our approach to hypotheses testing.

8.7 Classification and regression trees (CART)

Tree-based models are probably the most non-parametric kind of regression models one can use to describe the dependence of the response variable values on the values of the predictor variables. They are defined by a recursive binary partitioning of the data set into subgroups that are successively more and more homogeneous in the values of the response variable.[21] At each partitioning step, exactly one of the available predictors is used to define the binary split. The split that maximises the homogeneity and the difference between the resulting two subgroups is selected. When defining a split, either a quantitative or qualitative variable can be chosen (see Figure 16–16 for an example).

The response variable is either quantitative (in the case of **regression trees**) or qualitative (for **classification trees**). The results of the fitting are described by a 'tree' portraying the successive splits. Each branching is described by a specific splitting rule: this rule has the form of an inequality when a quantitative predictor is chosen (e.g. 'Variable X2 < *const*') or the form of an enumeration of the possible values for a chosen factor (e.g. 'Variable X4 has value a, c, or d'). The two subgroups defined by a split rule are further (recursively) subdivided, until they are too small or sufficiently homogeneous in the values of the response. The terminal groups (leaves) are then identified by a predicted value of the response value (if this is a quantitative variable in a regression tree) or by

---

[19] The restriction of permutations does not result in a different pseudo-F statistic, but rather in a different estimated distribution of the statistic values from the permutations under the null model.

[20] Permutation-based tests are offered only when a GLM is fitted as a part of an analysis, e.g. by means of the *Generalized-linear-model* template in the *Specialized Analyses* folder of the New Analysis Wizard.

[21] The division is similar to recursive binary splitting done by TWINSPAN, but in TWINSPAN (a) there is no distinction of predictor and response variables, (b) a multivariate data set is split, and (c) also the splitting criterion is different.

a prediction of the object membership in a class (if the response variable is a factor in a classification tree).

When we fit a tree-based model to a data set, we typically create ('grow') an over-complicated tree based on our data. Then we try to find an optimum size for the tree to predict response values. A **cross-validation procedure** is used to determine the optimum size of the tree. In this procedure, a series of progressively reduced ('pruned') trees is created using only a subset of the available data and then the remaining part of the data set is used to assess the performance of the created tree: these observations are passed through the hierarchical system of the splitting rules and the value of the response variable predicted by the tree is compared with its known value. For each size ('complexity') of the tree model, this is done several times. Typically, the data set is split into 10 parts of approximately the same size and, for each of these parts, a tree model of given complexity is fitted using the remaining nine parts and the selected part is used for performance assessment. A graph of the dependency of the tree model 'quality' on its complexity (model size) has typically a minimum corresponding to the optimum model size. If we use a larger tree, the model is overfitted – it provides a closer approximation to the data, but a biased description of the sampled population.

Canoco 5 does not implement CART models at all – the *rpart* package for the R software is recommended as a starting tool for fitting these models. Interestingly, a version of the regression tree method for the analysis of multivariate response data (called multivariate regression trees, see De'ath 2002, also implemented in R software in the *mvpart* package) was developed, and it can be seen as a counterpart of constrained ordination in the classificatory approach to data analysis. Unlike the (constrained) ordination, it cannot help us to find (and test) continuous gradients of compositional change, but it is able to describe in a very natural way the situations where different environmental factors affect community composition in different data subsets or non-trivial interactions among different environmental descriptors.

8.8 Modelling species response curves with Canoco

This section is a tutorial for fitting the various regression models that describe relation between a quantity of a particular species and the environmental gradients or gradients of community variation. With a single predictor variable, the resulting fitted curve is usually called a species response curve. The process of fitting various regression models (generalized linear models, generalized additive models and loess smoother) will be illustrated using the tools available in Canoco 5 software.

In this tutorial, you will study the change in abundance of six selected sedge species in wet meadow communities, along the compositional gradient represented by the first ordination axis of DCA. This sample study uses the same data set that is used in Case study 2 (Chapter 13), and you should check that chapter for a more detailed description of the data set. You can also check the *regress.c5p* example project, with *DCA* analysis, where all the models were fitted and visualised. Selected sedge species belong to the more frequently occurring ones in the collected data and their optima (approximated by

8.8 Modelling species response curves with Canoco

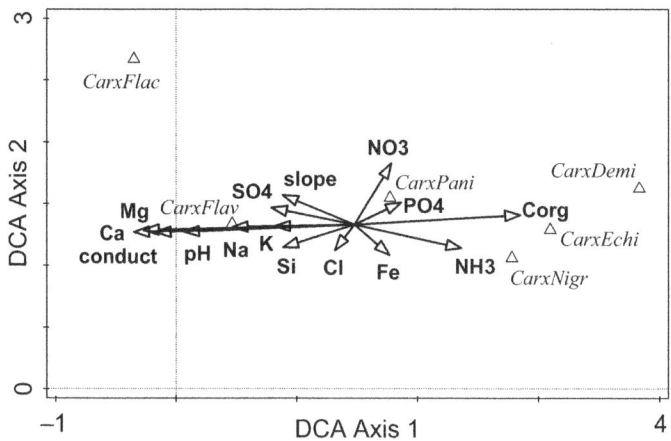

Figure 8–1 Biplot of environmental variables and the selected six *Carex* species, based on a DCA.

their scores in the ordination space) span the whole range of compositional variation in the sampled community types, as can be seen from Figure 8–1.

The first ordination axis represents a gradient of increasing availability of nutrients in underground water (from left to right in the diagram) and decreasing concentration of Ca, Mg and Na ions. In your regression models describing the change of species abundances (the 'species response curves') you will use the scores of individual cases on the first DCA axis as the explanatory variable. The results of the detrended correspondence analysis are stored in a Canoco 5 project named *regress.c5p*, in the analysis named *DCA*. Open the project in Canoco if you want to work through this tutorial.

Start with building a regression model for *Carex panicea* (labelled as *CarxPani* in the sample project). To get your first impression about the change of abundance of this species along the first ordination axis, create an XY-scatterplot and supplement it with the non-parametric loess model. To do so, use the *Graph | Attribute plots | XY(Z) diagram* command and in the dialog box select *Analysis 'DCA' / Case scores / Axis 1* in the top left list (for X variable). In the top right list select *Plant Species / CarxPani* item. In the upper right corner select the *Loess model* option and also make sure that the *Add points to model* option is checked in the lower right-hand area. The desired final state of the dialog is illustrated in Figure 8–2.

Often, you will be interested in species response to measured environmental descriptors, rather than to compositional gradients (represented by the ordination axes). In such case, you should select in the top left area the required variable in one of the folders representing a data table (e.g. here the *Soil Properties* folder).

Before leaving the dialog, you must set the options for the loess smoother. To do so, click the *Model Options* button and in the dialog box that appears set *Span* value to 0.67, *Degree* to *local linear model* and make sure that the *Robust fitting algorithm* option is checked. Then close the *Loess Options* dialog with *OK* and do the same with the *Create XY(Z) Plot* dialog. This will fit the model and create the diagram. During the process, you might see one or multiple dialogs (depending on currently set Canoco 5 options)

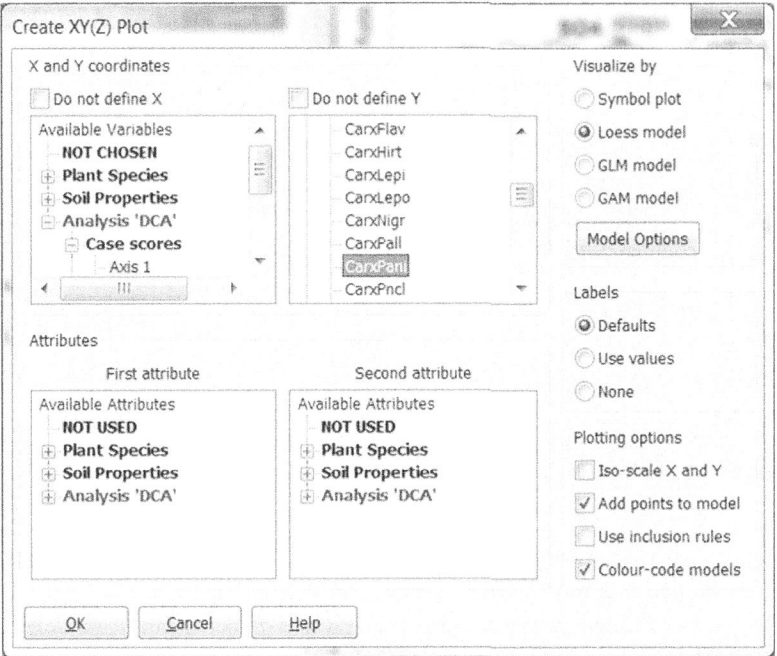

Figure 8–2 Selections to be made in the Create XY(Z) Plot dialog when fitting the loess model.

offering choices for the graph or summarising the fitted model, and finally a diagram similar to that in Figure 8–3 is created.

In your diagram, the individual data points are labelled. To remove the labels, select any one of them by clicking it with the left mouse button and then press *Ctrl-H* to select the other labels. Remove the labels by pressing the *Delete* key on the keyboard. You can see in the diagram that the vertical coordinates of plotted points are surprisingly regularly spaced. This is because the abundance of individual species in this data set was recorded on a semi-quantitative estimation scale with values ranging from 0 to 7.

The curve estimated by the loess smoother suggests a unimodal shape of the response. Therefore, you will now fit a classical unimodal response curve ('Gaussian model', see section 3.8.3 in the Canoco 5 manual) to the same data, i.e. a response curve describing the change of *CarxPani* values along the first ordination axis. Repeat the steps used in the previous fitting, including the selection of variables in the *Create XY(Z) Plot* dialog, as illustrated in Figure 8–2. In this dialog, the following options must be set differently: use the *GLM model* choice instead of *Loess model* in the *Visualize by* area, change in the middle right part of the dialog the *Labels* option value from *Defaults* to *None* (to remove case labels more easily than for the previous diagram), and then set options for GLM by clicking the *Model Options* dialog. The required settings are illustrated in Figure 8–4.

You are fitting a generalized linear model with the predictor variable (case scores on the first ordination axis) used in the form of a second-order polynomial (i.e. a *quadratic* model). In addition, you specify the *Poisson* distribution, with an implied *log* link function. We **do not** assume that the response variable values really have a Poisson

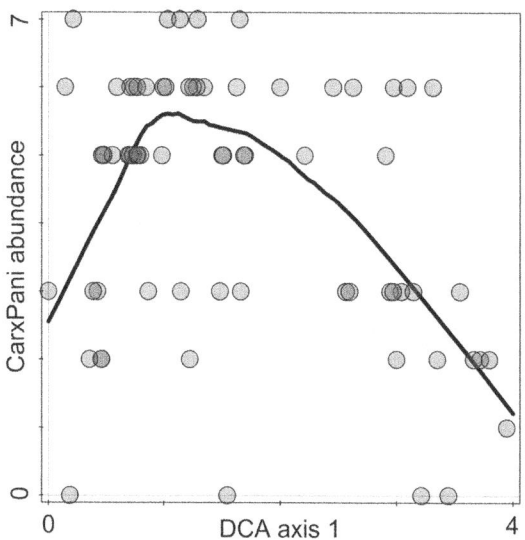

Figure 8–3 Abundance of *Carex panicea* plotted against the first DCA axis together with a fitted loess model.

distribution. But we find this is the best choice amongst the available options, particularly because the (implied) log link function is needed to fit the Gaussian-shaped response curve (with the tails approaching zero) rather than the plain second-order polynomial parabolic curve (which would predict negative values in part of the predictor range). Additionally, the choice of Poisson distribution implies that the variance is proportional to the mean (see Table 8–2), and this does not seem unreasonable. After you close this and preceding dialog with the *OK* buttons, the regression model is fitted and the results are summarised in another dialog box, shown in Figure 8–5.

Information in the upper part of the dialog summarises the model properties, which you have specified in the preceding dialogs. The fitted model is compared with a null model (EY = const) and the dialog shows both the raw deviance values for the fitted and null models as well as a deviance-based F test (labelled *Model test*) comparing the difference between the deviances of the null model and the fitted model (in the numerator of the F statistic ratio) with the deviance of the null model (in the denominator). Both terms are adjusted for the number of degrees of freedom, so they correspond to the mean-square terms in the traditional F statistic used in ANOVA models.

In the white area at the bottom of this summary dialog, the estimated values of individual regression coefficients, as well as the standard errors of the estimates are shown. The last column (*T value*) gives the ratio of the estimate and its standard error. In classical regression models these ratios are used for testing the partial hypotheses about individual regression coefficients, i.e. H_0: $\beta_j = 0$. Under the null hypothesis (and fulfilled assumptions about the distributional properties of the response variable and the independence of individual observations), these statistics have the t (Student) distribution. Note, however, that this approximation does not work very well for the

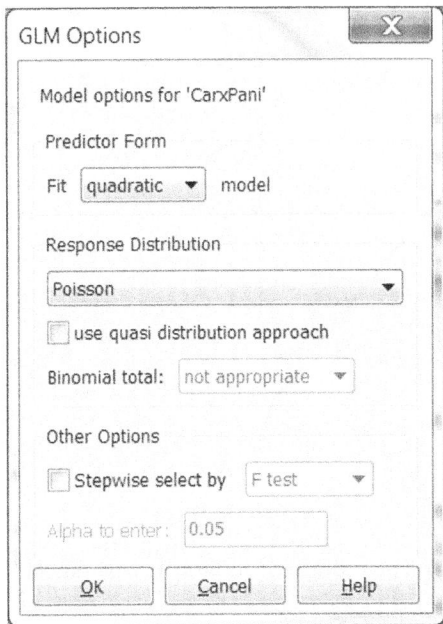

Figure 8–4 GLM Options dialog: options for fitting a unimodal response curve.

generalized linear models, so comparison of the values with the critical values of the *t* distribution is not recommended (and hence the implied significances are not shown).

The area above the list of coefficient estimates (named *Unimodal response curve*) in Figure 8–5 is used only if a second-order polynomial model was fitted, using either the *log* or *logit* link function (i.e. the *Distribution* type was specified as *Binomial*, *Poisson*, or *Gamma with log link*). In such cases, Canoco attempts to estimate two parameters of the species response curve. The first parameter is the position of the curve maximum on the gradient represented by the predictor variable (*CaseR.1* in your case), called *Optimum*. The second parameter is the 'width' of the fitted curve, named *Tolerance* in this dialog box. Depending on the meaning of the predictor variable, the tolerance parameter can be sometimes interpreted as the width of the species niche with respect to a particular resource gradient ('niche dimension'). Depending on the quality of the fitted unimodal model, Canoco can estimate the variability of the two parameters (the *s.e.* fields) and, in the case of the optimum, calculate the 95 percent confidence interval for its estimate.

The estimated values of the tolerances and optima of individual species can be used for further data exploration or for testing of hypotheses concerning the set of species. Please note that you must be careful to select an appropriate predictor for such models. If you use the case scores on the first DCA axis as the predictor variable, you can use the resulting models to illustrate in much detail the meaning of the ordination results, but you cannot draw any independent conclusions from the positions of species optima (or change in species tolerances) along the DCA axis. This is because the dispersion of the

8.8 Modelling species response curves with Canoco

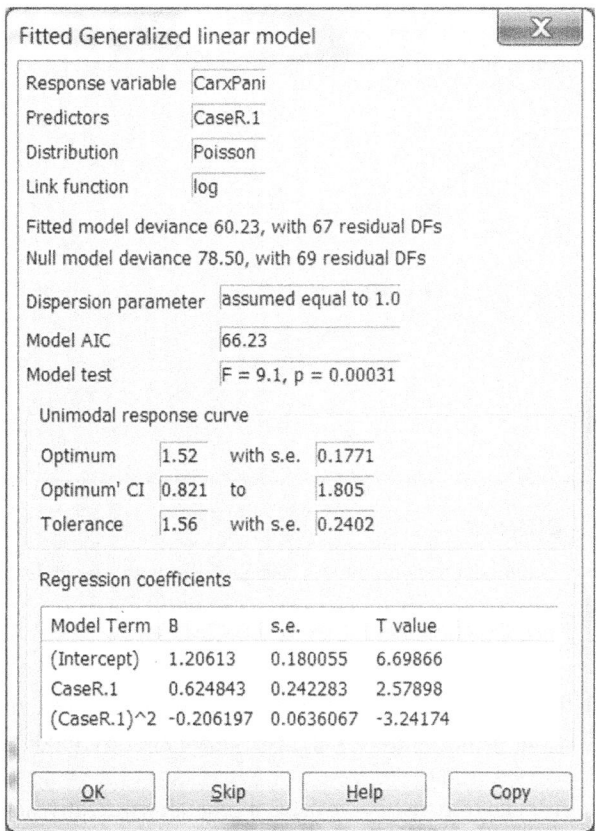

Figure 8–5 Summary of a fitted unimodal response curve.

response variable (species) scores, as well as the variation of the case scores (affecting the width of the fitted response curves) are systematically modified by the weighted averaging (and also the detrending) algorithm of unimodal methods.

More generally, in this and latter models of this section, you are estimating the relation of one species (response variable) to a synthetic predictor (case scores on the DCA axis) that is calculated from the values of all species in the data set, including the one used as a response here. Significant results of the performed tests should be therefore taken only as indicators of the relative contribution of the particular species to the definition of the DCA axis. To obtain valid tests of differences among the species in relation to resource gradients, you would need to specify independently measured explanatory variables as your predictors. In that case, the ordination performed with Canoco is not involved at all, but it still provides a useful 'framework', representing a rich source of ideas about which environmental gradients can be important for variation in community composition and which species respond to which factors.

The fitted regression model for the unimodal response curve is shown in Figure 8–6.

The second-order polynomial is a very strict specification for the shape of the response curve. The response of species often takes more complicated shapes and it is difficult to

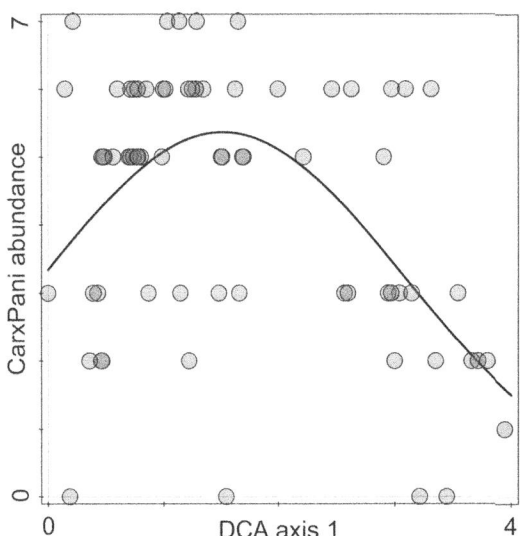

Figure 8–6 Fitted species unimodal response model, using a GLM with a second-order polynomial predictor.

describe them with the more complicated polynomial terms. Therefore, another family of regression models can be useful here, e.g. generalized additive models or the loess smoother.

You will now estimate the response curve using a generalized additive model (GAM). To fit the GAM with an amount of complexity comparable to the GLM fitted earlier, you would need to specify the smooth term's complexity as 2 DF. But as you are trying – in general terms – to find the existence and eventual shape of species response to (indirect) environmental gradients, it is better to let Canoco compare models of differing complexity (but with an identical predictor, i.e. the case scores on the first DCA axis) and select the best one. Canoco will measure the performance of individual candidate models using the *Akaike Information Criterion (AIC)*. This criterion attempts to measure model 'parsimony', so it penalises a candidate model for its complexity, measured using the number of model degrees of freedom (see Hastie & Tibshirani 1990, or Chambers & Hastie 1992, for additional discussion).

So, start your fitting with the same dialog box (*Create XY(Z) Plot*), but specify *GAM model* option in the upper right corner of this dialog. Then click the *Model Options* button and make sure you change the default settings to match those shown in Figure 8–7.

You specify the value 4 in the *Term Smoothness* field. If the *Stepwise selection using AIC* option were not selected, a generalized additive model with a smooth term for *CaseR.1* (i.e. the case scores on the first ordination axis) with a complexity of four degrees of freedom would be fitted. But the model selection option is enabled (checked), so Canoco will fit a null model (which is, in fact, identical to the null model we

8.8 Modelling species response curves with Canoco

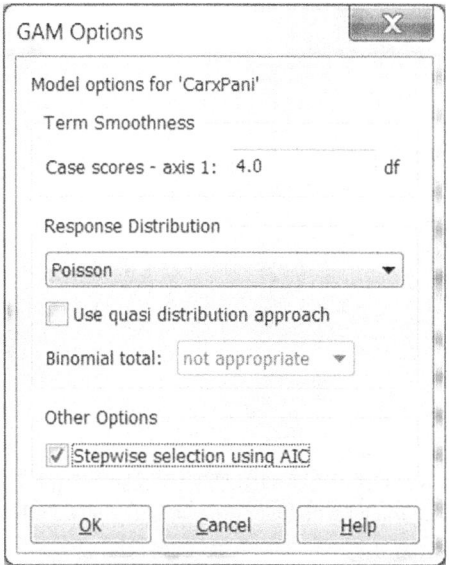

Figure 8–7 Settings for a generalized additive model with model complexity selection.

mentioned when fitting GLM) and four alternative generalized additive models with increasing complexity (*DF* equal to 1, 2, 3, 4). If you specify a non-integer number in the smoothness field, Canoco fits models with the integer sequence of DFs up to the integer immediately below the specified value and then a model with the specified complexity. For example, if the field value were *3.5*, Canoco would make a comparison between a null model and models with 1, 2, 3, and 3.5 DFs. The distribution specification is identical to the one you used when fitting the generalized linear model.

After you press the *OK* button (in two dialogs), Canoco selects the model complexity and informs you about it with the *Results of Model Stepwise Selection* dialog, as shown in Figure 8–8.

As you can see, the model with three degrees of freedom (i.e. describing the predictor effect as a smooth term with 2 DFs) has the lowest AIC value (the highest parsimony). This is the one which Canoco 5 accepts, summarises in a dialog box, and plots in the diagram illustrated in Figure 8–9.

The curve has a slightly asymmetrical shape, but in general corresponds well to the loess smoother model and the generalized linear model you have fitted before.

Finally, we illustrate how to fit and plot the response models for multiple species at the same time, determining the appropriate model complexity for each species (response variable) independently. You should return to the beginning of this section and study the diagram in Figure 8–1, containing the six selected sedge species, which you will now compare with respect to their distribution along the main gradient of meadow community variation, represented by the first DCA axis. Because it is quite possible that you may need to refer to this species group in several diagrams, you will define the

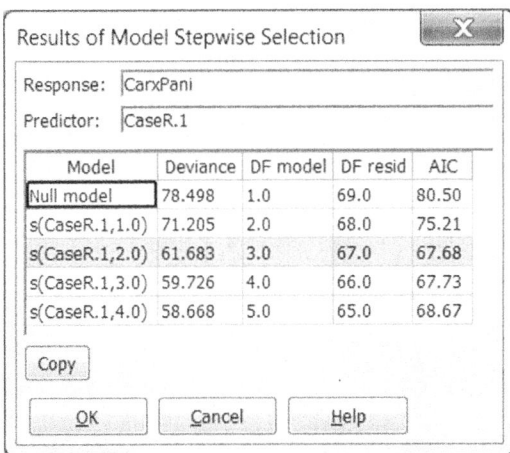

Figure 8–8 Report on stepwise selection of a generalized additive model's complexity. The emphasised line represents the selected model with highest parsimony (lowest AIC value).

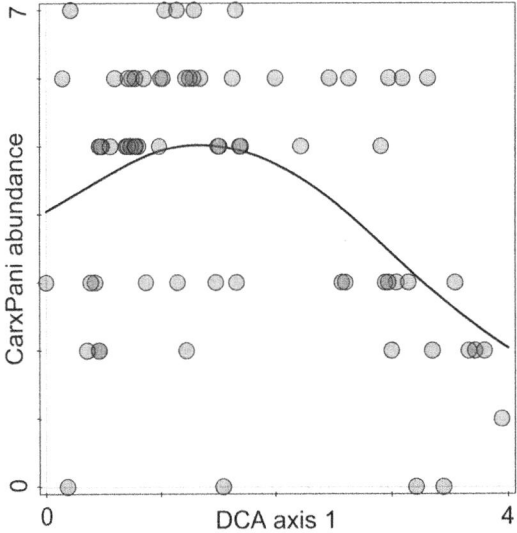

Figure 8–9 Species response curve fitted with a generalized additive model with df = 3.

group explicitly within the project. Go to the *Project* menu and select the *Groups | of plant species* menu command. A new dialog (named *Group Manager*) appears and you should click the *By Selection* button in the *Create* area. Another dialog appears and you should select the appropriate species names in the right list and move them to the left list by clicking the ≪ button. The desired final state is illustrated in Figure 8–10.

After you close this dialog with the *OK* button, you can change the default group name (*Plant Species group 1*) using the *Rename* button in the group manager dialog. Leave the dialog using the *OK* button.

8.8 Modelling species response curves with Canoco

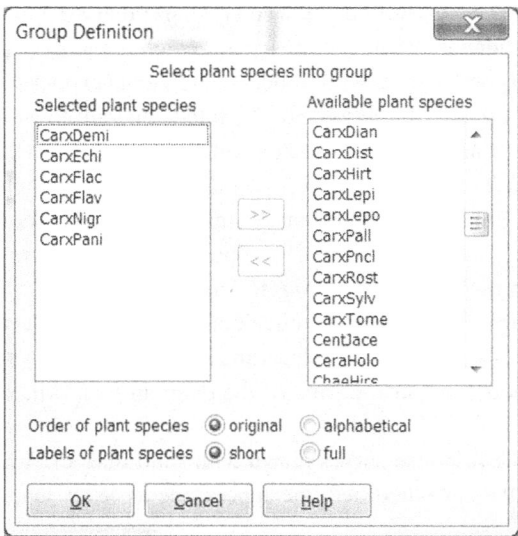

Figure 8–10 Definition of a group of sedge species.

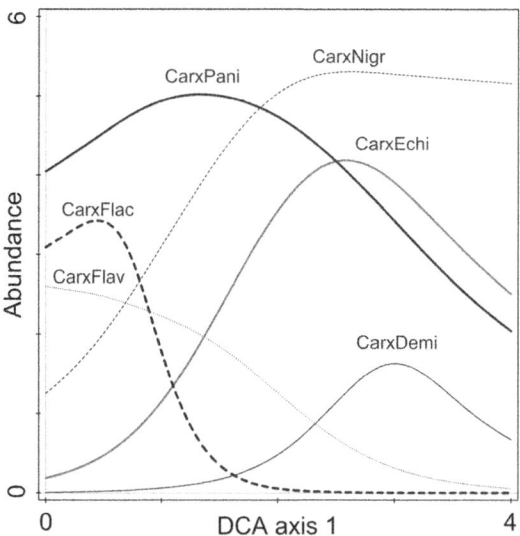

Figure 8–11 Response curves of six sedge species, fitted using generalized additive models.

To create a diagram with the response curves of multiple sedge species, you will use the *Graph | Attribute plots | Plant species response curves* command.[22] In the dialog box, which appears after you select the command, choose the *GAM* option in the *Model type* area. In the lower right area, select the name of the group you just created. You will

---

[22] The beginning part of the command name (here 'Plant species') of course varies depending on response variable terms used in your project.

see that the group members were selected in the central list (titled *Response variables*). Now you must select a predictor variable at the left side – click the *Axis 1* choice. You can alternatively fit the response curves with respect to individual explanatory variables.

Before leaving this dialog, specify options for the fitted generalized additive models in the dialog shown by the *Model Options* button. In the dialog you will specify the options for all the response variables (species) at once. To compare multiple species models, you probably want to keep the fitted curves simple. Specify, therefore, value 3.0 in the *Case scores – axis 1* field. Use the *Poisson* option in the *Response Distribution* area and make sure the *Stepwise selection using AIC* option is checked.

After you close this and preceding dialogs with *OK* buttons, Canoco selects 'optimum' model complexity for each of the specified species and reports about the model selection and about the selected model properties. Finally, the diagram is drawn, similar to the one in Figure 8–11.

Your diagram has all the response curves drawn with solid lines of different colour, but with uniform width. We have modified line appearance to better support the black and white reproduction.

We will remind you here that the roughly unimodal shape of the response curves for all the species is not an artefact of the specified model type (unlike the situation where you would fit second-order polynomial GLMs without model complexity selection). Generalized additive models are well able to fit monotonically changing or bimodal response curves, even with the limit of three degrees of freedom imposed in our example.

9 Interpreting community composition with functional traits

The composition of ecological communities changes along environmental gradients and such change is often predictable and repeatable across various biogeographical areas when expressed in terms of prevailing species trait composition (Garnier et al. 2004). We can often predict over various biogeographical regions what will be the traits of dominant species in a certain environment and which traits will determine species response to an environmental change. Functional groups (or functional types) defined in various ways are used also in various simulation models predicting e.g. the dynamics of vegetation facing environmental or land use changes, or in the description of these changes.

Functional types used in large-scale models of vegetation dynamics are usually broad categories of species based on their morphological and physiological traits or on large phylogenetical groups, or these two types of criteria can be combined. For example in temperate grasslands, four groups are often used – grasses (further distinguished into C3 and C4 groups in North America), sedges (and rushes), legumes, and (non-legume) forbs. In large-scale dynamic simulation models, the overwhelming diversity of species does not allow for modelling directly the changes in species composition, and so we need some broader groups of similarly functioning species; nevertheless, it is clear that rough classifications such as the one mentioned above are often not sufficient for this purpose. Detailed (and usually spatially limited) studies, where we can evaluate the species composition response to environmental gradients and we also have the data on traits of individual species available, provide a good lead for designing functional classification of species that can be subsequently used in large-scale modelling.

To create truly functional classification, we need first to define the actual purpose of grouping species together. One goal often required is to know which traits (and trait combinations) are responsive to changes in environmental conditions and if there are groups of species that respond similarly to these conditions ('functional response groups'). But the response of trait composition to environmental gradients is also interesting independently of the need to define functional groups. The ecophysiological correlates of individual traits are known and so the analyses of changes of trait composition along environmental gradients are used to test mechanistic hypotheses about ecological forces and the constraints controlling the community assembly.

Many methods were suggested to analyse the relationships between the species traits, species composition in individual communities, and properties of environment, and eventually define functional groups. See Kleyer et al. (2012) for comparison of selected methods, their detailed description, and R code for their execution. Below we discuss how to implement the analyses that detect the changes in trait composition along environmental gradients, using Canoco 5.

9.1 Required data

To assess the variation in trait composition along environmental gradients, we need three data tables: the species x traits table, characterising the traits (also called life history attributes) of individual species, the cases x species table, characterising species composition in individual community sampling units (e.g. classical relevés or any (quantified) list of species in each unit), and finally the cases x environmental variables table. As the topic of this task is an issue specific to community ecology, we will use the term *community* to refer to response data values in a particular case; *trait* will be any measured property of species; and *environmental variable* is generally any explanatory variable affecting the community composition – typically soil or climatic variable, but also the type of management or time in successional studies. The data could be based on an observational study, but also on a manipulative experiment. In the latter case, the 'environmental variables' reflect the experimental design (including e.g. the structure of repeated observations).

In this section we focus on the species x trait table. As noted above, the **species traits** are considered here in a very wide sense – for example, the relationship of species geographical range of distribution (that can be used as species 'trait' in the analyses) with its habitat preferences was demonstrated by Spitzer et al. (1993). A wide variety of species properties can be used, yet they should be obtained independently from the data used to infer species response to environmental gradients.

From this point of view, we should be extremely careful when using species strategies or Ellenberg indicator values as 'traits'. Whereas use of strategies derived only from measured traits (as in the LHS system of Westoby 1998) is fine, the assignment of plant species to Grime's C-S-R strategies (Grime et al. 1988), as present in various databases is likely based also on the species ecological distribution, i.e. on species habitat preferences. This can make the C-S-R strategy a powerful predictor, which is useful when our goal is to predict species response to environment. But the results based on C-S-R strategies used as traits cannot make a contribution to ecological theory. We just demonstrate that e.g. the species that others have observed in disturbed places (and consequently assigned them to the R-strategy) really grow in disturbed places. This is fundamentally different from a conclusion that species with, e.g., many small seeds grow in disturbed places. The same problem concerns also various indicator values, such as those of Ellenberg.

The use of various statistical methods based on the trait data increased dramatically with the availability of species trait databases, such as LEDA (Kleyer et al. 2008), BiolFlor (Kühn et al. 2004), or TRY (Kattge et al. 2011). Within such databases, the traits are treated as fixed properties of individual species (rather than properties specific to a particular site where a trait was measured) and their use in other sites is (implicitly) based on the assumption that the within species variability is smaller than the between species variability (or even negligible).[1] Nevertheless, these databases contain data from various sources and from various regions, and so we should pay attention to the selection of an appropriate value. Further, some taxonomic expertise is needed to assign correct values to particular species, because the databases sometimes store values for more broadly or differently defined taxa. If we measure the trait data as part of our study, instead of using database values, we know exactly which species we have measured, and we can also measure separate values for different environmental conditions. The situation is further complicated by the fact that while some species traits are quite stable (typically the life form or less, but still rather stable seed weight), other traits are very plastic (e.g. plant height or number of flowers). Consequently, for the latter traits, we might use different values of a trait for the same species, depending on the conditions.

Whatever data we have, however, the trait data table will contain a number of missing values, very probably the highest proportion of missing values among the three tables considered here. Species x plots data are usually without missing values and even in the environmental data, missing values can be often avoided. But we often do not have trait data for all the species, irrespectively of the way we obtain them. Even when recording the traits as part of our study, we are frequently not able to measure them for all species and particularly not the traits of all the species in all the recognised types of environment, if we consider trait variability.

There are then several possibilities how to handle the missing trait values and their use depends on the trait type and on the statistical methods used. For example, when calculating the community weighted means (see the next section), just the species with trait data available are used: Pakeman and Quested (2007) suggested that the community weighted means might be reliable when based on a set of species representing at least 80 percent of the total vegetation cover or biomass. Nevertheless, they also noted that this recommendation is valid only if the traits do not vary greatly among species. They warn that if this variability is high and the trait values are likely to be correlated with the species abundance, then a greater effort in sampling species for traits is required.

Another possibility for handling missing values is to use some imputation technique; in this context, we have probably better possibilities than the simple imputation using mean or median, as provided for general data tables in Canoco 5.[2] Typically, non-measured trait values can be amended with values obtained from various databases. If doing so, however, one should perform some checks using the species that were both measured and are available in a database, to see to what extent the values correspond and whether one source is not systematically shifted against the other. Also, in some cases, only particular life stages are present in the sampled communities. Typically, in mown meadows in wooded landscapes, seedlings of woody species are often present, and sometimes even abundant. If we then mechanically apply the plant height values

---

[1] Alternatively, when the within species variability of traits is high, we can take the individual plant as a unit of analysis and measure both its trait values and the environmental conditions at which it grows. This gives then two data tables (species x traits and species x environment data) where the latter may include design factors such as plot or location. Such data can be then analysed by RDA with customised permutation tests.
[2] The R package *mice* can be of help here.

obtained from a database, height of mature trees is assigned to seedlings, which can lead to rather strange results. If we use the database trait data for our own small-scale project (comprising tens or several hundreds of records at maximum), we will probably notice these problems, and we will try to solve them. But this might became difficult for large-scale projects such as those connecting plant trait databases with large databases of vegetation composition records.

In databases, the easy-to-measure (soft) traits are available for more species than the traits that are difficult to measure, and more traits are available for common species than for the rare ones. Moreover, phylogeny of species is gradually better resolved, and because many traits are phylogenetically conservative, we can use this information similarly as we can use the fact that traits are usually correlated. An intuitive approach might be either to use the mean or median value of a genus, or to infer the data from the known correlation structure among traits, or to combine these two.[3]

Recently, sophisticated formalised algorithms were developed (Shan et al. 2012) that can combine these intuitive approaches, i.e., take into account the best resolved available phylogeny and trait correlation structure. Such methods will be particularly useful for large-scale comparisons, such as those covering multiple continents, with large trait variation among biomes, where the difference between closely related species should be rather small in comparison with the total trait variation. But whenever we use these imputed trait values, we should very carefully consider whether they will not introduce some artefacts into results of our analyses – for example, if the traits are imputed according to phylogenetic relatedness with other taxa present in the data set, this will artificially increase the phylogenetic conservatism in our data, which might in turn affect any analysis taking the phylogenetic relations into account.

Whatever source of traits we use and whatever method for handling the missing values, we should be aware that the selection of traits for particular study is a decision of the researcher, affected both by scientific criteria (i.e. which traits are relevant for the questions studied) and by purely pragmatic factors (which traits are available in databases or are feasible for direct measurement). We will be never sure that all (or at least the most relevant) traits have been measured and this fact should be always kept in mind when interpreting results.

9.2 Two approaches in traits – environment studies

In general, there are two basic questions that we can ask in this context: (1) 'How is the trait composition of a community affected by environmental conditions?' and (2) 'How is the species response to environmental gradients determined by its functional traits?' These two questions are matched by two basic approaches (with many variants) to the problem of relating species traits to environment.

---

[3] Again, these approaches should be used with caution, and their results carefully checked. For example, it would not be the best idea to use a genus mean for the height of dwarf willow (*Salix herbacea*), but similar mistakes can be usually avoided using 'ecologist's common-sense'.

9.2 Two approaches in traits – environment studies

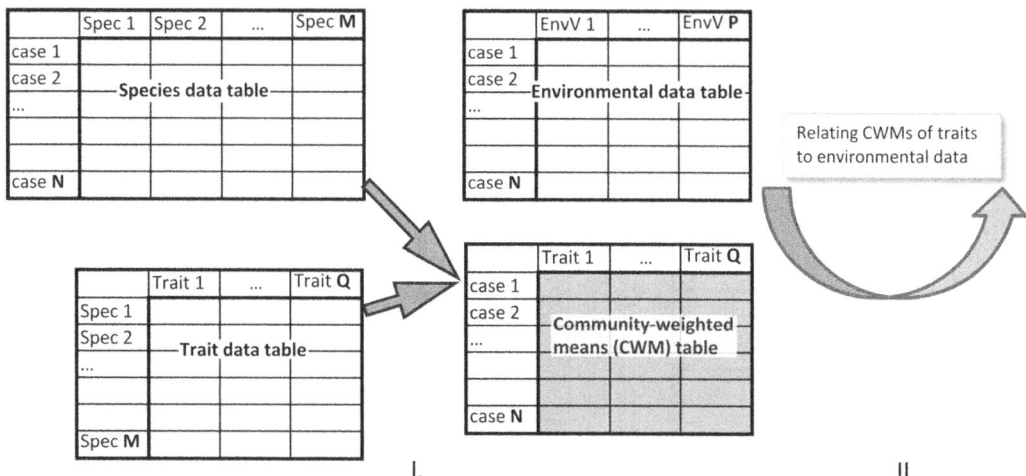

Figure 9–1 Schematic diagram of the two steps for community-based approach of relating the functional traits to environment. CWMs of traits are most often related to environmental data using a constrained ordination.

In the first, **community-based approach** (see Figure 9–1), we first characterise the cases by summaries of trait composition, and then use the environmental values as predictors (explanatory variables) in statistical analyses (e.g. Garnier et al. 2004). The basic unit in these analyses is a case, representing a particular community. The basic research question is how the community composition characterised by its prevailing species traits changes on environmental gradients, i.e. question (1) above. The composite trait characteristics of a community are the response, predicted by environmental characteristics: most often, each case is characterised by the (weighted) average of trait values, usually termed **community weighted mean** (CWM). For categorical trait variables, it will be a proportion of given category, which is in fact the (weighted) average of zeros and ones, if categories are coded as a set of dummy variables (see the end of Section 2.2).

We might be interested not only in averages, but also in trait variability (or, generally, in functional or morphological diversity), and then we can use such trait variation statistics as response variables (de Bello et al. 2006; Lepš et al. 2006). Because each species is considered separately in each case, community weighted means can be calculated either using general fixed trait values over all the plots, or using case-specific trait values, and even the within-species within-case variability can be reflected.

In the second, **species-based approach** (see Figure 9–2), we first quantify the species response to individual environmental variables (or to composite environmental gradients), and then we construct a model predicting species responses on the basis of their traits (e.g. de Bello et al., 2005). In this approach, the basic unit of the central, second step is a species. There are other approaches possible. For example section 6.3.6.2 of the Canoco 5 manual suggests an analysis where the responses to individual gradients are calculated first (similarly to our approach), but they are subsequently used as predictors and the species trait values are considered the response. This approach might be quite

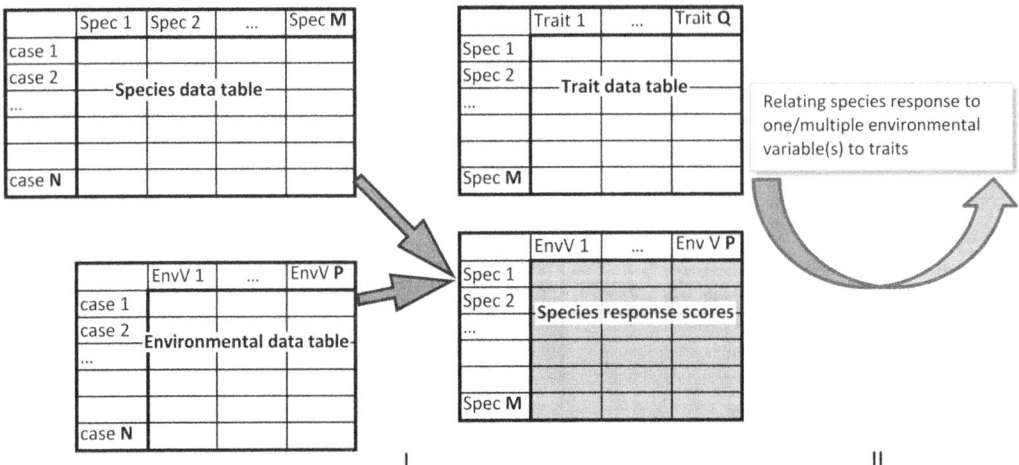

Figure 9–2 Schematic diagram of the two steps for species-based approach of relating the functional traits to environment. The species scores expressing their response to particular environmental variables usually come from separate constrained ordinations for each environmental variable and they are then often (but not necessarily) used as a response variable in a regression model, with trait variables acting as predictors.

useful particularly when we have many traits, and so a high dimensionality of the trait matrix, which might cause some problems on the predictor side (too many predictors can decrease test power).

In the following text, we will stick to the idea that species traits determine the ecological distribution of species, and so we use them as predictors, and the species response to a gradient as a response variable. If there are too many traits, we try to find the ones that have predictive power. This means that the basic aim in this approach is to find a rule (or a set of rules), which would enable us to predict species response to environmental gradients on the basis of known species traits. For example, we might try to predict which species would be endangered, or which species will become dominant under various land use change scenarios, or which species are potentially dangerous invaders.

In the species-based approach, each species must be characterised by a fixed value of each trait. The trait values should be general species characteristics (e.g. potential plant height), and not the characteristics of the species in individual investigated sites. But if known a priori, characteristics of species plasticity can also be used as predictors, as a trait itself. Using traits as predictors of species response enables us to take explicitly into account trait interactions, i.e. non-additivity of their effects: for example, early flowering time might be important for annuals, but not for perennials, in response to grazing.[4]

In statistical analyses, each observation unit is normally considered independent, but this might not be an appropriate assumption when species are used as sampling

---

[4] The annuals are crucially dependent on seed production and so need to finish their life cycle before the start of the intensive grazing period, while perennials are able to survive and often to spread by vegetative multiplication.

units. Consequently, in some cases we will have to consider the problem of phylogenetic dependence among the species as in other comparative studies (Harvey & Pagel 1991).

In both approaches introduced above (Figures 9–1 and 9–2), we can keep the multivariate character for all the variables throughout the whole analysis, or we can focus on individual variables. In the latter case, we can look at each trait separately or we can separate species responses to individual environmental variables. The univariate approach will be particularly useful when we work with a limited number of traits, e.g. when we stick with the Westoby's (1998) LHS strategy system or in manipulative experiments, where a limited number of environmental variables was manipulated.

Other examples are the observational studies where the sampling was designed to reflect some pre-selected gradients: for example, de Bello et al. (2005) selected plots along climatic gradient represented by elevation and along the gradient of grazing pressure. Focusing on single traits will be also useful, if we would like to disentangle the contribution of species turnover and trait plasticity to community weighted mean response to environmental gradients (see Lepš et al. 2011). Although the multivariate approach is also possible in this context, in our view the trait-by-trait analysis is in principle more informative, because the plasticity contribution is very different for different traits, and compounding the data about all traits together could obscure important ecological information.

The community-based and species-based approaches usually provide similar results (Kleyer et al. 2012). This is not surprising – if e.g. the tall plants are favoured in a more fertile environment, then the plant potential height will be a good predictor of positive response to increased fertility and, vice versa, an increase in soil fertility will predict the increase in CWM for plant height. Yet there are some differences between the two approaches, which reflect the questions answered by them. In the species-based approach, the trait value is the characteristic of a species (i.e. has a fixed value for given species over all the sampling units). If you predict the species response to environment by trait values, these values must be known a priori (species success in fertile environment is predicted by its potential ability to grow tall). Using the traits as predictors in the species-based approach enables us to model explicitly their interactions (i.e. the non-additivity of trait effects on species ecological distribution).

Also, some species survive in a community because they are very different from the dominants and so they are not competitively excluded by them. Typically, some herbs of deciduous forests are low, finishing their life cycle before the budding of dominant trees. Whereas these herbs will probably not affect too much the CWM values, the species-based approaches can be more efficient to identify that 'to be small and early flowering' is a successful strategy of survival in deciduous forests.

In the community-based approach, we can (but need not) use different trait values for a species in different sampling units and we can therefore explicitly reflect the trait value response to environmental differences. For example, Lepš et al. (2011) demonstrated that the plants in a fertile environment are on average taller than the plants in an infertile environment and they have subsequently shown to what extent this is caused by species turnover and to what extent by the fact that the same species grow taller in fertile conditions. Finally, the community-based approach enables us to model not only the changes in CWMs, but also in the trait variation.

9.3 Community-based approach

Using this approach, we try to relate the average trait values of individual plots to environmental characteristics measured at individual plots, usually with a linear constrained ordination.[5] You can find a simple example of this approach in Case study 3 (Section 14.4) and a more detailed example in Case study 5 (Section 16.8.1). In this analysis (also called *RDA of community weighted means – CWM RDA* – by Kleyer et al. 2012), we first calculate community weighted means (CWM) of all the traits and for all cases. We use CWM values subsequently as response variables, characterising the plots and predicted by environmental variables using a linear constrained ordination (RDA). The categorical traits are represented by as many 'indicator traits' as there are categories. Fuzzy coding can also be used on the original table with traits, for example for pollination categories, such as when a plant species is partially wind- and partially insect-pollinated. The table with CWMs of traits can be calculated in Canoco 5 using the *Data | Add new table(s) | Trait averages* menu command.

When we test the significance of the trait CWM–environment relationship, we should pay attention to what the null hypothesis means biologically, how to interpret it, and what permutation type is appropriate for our null model. The classical permutation test in RDA is valid for the null hypothesis that in a large (potentially infinite) number of experimental units, the trait composition is independent of the environmental variables. This means that our null model assumes that there is no change in species composition across the studied environmental gradient caused by the traits of the species. It does not however tell anything about whether all present species respond in a concordant way according to their traits. Our null model is that species do not respond to environmental variables or, if they do respond, then in a way where the trait averages remain the same for all traits under all environmental conditions.

As a consequence, if e.g. dandelions would completely prevail in fertilised plots at the expense of any species with flower colour other than yellow present in both fertilised and unfertilised plots (so at least one species responds to the treatments), we can get a highly positive response of the 'yellow flower colour' trait to fertilisation (which is correct, because in our 'universe', all the fertilised plots have a greater abundance of yellow flowers). Nevertheless, we should be aware that such a response might be caused even by a single responding species.

The other possibility would be to randomise trait assignment to species when calculating the CWMs and for each randomisation to calculate RDA with environmental variables and use the resulting pseudo-F statistic as a reference distribution under the

---

[5] Because individual traits are measured in different units, it is absolutely necessary to use the option *Center and standardize* by variables, i.e. to calculate RDA on the correlation matrix. For the same reason, the unimodal constrained method (CCA) cannot be used.

null model. Under this null model, the species might respond to environmental variables, but in a way that is independent of their traits. To get a complete picture, both tests could be performed – unfortunately, this latter test cannot be carried out in the present version of Canoco and we would need to write our own R-script for it to further explore potential species responses of CWM to environmental factors.

What can be done in Canoco 5 is to relate the species x case data table to species x trait data table by CCA or RDA and test their relationship by randomly permuting the species. This establishes the link between the species x case and the species x trait data tables (disregarding the environmental table), whereas CWM-RDA establishes the link between the species x case data and the environment x case data (taking into account the trait data table). Both of these two links need to be significant to demonstrate trait–environment relationships (see Dray & Legendre 2008 and ter Braak et al. 2012 for additional discussion and solutions).

If we are interested in the response of trait variability ('functional diversity') instead of community weighted means, we have basically two possibilities of how to proceed. We can either (1) calculate for each case some measure of functional diversity (FD, usually the Rao index is used), which will be based on all the traits we are interested in (see e.g. Lepš et al. 2006 for discussion of the problems with combining multiple traits into one FD index); or (2) we can calculate the variability of each trait separately. In the first case, we have a single characteristic for each case and we can evaluate it using classical methods of univariate statistics (see, e.g. de Bello et al. 2006). In the second case, we use the variability of individual traits as a multivariate response in RDA, with available environmental data as predictors. This approach will be quite analogical to CWM-RDA: just as the CWM-RDA asks how the weighted means of individual traits respond to environment, here we ask how the variability of individual traits changes along the environmental gradients.

One of the most widely used indices of FD is the Rao coefficient (Rao 1982, Lepš et al. 2006). In fact, it is a generalised form of the Simpson index of diversity (see Section 10.5). If the proportion of the i-th species in a community is p_i and the (functional) dissimilarity of species i and j is d_{ij}, then FD expressed by the Rao coefficient has the following form:

$$FD = \sum_{i=1}^{s}\sum_{j=1}^{s} d_{ij} p_i p_j$$

where s is the number of species in the community and $d_{ii} = 0$, i.e., dissimilarity of each species to itself is zero. If p_i is the proportion of individuals of species i in an infinitely large community, then FD is the expectation of dissimilarity of two individuals, randomly selected from the community. If $d_{ij} = 1$ for any pair of species (so each pair of species is considered completely different), then FD is the Simpson index of diversity expressed as 1 minus Simpson index of dominance D (see Section 10.5). The Rao index can be also seen as a measure of trait variance around the CWMs (see below).

The Rao coefficient can be used with various dissimilarity measures (see Chapter 6). The main methodical decisions are mainly how to measure the species dissimilarity (which can be a tricky task, particularly when we combine traits measured on various scales) and how to characterise the proportion of a species in the community (see Section 6.4 and Lepš et al. 2006, de Bello et al. 2013 for detailed discussion). The Rao coefficient can be used with both single and multiple traits. If used with a single trait and the dissimilarity measure is squared Euclidean distance, it is proportional to (weighted) variance of trait values. Trait diversity based on the Rao coefficient can be computed in Canoco 5 using the *Data | Add new table(s) | Functional diversity* menu command.

Further, the FD identified at various spatial scales (and under various environmental conditions) can be decomposed into its alpha and beta components, and this decomposition can be used to test various hypotheses about trait convergence/divergence. This approach can reveal possible mechanisms of species coexistence and community assembly (de Bello et al. 2009; Götzenberger et al. 2012).

Whereas using measures of FD based on multiple traits could make the analyses more straightforward,[6] our experience shows that the variability of different traits responds differently to environmental gradients, and so keeping the variability of individual traits separated can provide better understanding of ecological processes. Similarly to calculation of CWM, we face the problem of missing trait values for some species when calculating functional diversity. Whereas for the CWM, at least one study evaluated the effect of missing traits and provided some recommendations (Pakeman & Quested 2007), we are not aware of any such study for the functional diversity. Moreover, while we can expect that, provided the species with missing values are a random subsample of all the species, with the increasing number of missing trait values the estimates of CWMs will be less precise, but not biased (and the same is true if we replace the missing values with trait means), the missing traits will probably decrease the estimates of trait variation and this will also happen when we replace the missing values with trait means.

These two approaches (i.e. focusing on the trait weighted means and on the trait variability) are complementary and should be ideally used together (Ricotta & Moretti 2011). For many communities, the coexistence of species differing in their traits is very typical, and the use of sole trait means could be rather misleading. For example, average plant height in forests with a rich understorey is often not very informative due to typical differentiation in plant heights between the understorey and canopy species. Depending on the weighting used, the CWM might be a value between the tree and herb layer heights, not typical for either of them.

This example also shows that we must pay attention to the 'abundance variable' which is used for weighting in community weighted means calculations (Pakeman et al. 2008). If we use the biomass, the CWM for height will correspond to the tree layer (because trees have considerably higher biomass than the herbs), but if we use the cover, the height CWM would be somewhere between herb and tree layer, and if we use a non-weighted mean, the CWM can be close to the herb layer values, if the number of species in the herb layer is considerably higher than the number of tree species.

---

[6] You might have some general theoretical predictions how the functional diversity changes on gradients of disturbance and/or environmental stress, and under which conditions we can expect trait convergence/divergence.

Community-based methods also enable us to take into account the intraspecific trait plasticity. By species trait plasticity we understand here any variability in the traits of a species, connected with environmental variability.[7] There are (at least) two levels of intraspecific trait variability in community-based studies: within-case variability, which we usually consider a random one, because we expect that the environmental conditions are constant within a single case (even though many of the differences might be caused by microscale environmental variability), and the variability represented by systematic among-case differences that we can ascribe to environmental variability, and which we will explain by measured environmental descriptors.

For a single species, we can define a general(ized) linear mixed-effect model (LMM or GLMM) analysis, which would disentangle the individual components of the variability. Such a model should include the case identity as a random factor and we will get three components of variation: within-case variability and the between-case variability – explained and unexplained by measured environmental factors. But such a model would reflect neither the response of species abundance to environmental variables, nor the fact that not only the abundance, but also species presence reflects the community response to environmental gradients. The mass ratio hypothesis (Grime 1998) suggests that the community is characterised by the dominant strategies, i.e. by prevailing traits.

Consequently, we have proposed an approach that reflects both the intraspecific trait variability and trait composition changes due to changes in species composition (Lepš et al. 2011). The method was suggested and implemented for single traits and it is based on the sum of squares decomposition in ANOVA, but can be used in an analogous way with RDA for simultaneous analysis of multiple traits. The method is based on community weighted averages of the traits. When the traits are measured in each case separately, case-specific trait values can be used to calculate CWM (*specific CWM*). The difference between two plots in the specific CWM can be then caused either by a change in species composition or by trait plasticity.

For example, in more fertile plots, the specific CWM for height is higher either because small species were replaced by the tall ones (e.g. small sedges were replaced by reed), or due to species plasticity (even the small sedge species grow taller when in fertile conditions), but usually by a combination of both. If we use trait values fixed for the whole study (*fixed CWM*), then any change can be caused only by a change in species composition. Simultaneous analysis (in fact a decomposition of the sum of squares) of fixed and specific CWMs then enables us to disentangle the contribution of species turnover and species plasticity for the variability of specific CWM (see Lepš et al. 2011, who provide also an R-script for execution of the method for single traits).

---

[7] For example, the fact that plant individuals of the same species are taller in fertile habitats than in unfertile ones can be caused simply by the fact that they have more nutrients in the former, but they can be also under different directions of selection pressure in the two habitat types. Denser vegetation in fertile habitat could select against lower individuals and the final result can be genetic differentiation among habitat types. Having measured just trait values under various conditions, we are not able to distinguish the phenotypic plasticity from genetically based differences (Silvertown & Lovett Doust 1993) and consequently here we subsume both under the term 'plasticity'.

As noted above, it is not feasible to measure individuals of all the species in all the plots. In a factorial experiment resulting in a model with categorical predictors (Lepš et al. 2011), we have ignored the inter-case differences and simply measured (about) ten individuals for each combination of factor levels. If an environmental predictor is continuous (typically pH in observational studies), it will be necessary first to measure the trait values only for selected (ideally evenly distributed) values of environmental characteristics, then to build a regression model predicting the trait values, and finally to use these predicted values to calculate specific CWM. When using this approach, one should be very careful in interpretations, because by this procedure, we deliberately decrease the variation among individual plots (as happened also to us, see Lepš et al. 2011).

9.4 Species-based approach

The general goal in this approach is to predict the species response to environmental variation using the traits of individual species.[8] In the first step, we calculate a characteristic of species response to environmental variation and then we predict this response by species traits in the second step (see Figure 9–2). A simple illustration of this approach is provided in Section 14.4 (Case study 3) and a more detailed example can be found in Case study 5 (Section 16.8.2).

As the traits are used as predictors, they have to be general characteristics of species, not the characteristics of species in individual sites. This does not mean that these can be only fixed trait values, obtained from some database, as trait plasticity can be considered a trait itself.[9] Because species are basic units of the analysis, we should take into account the following two facts:

1. The response to environmental gradients is not determined for all the species with the same precision. In particular, for the species with lower frequency the estimate of the response could be more imprecise. Consequently, we will need some threshold for the species to be included in the final analysis or, alternatively, to weight the species by their frequency and/or by the sum of their abundances (as in unimodal ordination or similarly to CWM calculation).[10]
2. Species are not fully independent observations because of their phylogenetic relationships – consequently, we should consider taking this dependence into account: we can use one of the methods of phylogenetic corrections (Harvey & Pagel 1991) or we can alternatively separate the variability in response explained by phylogenetic

---

[8] You can find a detailed example of the methods discussed in this section in Case study 5 (section 16.8.2).
[9] If e.g. independently assessed degree of trait plasticity is known for the species in the project (which is usually not the case), it can be used as predictor also. For example, in a study of species response to grazing in a field experiment (Benot et al. 2013), the authors used trait plasticity experimentally assessed in a pot experiment as the predictors of species behaviour in the field.
[10] The weighted-analysis approach introduces, however, some imprecision into the permutation test in Canoco, because the explicit weights of permuted units (here the species) do not 'travel' with them during their permutations, so its results are unreliable.

relatedness from the variation explained by functional traits by variation partitioning (using the method of Desdevises et al. 2003 – see also Canoco 5 manual, section 6.3.7, for the use of this method with Canoco 5).

There are many possible ways to calculate species response in the first step and also how to combine the two steps. If we have multiple environmental variables, the first decision will be whether we want to predict the response to each of environmental variables separately or whether we want to get a general picture of trait correlations with all the measured environmental variables. For the interpretation of results, it is also important to take into account whether the environmental factors were experimentally manipulated or not and if not, how much they are correlated.

Finally, there is a question of statistical testing and appropriate null models. The characteristics of species response will be considered in the second step probably only if we are able to demonstrate their significance in the first step – i.e. if the independent (marginal) or partial effect of the corresponding environmental variable on the community composition is significant. This step involves permutation of the plots. The null model tested in the second step, i.e. when predicting the species response from traits, is that of the independence of species response **as identified by the first step** from the species traits. So this analysis takes the species response values as granted and also considers the species to be independent observations – consequently, in each permutation, the species are permuted without any restriction.

Note that if in the first step none of the environmental variables is significant (e.g. as judged by a single overall test of an RDA/CCA of community against environmental variables), then the second test is not performed. The test is therefore very similar to the sequential test that ter Braak et al. (2012) developed for RLQ analysis. They showed that the final P-value of the two tests joined together in this way is the maximum of the individual P-values.

The null model underlying such an approach can therefore include also the phylogenetic relatedness of species. Similarly as in the case of the similar approach for the community-based analyses, we would need to write our own script (in R) for such a permutation test. As in the community-based approach, the statistical test of trait-environment relationships thus involves two tests.

9.4.1 Selection of species

Whereas for the community-based methods rare species have lower weight on CWMs (and for this reason they are often neglected in calculations), in species-based methods all the species have the same weight, unless we explicitly specify otherwise (see footnote 10 about explicit weighting). We can, however, omit some of the species. There will be two reasons to do so – first, to omit the most infrequent species with unreliably estimated response along environmental gradients, and second to omit the species with missing trait information. There is no general rule concerning the frequency value below which the response estimate is not reliable, because the decision depends not only on the total number of plots in the study, but also on the study design.

For example, if we have a repeated measures design of a field manipulative experiment (say, studying effect of fertilisation), we can expect that in the first years, the species biomass will change according to treatments, but very few species will completely disappear or newly invade the plots. If a species is missing at the baseline time (i.e. before the treatments were imposed) from the plots designed to be fertilised, its response to fertilisation will very probably be estimated as zero, simply because it was absent before and remained absent after the fertilisation. But we might expect that if a species is present in the plots at the start, it will respond to fertilisation in some way. Also a species that was completely absent at the baseline time, but after treatments invaded vigorously many plots will be quite correctly estimated as an increaser. Consequently, the decision can be based on a combination of total frequency and the frequency at the beginning of study.

9.4.2 Predicting response to each environmental variable separately

We can calculate species response to individual environmental factors in many ways, from very simple ones to more sophisticated approaches. Their use will depend on many factors, the main one being the sampling/experimental design of the study, including the number of environmental variables in study and their possible correlation. Some measures can be very simple (e.g. simple correlation coefficient between species cover and an environmental variable). Dem et al. (2013) characterised preferences of species for late successional stages by the ratio of basal area of the species in a primary forest to the sum of its basal areas in primary and secondary forest (same total area in both forest types).

Here we focus on the use of constrained ordination techniques for this purpose, which enables us to perform a global test of species composition response to the environmental variable investigated, and also the separation of responses to multiple, correlated environmental variables. In general, we can consider the species score on a constrained (canonical) axis as the characteristic of species response to the explanatory variable defining the axis (Lepš 1999). In this case, there should be a single explanatory variable in the analysis, but we can have one or more covariates. The use of covariates enables us to separate the partial (conditional) effect from simple (marginal) ones in observational studies with correlated predictors, or to use as a response the species dynamic response to treatments in repeated observation experiments (see Section 16.8.2 for an example).

In observational studies, the predictors are often correlated. Typically, pH and the concentrations of Mg or Ca are usually positively correlated in the field (as in the spring meadow case study, Chapter 13). We can therefore expect that the species scores on axes, where either of the three variables is used as the sole predictor (i.e. focusing on simple effects) will be also highly correlated. Nevertheless, we can still ask which traits predict best how the species depend in the field on Mg concentration in soil, and then the species scores on a constrained axis defined solely by Mg concentration are the appropriate characteristic of species response. On the contrary, we can be interested in how the species respond to the Mg concentration, when the effect of Ca and pH is first partialled out and which traits predict this response. This information will probably not

9.4 Species-based approach

help to predict the species response in the field, where Mg is always correlated with the other two variables, but could contribute to the understanding of eco-physiological mechanisms behind the species response better than the earlier approach. None of the two approaches has a logical precedence, use of one or another method depends on the questions we ask.

In the case of designed experiments with repeated observations of experimental units (usually plots), where we have baseline data available (i.e. we know the state of the plots before the treatments were introduced), the community response is best characterised by a constrained axis defined as an interaction between treatment and time. In this case, the time should be coded so that the baseline time is zero and the covariate structure should reflect the repeated observation and experimental design (see our case study in Chapter 16).

We usually do not want the species response to be dependent on total species quantity. This is not an issue in unimodal (weighted averaging) methods, where the species scores are weighted averages of case score/explanatory variable values (and so, the standardisation by species is implicit), but it might be a problem in linear methods, when we centre the individual variables (species), but do not standardise to unit variation. In this case, we must use the post-transformation of species scores (i.e. divide by standard deviation). If we do not post-transform, the species with a high total quantity will have on average higher absolute values of their response to an environmental factor (i.e. it will be highly positive or highly negative) than the species with a low total.

Any statistical model can be used to predict the species response by its traits. Usually, we will have several traits available, so we will need some procedure simplifying the model, i.e. selecting the subset of traits with most parsimonious capacity to explain the species response. In Chapter 16, we demonstrate the use of two of them, multiple regression with forward selection of predictors and regression trees. The regression tree method is a non-parametric regression that produces a binary tree by a binary recursive partitioning (se Section 8.7). In this way, it allows for traits having a positive effect in one group but e.g. no effect in other groups (for example, de Bello et al. 2005 demonstrated that early flowering is important for the response to grazing of annuals, but not of perennials). By building the binary recursive partitioning, the regression tree method also provides groups of species with similar trait values, which are predicted to have the same response to environmental gradient, and so can be considered a functional type (nevertheless, the primary goal of regression tree is not forming groups, but a prediction based on a set of trait combinations).

There are several additional possibilities of how to use the traits to interpret results of ordination diagrams. When we have the ordination of species, we can passively overlay values of a trait over the diagram.[11] Because the traits are usually multivariate and often correlated (Verberk et al. 2013), it might be also useful to perform a reverse analysis (see Section 4.14), where the response of species are predictors, predicting the traits of

---

[11] This can be done in Canoco 5 using the *Graph | Attribute plots | XY(Z) diagram* menu command, by choosing the *Response variable scores* from an analysis to define X and Y axis (checking the *Iso-scale X and Y* option) and one or two traits as attributes.

individual species. Such analyses are offered in Canoco 5 as four analysis templates in the *Response Trait Analyses/Summarizing response traits* group in the New Analysis Wizard.[12]

Further, there are analyses that join the two steps present in both community-based and species-based approaches into one, i.e. combining the three data tables (with species composition, environmental variables, and trait values, respectively) in a single analysis that directly relates the traits to environment. Among the available methods (see Kleyer et al. 2012), the RLQ analysis is most prominent (Doledec et al. 1996), but currently implemented only in the *ade4* package of the R program.[13] The proper statistical test to demonstrate trait–environment relationships in RLQ consists of two tests that are executed sequentially (ter Braak et al. 2012), very much in the same way as the two tests mentioned for both the community-based and the species-based approaches in this chapter. Their main difference from RLQ is that RLQ uses the same test statistic (involving all three data tables) in both tests, whereas the test statistics differ among the two steps in both the community-based and the species-based approaches.

Finally, a completely different approach was suggested by Kleyer et al. (2012), as their cluster regression method. The idea of the method is to start with forming clusters of species, based solely on their trait similarity (using single traits and all their possible combinations) and then to select the best classification as the one best fitted by the underlying environmental gradients.

---

[12] The wizard offers them, however, only when you select the table with the trait data as focal in the second wizard page.

[13] In fact, RLQ analysis can be executed in Canoco 5 (see Canoco 5 manual, section 6.3.6.4), but the proper test of significance of the trait–environment relations via RLQ is not yet available there.

10 Advanced use of ordination

This chapter introduces six more advanced techniques, which build on the foundations of constrained ordination methods and can be used with the Canoco software. Principal response curves (PRC) (Section 10.1, illustrated in Section 16.6) are useful when comparing the development of biotic communities in time under different conditions. Principal coordinates of neighbour matrices (PCNM, also known as dbMEM) technique (Section 10.2, illustrated in Chapter 19) allows us to study the spatial structure present in data and compare it with (separate it from) the effects of environment or experimental manipulation. Linear discriminant analysis (LDA, also known as CVA) is a traditional method of taxonomical studies, but it can be also useful in the field of population or community biology (Section 10.3, illustrated in Section 18.3). Section 10.4 demonstrates the hierarchical decomposition of community composition variation and Chapter 17 presents a case study using this method. Decomposition of the total biotic diversity into its alpha- and beta-diversity components using ordination methods is explained in Section 10.5 and illustrated with a practical example in Section 12.4. Finally, Section 10.6 briefly summarises how the community composition can be predicted for a given combination of environmental conditions.

10.1 Principal response curves (PRC)

When we experimentally manipulate whole communities, we often evaluate the effect of the experimental treatments over a longer period, because the response to disturbance or to nutrient addition has a strong temporal aspect. When we need to compare the community composition at sites differing in experimental treatment with the control (unmodified) sites at different sampling times, it is quite difficult to do so using an ordination diagram from a standard constrained ordination. The temporal trajectory of each site is usually not a straight line going through the ordination diagram, but rather a complex, wiggling path (see Figure 16–8 for an example). It is therefore difficult to evaluate the extent of differences among the individual time steps and – more importantly – among the individual treatments, across the time. Van den Brink and Ter Braak (1998, 1999) developed a new method called principal response curves (PRC), which focuses exactly on this aspect of the data.

The primary result of the PRC method is one or several sets of response curves, representing temporal trajectories of community composition for each of the experimental

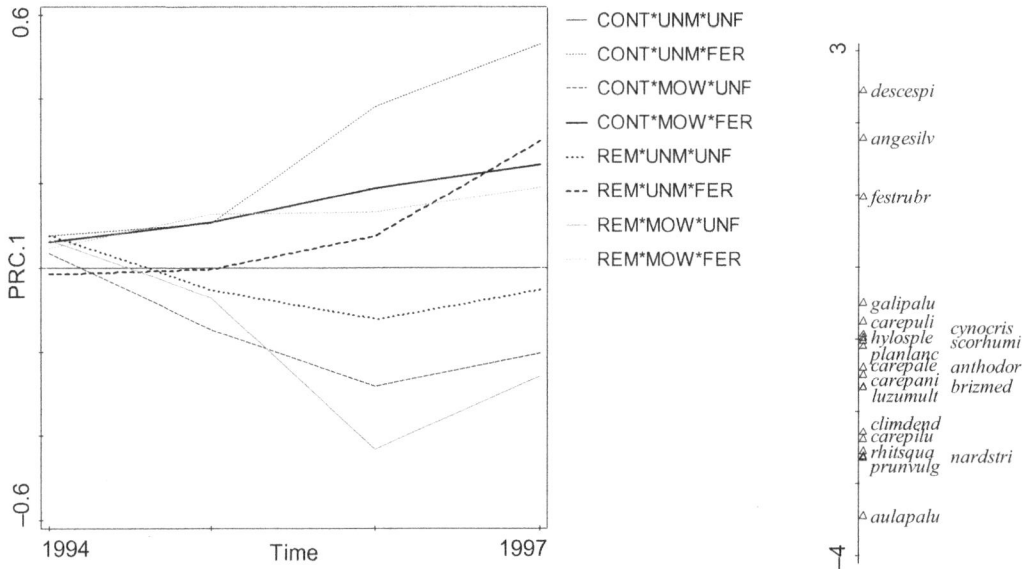

Figure 10–1 Diagram with principal response curves, from Section 16.6. *MOW/UNM* is for mowing the plots (or lack of it), *FER/UNF* for their fertilisation, *REM/CONT* for the experimental removal of the dominant plant species (*Molinia caerulea*).

treatments. An example of a set of eight principal response curves (corresponding to eight experimental treatments applied to grassland vegetation) is shown in Figure 10–1.

The diagram displaying PRC can be usefully supplemented by a one-dimensional diagram showing the response variable (typically species) scores on the corresponding RDA axis, as illustrated at the right side of Figure 10–1. We can combine a value read from PRC for a particular treatment and time with the response variable (species) score: if we calculate an exponential transformation of their product, the result predicts the relative size of that species abundance in comparison to its abundance in the cases with the control treatment at the same time. For example, the species *Scorzonera humilis* (*scorhumi*) has its score equal to −1.0 on the first RDA axis. We can predict from the diagram that in the third year its cover will be 32% **lower** on the fertilised-only plots, compared with the control plots. This is because the *CONT*UNM*FER* curve value in 1996 is (approximately) 0.38, so we predict the relative species cover as exp(−1.0 * 0.38) = 0.68. Such a quantitative interpretation rule requires a log-transformation of the original response data[1] and it was derived with the assumption of count data. But the interpretation carries on to the log–transformed cover estimates used in the study from which Figure 10–1 originates.

The PRC method is based on a partial redundancy analysis (RDA). If we have K treatment levels, coded by a factor variable Z in a Canoco data table, and we measure

---

[1] These predictions can be, however, biased when you need to use a non-zero C constant in the general log-transformation formula (see Section 1.3.1), as you then do not transform perfectly the multiplicative comparisons into additive ones.

the community composition at permanent sites multiple times, with time coded as a factor T, then we should set up the redundancy analysis model where the factor T is used as a covariate, and the interaction of the Z and T factors is used as the explanatory variable. Because the time is used as a covariate, the plotted response curves portray the overall differences among the treatments and their change with time. As the partial RDA, providing input data for PRC curves, uses time as a factor variable, the PRC can be used even with data where the sampling times are irregular.

Such an analysis can be easily set up in Canoco 5, where PRC has its own template.[2] When you use such template, Graph Wizard is able to construct the specialised PRC diagram without the effort needed in earlier CANOCO versions. There are as many response curves originating from the first RDA axis as there are treatment levels. Because the effect of experimental treatments at different sampling times cannot be usually summarised with only one constrained axis, we can construct a second or an even higher set of PRCs. The effect represented by the first-order and the additional response curves should be ascertained using a permutation test of the corresponding constrained axes. Correct setup of the permutation test (based on a split-plot nature of the repeated-measures data analysed by the PRC) is arranged for by the Canoco Adviser.

The construction of a PRC diagram and testing of the corresponding effects is illustrated in Case study 5, Chapter 16. Since its invention (Van den Brink & Ter Braak 1998 and 1999), hundreds of papers utilised PRC not only in its original domain of ecotoxicology (see e.g. Trekels et al. 2011 as an interesting application), but also in restoration ecology (e.g. Maccherini & Santi 2012) or in general ecological research (e.g. Veen & Olff 2011).

10.2 Separating spatial variation

The composition of biotic communities varies across space for various reasons. One of them is the spatial variation of environmental and other external factors, sometimes acting in the past and no longer measurable under present conditions, called 'induced spatial dependence' by Legendre and Legendre (2012). Other reasons can be internal to the studied community, such as dispersal limitation of the organisms involved or their competition, creating spatial autodependence in the community (response data) itself. The spatial variation leads to dependency among observations made at both closer and more distant locations and this phenomenon is usually called the spatial autocorrelation (Legendre, 1993). Further, an independently generated spatial variation in the community composition and in the environmental properties might generate an 'apparent' correlation between the community and the environment. Under such circumstances, the spatial variation acts as a 'spatial nuisance' (Peres-Neto & Legendre 2010).

If we are able to describe spatial variation in response (typically community) data using a set of spatial descriptors, we can separate – to a certain extent – the effects of spatial

---

[2] Two templates, in fact.

heterogeneity from the effects of environmental factors, using variation partitioning (see Section 5.7, Peres-Neto & Legendre 2010, or Legendre & Legendre 2012, pp. 877–881). Initially, the spatial descriptors were based on spatial coordinates of sampling points (e.g. latitude and longitude) but usually extended with power terms calculated from those coordinates and forming so-called trend-surface polynomials (Legendre & Legendre 2012, pp. 822–825). But this method is able to describe only gross spatial patterns, with a broad-scale change of response variables across the whole sampling domain, and it is not appropriate at all for localised patterns with small patches and/or medium-scale changes. Moreover, the trend-surface polynomials were found not to perform well in controlling the Type I error in the test of environmental effects when the spatial variation acts as a nuisance factor (Peres-Neto & Legendre 2010).

To model fine scale patterns (but also gross spatial patterns), so-called spatial eigenfunctions can be used as predictors in multivariate methods and they can represent the spatial variation in the variation partitioning procedure. The initially generated set of spatial eigenfunctions represents all spatial scales covered by the sampling design,[3] but we must usually select among those potential spatial predictors only the ones that are important for our response data. This is done with a stepwise selection of predictors in a separate constrained ordination, preceding the variation partitioning.[4] Because the standard forward selection procedure of spatial eigenfunction predictors tends to produce biased Type I error estimates and select too large a number of predictors (Peres-Neto & Legendre 2010, p. 178), it is important to protect the selection procedure by an initial test of significance using all spatial predictors and by the adjusted coefficient of determination (sensu Blanchet et al. 2008).

The initial set of spatial eigenfunction predictors can be generated using the PCNM (principal coordinates of neighbour matrices) method, initially described by Borcard and Legendre (2002), and further improved by Dray et al. (2006) who also identified this method with the distance-based MEM (Moran's eigenvector maps). Details of the computational procedure for the PCNM (dbMEM) can be found in Dray et al. (2006) or in Legendre and Legendre (2012, pp. 861–881), yet we provide a simplified description here.

Initially, a matrix of Euclidean distances among all sampling locations is computed and a threshold value estimated. All distances above this threshold are replaced with an identical, large distance value (its minimum is four times the threshold value), but smaller distances are not changed.[5] The distance matrix adjusted in this way is then processed

---

[3] I.e. not the scale below the sampling resolution or above the size of sampled area.

[4] This stepwise selection step normally uses community composition as the response data, so that all spatial pattern reflected in them is modelled. If you, however, want to remove the induced spatial dependence and to use just the inherent spatial dependence (i.e. Model 2 sensu Legendre & Legendre 2012, pp. 11–12), you can use the environmental descriptor data as the response variables to select spatial eigenfunctions describing the spatial dependence in environmental properties and then use the selected eigenfunctions as a priori covariates in remaining analyses, including the selection of additional spatial eigenfunction with the community data table as the response data (see p. 343).

[5] Distances of the locations to themselves (i.e. on the main diagonal of the distance matrix) are also replaced in the improved PCNM procedure, reflecting the view that the observations are not auto-correlated with themselves.

10.2 Separating spatial variation

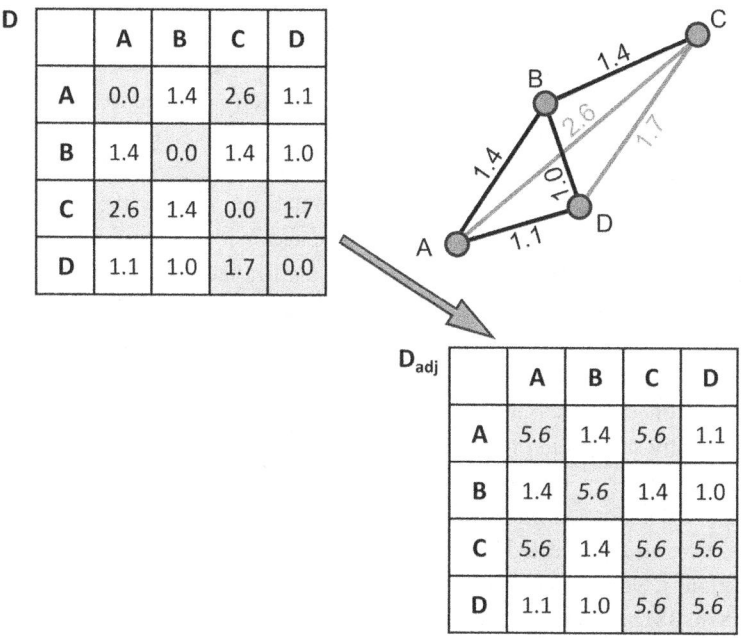

Figure 10–2 Calculating the adjusted distance matrix, using the default algorithm for threshold distance value: location C is most isolated from the others (see matrix D), so its minimum distance (to location B) becomes the threshold. Values above this threshold become replaced with (4*1.4=) 5.6 value.

with principal coordinates analysis (PCoA), also known as metric multidimensional scaling (see Section 6.5). The case scores on individual PCoA axes are then treated as independent (uncorrelated) spatial predictors, called Moran's eigenvector or, more generally, **spatial eigenfunctions** in Legendre and Legendre (2012). The threshold value can be chosen on a priori grounds, but as a default approach, the minimum distance needed to connect all locations together (i.e. the maximum edge length of a minimum spanning tree) is recommended.[6] This choice and the calculation of adjusted distance matrix (D_{adj}) are illustrated in Figure 10–2.

The PCoA applied to the adjusted matrix of distances yields not only axes with positive eigenvalues: about half of the computed axes have negative eigenvalues. The eigenvalues in the PCNM method are related to Moran's I coefficients of spatial correlation at the spatial scale captured by the corresponding eigenvector (predictor). When selecting spatial predictors to describe the spatial structure in data, most studies focus on the predictors with positive eigenvalues, corresponding to patterns of positive spatial correlation, but

---

[6] Here we assume that the sampling locations are all placed in a single continuous area. If there are multiple, distant sampling areas (the distances among them far exceeding the size of individual areas), it is usually better to compute the spatial predictors separately for each area and combine them later in a single matrix of predictors, with the predictor values set to 0 for the 'foreign' areas.

for some studies, the negative correlation is also of interest (see Legendre & Legendre 2012, p. 863).

Further, the axis order reflects the spatial scale: the first few axes capture the gross (overall) spatial patterns (of positive autocorrelation), often representing simple spatial trends if present in the data,[7] while the later axes (with positive eigenvalues) correspond to gradually finer spatial scales, and this property can be used to divide the spatial predictors into several distinct groups, interpreted separately in the variation partitioning procedure (see e.g. Borcard et al. 2004). The axes with negative eigenvalues (reflecting negative autocorrelation) usually represent small-range spatial scales.

Based on the results of variation partitioning using spatial predictors as one group (or multiple groups, see above), we can enhance the interpretation of data by separating the environmental effects that are not related to spatial variation (the partial effects of environmental descriptors after accounting for the effect of spatial predictors) from the spatially structured effects of environmental variables. We can view the latter type of effects either as a consequence of indirect (false) correlation between community composition and environment (if we believe that both the community and the environment are causally affected by external factors) or as a direct effect of the environmental variation, which on its own has a spatial heterogeneity (Legendre & Legendre 2012, pp. 878–881). In practice, however, the causal background is usually impossible to disentangle.

In many studies, it might be also meaningful to test and visualise the unique spatial variation in the community composition, not accounted for by the environmental properties i.e. the results of a partial constrained ordination using environmental predictors as covariates and selected spatial eigenfunction predictors as explanatory variables. This part of community variation might include the effects of non-measured environmental descriptors, shadow effects of past events or the consequences of biotic processes within the ecosystem (competition, predation, herbivory, dispersal, etc.).

Beside the PCNM corresponding to the MEM based on spatial distances, we can also use a more general type of MEMs, where the 'distance' matrix reflects the connectivity[8] among the sampling locations. Such a matrix might have a hierarchical nature (e.g. when describing connectivity among river branches forming a river network) and it can be as simple as having 0s for pairs of connected points and 1s for disconnected pairs. Canoco 5 implementation supports this extended approach (referred to as a general form of MEM in Legendre & Legendre 2012, pp. 881–893), with the distance matrix included in the analysis as a data table.

Case study 8 in Chapter 19 demonstrates the use of PCNM with variation partitioning in Canoco 5, but there are many published studies using the same method including, among others, studies focusing on the comparison of niche separation effects

---

[7] But Legendre and Legendre (2012, p. 868) recommend to model such simple trends separately (using X and Y coordinates or even the response surface polynomials), i.e. to detrend the data before PCNM analysis. Canoco 5 supports this using the PCNM analysis template with a priori covariates.

[8] In a broad sense, including the connectivity as defined in meta-population studies as a special case.

with neutral-theory explanation for beta diversity patterns of damselfly assemblages (Siepelski & McPeek 2013), Quaternary and pre-Quaternary history effects on world wide distribution of palms (Kissling et al. 2012), or the role of environmental and spatial processes in forming fish communities in complex North American lake systems (Sharma et al. 2011).

An approach similar to PCNM was also suggested for the analysis of data in comparative studies, where individual cases represent biotic taxa (e.g. species) and where the spatial correlation among cases is replaced by dependency among taxa, due to their shared evolutionary history (Desdevises et al. 2003). Instead of starting from a matrix of geographical distances, so-called patristic distances among taxa are calculated from a phylogenetic tree, quantifying the extent of the evolutionary past that the individual taxa pairs share. The Canoco 5 manual has an example in section 6.3.7, illustrating the use of Desdevises method.

10.3 Linear discriminant analysis

In some situations, we have an a priori classification of the studied objects (individuals of various species, sampling plots assigned to vegetation types, etc.) and we want to find a quantitative classification rule which uses the values of measured (explanatory) variables to predict the membership of an object in one of the a priori classes. This is a task for classical discriminant analysis. We will focus here on presenting the discriminant analysis as an ordination method that best reflects group membership.

Fisher's linear discriminant analysis (LDA), also called canonical variate analysis (CVA, e.g. in the Canoco 5 manual), is a method that allows us to find the scores for the classified objects (i.e. the cases in Canoco terminology). These scores are expressed as linear combinations of the explanatory variables that optimally separate the a priori defined groups (classes). This method (described in more detail in Legendre & Legendre 2012, pp. 673–690) is available in Canoco 5 with additional features not available in standard implementations of this method.

To perform LDA in Canoco 5, a data table with the classification of cases encoded by a single factor variable must be used as the response data. This is the appropriate coding for classical discriminant analysis, where the classification is 'crisp'.

The variables we want to use for the discrimination enter the analysis in Canoco as explanatory variables. We then use a canonical correspondence analysis (CCA) using Hill's scaling with a focus on the response distances (see Section 11.2). This is arranged for by the *Discriminant-analysis* template, offered by the Canoco Adviser when a data table with classifying factor is chosen as the focal table.

One distinct advantage of doing LDA in Canoco is that we might perform a **partial** discriminant analysis. In such an analysis, we can look for explanatory variables allowing us to discriminate between given classes in addition to other known discriminatory variables. Other useful features of LDA in Canoco are the ability to select a subset of the discriminating (explanatory) variables by means of forward selection of explanatory variables and the ability to test the discriminating power of the variables by the non-parametric Monte Carlo permutation test.

When plotting the results, the response variable scores represent the means (centroids) of the individual classes in the discriminatory space. Case scores (the *CaseE* scores) are the discriminant scores for the individual observations. A biplot diagram containing both response variable scores and the biplot scores of explanatory variables (*BipE*) portrays the table of averages of individual explanatory variables within individual classes, while the regression/canonical coefficients (*Regr* scores) of explanatory variables represent the loadings of the individual variables on the discriminant axes.

The lengths of the arrows for discriminating descriptors (plotted as the biplot scores of explanatory variables) must be adjusted to portray properly the discriminatory power of each variable, and the two *Discriminant-analysis* templates perform this adjustment automatically, as described in Canoco 5 manual, section 6.4.3.

10.4 Hierarchical analysis of community variation

If we study the variation in a biotic community at several spatial or temporal scales, we must take into account the hierarchical arrangement of the individual scale levels when analysing the data. If our sampling design allows us to do so, we may ask questions that involve the hierarchical nature of our data:

1. How large are the fractions of the total community variation that can be explained at the individual levels?
2. Is it possible to identify the levels that influence significantly the community composition and what are the differences at a particular level?

In this section, we will illustrate the required procedures using an imaginary example. An example of studying community variation on different spatial (landscape) levels is provided by Case study 6 in Chapter 17, using crayfish communities.

Let us start with the definitions of terms used in the following discussion. We will assume the response data (*Y*) were collected at three spatial levels. Let us work with a hypothetical project, for which we can imagine that three mountain ranges were sampled; in each of them three mountain ridges were selected (at random), and then three peaks were selected within each of the nine ridges. Response (community) data were collected using five randomly positioned plots on each of the 27 mountains (peaks). Figure 10–3 illustrates the sampling hierarchy for one of the three mountain ranges.

We will describe the spatial location of each plot by a factor variable representing mountain ranges (*Range*), by another factor variable identifying the ridges (*Ridge*), and another factor variable identifying the mountain peaks (*Peak*). The way of coding factors *Range*, *Ridge*, and *Peak* in the Canoco 5 data table is illustrated in Table 10–1.

The term 'variation explained by' in the following description refers to the classical variance when linear ordination (PCA, RDA) is used. When unimodal (weighted-averaging) ordination (CA, CCA) is used, it refers to inertia. For simplicity, we will

10.4 Hierarchical analysis of community variation

Table 10–1 Coding of the three spatial levels (ranges, ridges and peaks) using three factor variables. Please note that the numbering of lower hierarchical units (peak or ridge) must be continuous throughout the whole data set and cannot be restarted from 1 within each higher-level unit (ranges and ridges for peaks or ranges for ridges), see e.g. *plot 4b*.

Case	Range	Ridge	Peak
plot 1	range 1	ridge 1	peak 1
plot 2	range 1	ridge 1	peak 1
plot 3	range 1	ridge 1	peak 1
plot 4	range 1	ridge 1	peak 1
plot 5	range 1	ridge 1	peak 1
plot 6	range 1	ridge 1	peak 2
plot 7	range 1	ridge 1	peak 2
...	range 1	ridge 1	...
plot 15	range 1	ridge 1	peak 3
plot 16	range 1	ridge 2	peak 4
plot 17	range 1	ridge 2	peak 4
...	...	...	...
plot 4b	range 2	ridge 4	peak 10
...	...	...	...

Figure 10–3 Hierarchical sampling design used to separate variation at the different spatial levels (mountain ranges, mountain ridges and mountain peaks).

refer to linear methods when describing the required kind of analysis, but you can obtain a corresponding description in the context of unimodal ordination by replacing references to PCA by CA and those to RDA by CCA, when the unimodal methods are more appropriate for your data.

Total variation

The total variation in response data can be calculated using an unconstrained PCA, with no explanatory variables or covariates. We can describe the corresponding ordination model as:

$$Y = const$$

i.e. this is the null model, with no predictors.

The total variation in Y can be decomposed into four additive components: variation explained at the range level, variation explained at the ridge level, variation explained at the mountain peak level and the within-peak (or residual) variation.

Variation among ranges

To estimate the variation explained at the highest spatial level of ranges (*Range*), in an RDA we must use the *Range* variable as an explanatory variable and ignore the other two factor variables, i.e. the ordination model will be:

$$Y = \text{Range}$$

To test this effect (the differences among mountain ranges), we can use a Monte Carlo permutation test where we randomly reassign the membership in *Range* classes (i.e. which mountain range the data come from). But we should keep the other, lower-level spatial arrangement intact (not randomised), so that we test only the effect at this highest spatial level. Therefore, we must permute the membership of whole mountain ridges within the ranges, i.e. we must use a split-plot design (see Section 5.5) where the ridges (**not** the ranges) represent the whole-plots and the groups of 15 plots within each ridge represent the split-plots. The whole-plots will be permuted randomly (they will randomly change their membership in ranges), while the split-plots will not be permuted at all.

Variation among ridges

When estimating the variation explained at the intermediate spatial level of ridges (*Ridge*), we must use the *Ridge* factor as a predictor. This is not enough, however, because its simple (independent) effect includes the effect at the higher spatial level (*Range*), which we already estimated.[9] Therefore, the *Range* variable should be used as a covariate in the partial RDA. We can formally describe this partial constrained ordination as:

$$Y = \text{Ridge} \mid \text{Range}$$

i.e. we extract from Y the effect of *Ridge*, after we accounted for the effect at level of *Range*. In other words, we model here the variability **among** the ridges, but **within** the ranges.

---

[9] We have nine ridges and from the identity of a ridge we can tell to which mountain range a plot belongs.

When testing the *Ridge* effect (differences between the ridges), we must randomly (in an unrestricted way) permute the membership of mountain peaks within the ridges (with all five plots from each peak held together during the permutations), but we shall not permute across the mountain ranges (the *Range* levels). To achieve this, we must do restricted (split-plot type) permutations within blocks defined by the *Range* covariate.

Variation among peaks

The variation explained by the mountain peaks can be estimated (and tested) with a similar (partial RDA) set-up as used before, except we are now working at a lower spatial level. We will use the *Peak* factor as an explanatory variable and the *Ridge* factor as a covariate:[10]

$$Y = \text{Peak} \mid \text{Ridge}$$

Further, we will not use the split-plot design permutation restrictions here. Instead, we will randomly allocate mountain peak identity to the plots, but only within the permutation blocks defined by the *Ridge* covariate (i.e. within each of the nine mountain ridges).

Residual variation

We still have one hierarchical component to estimate. It is the variation in community composition among the individual plots collected at the individual peaks. This is the residual, lowest-level variation, so we cannot test it. To estimate its size, we will perform **partial unconstrained** ordination, i.e. a PCA with the *Peak* variable used as a covariate and with no explanatory variables.

Note also the following aspect of the hierarchical partitioning of community variation. As we move from the top level of the hierarchy towards the lower levels, the number of degrees of freedom for spatial levels increases quickly: we have only 3 mountain ranges, but 9 ridges and 27 peaks. Predictably, this fact alone will lead to the increasing size of variation explained by the progressively lower levels. The amount of explained variation (represented by the sum of canonical eigenvalues) corresponds to the sum-of-squares explained by a regression term in a linear regression, or by a factor in an ANOVA model. To compare the relative importance of individual hierarchical levels, we should divide the individual variation fractions by their respective degrees of freedom. The method for calculating the number of DFs for individual hierarchical levels is outlined in Section 17.3.

10.5 Partitioning diversity indices into alpha and beta components

Results of ordination methods are often combined with an independent calculation and visualisation of the diversity found at individual sampling units (i.e. their **α-diversity**

---

[10] Note that the use of *Ridge* as a covariate is fully sufficient, as its value implies the value of the higher level *Range*. In other words, the information about *Range* is fully embedded within *Ridge*.

values) and the diversity among the sampling units (i.e. the β **diversity** of the data set). The β- and α-diversity estimates intuitively combine to the total diversity in our data set, known as the γ **diversity**. We can further expect that the extent of β diversity is somehow related to the total variation in the response data, as reported by ordination methods, but the exact relation for the commonly used diversity measures, as well as the way to decompose the total diversity into its α and β components are not obvious.

In this section we explain how to modify data tables and existing ordination methods to perform a decomposition of the total diversity (measured by several diversity indices) into α and β diversity. As the notion of diversity is so strictly bound to the community ecology field, we will refer in this section to response variables as *species*.

Legendre et al. (2005) presented a framework for evaluating β diversity with constrained ordination, but the relation between the community variation – as reported by the ordination methods – and the usual indices measuring biotic diversity (species richness, Simpson diversity index, or Shannon–Wiener H index, see Legendre & Legendre 2012, pp. 250–255) was not explained. On the other hand, Pélissier et al. (2003) clearly define the links between classical diversity indices (including their additive partitioning into α and β components) and the ordination methods, but the terminology they use is closely bound to the *ade4* software and so it leaves many readers wondering how applicable the suggested methods are with other software for ordination, even after the paper of Pélissier and Couteron (2007) appeared. So we explain here the general principles of diversity decomposition using constrained ordination and relate the variation measures used in linear and unimodal ordination to the commonly adopted diversity indices. The last section of Chapter 12 provides a practical example of some of the procedures discussed here.

The additive model of community diversity decomposition, in which the total (γ) diversity represents the sum of the within-plot (α) and the among-plots (β) diversity (Lande 1996), can be reproduced by ordination methods only if these methods can quantify also the variation **within** the plots (cases). To achieve this, the standard response data table describing community composition for individual cases must be replaced by an **inflated data table** (Pélissier et al. 2003). In this table (illustrated in Figure 10–4(a)), each row represents a single species occurrence, i.e. one non-zero cell of the original species data table.

But the inflated species composition table shown in the lower left corner of Figure 10–4(a) does not fully correspond to the inflated table presented in Pélissier and Couteron (2007), in which each row represented a single individual. The individual-based table inflation is illustrated in Figure 10–4(b) and there the values in the non-inflated data table are interpreted as individual counts. The total number of rows in this type of inflated table is equal to the total number of individuals in the original data. When used in the ordination methods, both forms of inflated tables are equivalent[11] as long as the row weights (given as extra column, next to the inflated table) are used with the occurrence-based version of the inflated table (Figure 10–4(a)). In addition, only the occurrence-based inflation allows us to inflate tables for community composition data where the quantities differ from individual counts (e.g. percentage cover or biomass).

---

[11] With the exception of permutation tests, as discussed below.

10.5 Partitioning diversity into alpha and beta components

(a)

	Spc 1	Spc 2	Spc 3
plot 1	2	0	1
plot 2	3	0.5	2
plot 3	3	0	2
plot 4	2.8	0	1

	Spc 1	Spc 2	Spc 3	Weight
plot 1.Spc 1	1	0	0	2
plot 1.Spc 3	0	0	1	1
plot 2.Spc 1	1	0	0	3
plot 2.Spc 2	0	1	0	0.5
plot 2.Spc 3	0	0	1	2
plot 3.Spc 1	1	0	0	3
plot 3.Spc 3	0	0	1	2
plot 4.Spc 1	1	0	0	2.8
plot 4.Spc 3	0	0	1	1
			Sum =	17.3

	Plot
plot 1.Spc 1	plot 1
plot 1.Spc 3	plot 1
plot 2.Spc 1	plot 2
plot 2.Spc 2	plot 2
plot 2.Spc 3	plot 2
plot 3.Spc 1	plot 3
plot 3.Spc 3	plot 3
plot 4.Spc 1	plot 4
plot 4.Spc 3	plot 4

(b)

	Spc 1	Spc 2	Spc 3
plot 1	2	0	1
plot 2	3	1	0
plot 3	2	0	2
plot 4	1	0	1

	Spc 1	Spc 2	Spc 3
plot 1.Spc 1 a	1	0	0
plot 1.Spc 1 b	1	0	0
plot 1.Spc 3	0	0	1
plot 2.Spc 1 a	1	0	0
plot 2.Spc 1 b	1	0	0
plot 2.Spc 1 c	1	0	0
plot 2.Spc 2	0	1	0
plot 3.Spc 1 a	1	0	0
plot 3.Spc 1 b	1	0	0
plot 3.Spc 3 a	0	0	1
plot 3.Spc 3 b	0	0	1
plot 4.Spc 1	1	0	0
plot 4.Spc 3	0	0	1

	Plot
plot 1.Spc 1 a	plot 1
plot 1.Spc 1 b	plot 1
plot 1.Spc 3	plot 1
plot 2.Spc 1 a	plot 2
plot 2.Spc 1 b	plot 2
plot 2.Spc 1 c	plot 2
plot 2.Spc 2	plot 2
plot 3.Spc 1 a	plot 3
plot 3.Spc 1 b	plot 3
plot 3.Spc 3 a	plot 3
plot 3.Spc 3 b	plot 3
plot 4.Spc 1	plot 4
plot 4.Spc 3	plot 4

Figure 10–4 Creating an inflated data table from the original species composition data table. Additional description is in the text.

Note also that the row weights accompanying the inflated table in Figure 10-4(a), represent the original quantity of each species in the original case and so their sum is equal to the sum of all values in the original data table. As a part of the inflation procedure[12], another data table is generated and shown in both parts of Figure 10-4, coding the relation of the inflated table rows to the cases (rows) of the original data set. This secondary data table is essential for partitioning the γ diversity into α and β components.

If we take an inflated species composition table and analyse it (its first three variables, in our simplified example) using a weighted[13] principal component analysis (PCA) based on a variance–covariance matrix (i.e. we specify only the centring by species in Canoco), the analysis results will show that the total variation in response data is equal to 8.46243. When we divide this value by the sum of row weights (17.3, as shown in Figure 10-4(a)), we obtain an estimate of the total (γ) diversity in the species composition data, measured by the Simpson diversity index:

$$8.46243/17.3 = 0.48916$$

There are two widely used formulas for Simpson diversity index, both starting from the sum of squared relative frequencies of present species (see Legendre & Legendre 2012, p. 254, where λ is used instead of D):

$$D = \sum_{j=1}^{s} f_j^2$$

The two formulas are $1 - D$ and $1/D$.[14] Only the index version computed as $(1 - D)$, also known as the Gini–Simpson index, can be additively decomposed into α and β components. To verify the index value calculated via weighted PCA, we must first calculate the relative frequencies for the three species in our example, with the resulting values 0.6243, 0.0289, and 0.3468, respectively.[15] The Simpson diversity value can be then calculated as

$$1 - (0.6243^2 + 0.0289^2 + 0.3468^2) = 0.48914.$$

As we have already stated, the γ diversity in this analytical approach is the sum of α (within-plot) and β (among-plot) diversity, so it is therefore hardly surprising that to separate these two components, we must use the plot membership (the *Plot* variable shown in Figure 10-4 for both types of inflation) as an explanatory variable in a constrained form of the above PCA, i.e. in a redundancy analysis (RDA). If we continue our original example (focusing on the analysis without explicit species weights, i.e. reproducing Simpson diversity index), an RDA with the inflated table and using the *Plot*

---

[12] As implemented in Canoco 5.
[13] Using the Weight variable in the fourth column of the occurrence-based inflated table; no case weights are needed when you start from an inflated table illustrated in Figure 10-4(b).
[14] We note that the $1/D$ form corresponds to N2 diversity index that appears in Canoco results (shown with case scores in the analysis notebook) as a measure of case diversity and this equivalent-number transformation of the diversity indices is presently recognised as the more intuitive representation of diversity concept (see Jost 2006 or Chao et al. 2012).
[15] For example for species Spc 1, this can be calculated as $(2+3+3+2.8)/17.3$.

10.5 Partitioning diversity into alpha and beta components

as the single constraining variable produces the same value of total variation (8.46243), but also shows that the two constrained axes explain 1.9 per cent of the total variation. This is, therefore, the relative contribution of the β (among-plot) component to the Simpson index calculated for the whole data set.

This additive decomposition of the Simpson index is the only one (out of the three diversity indices considered in this section) where the β component corresponds to a difference between the Simpson index calculated for whole data set and a (weighted) average of the Simpson indices computed for individual cases. This additive decomposition was criticised due to the dependence of the β-component on the α-component value (Jost 2007).

To test the null hypothesis that β diversity is nill ($\beta = 0$), i.e. to test whether there is a significant amount of β diversity (rather than a random distribution of relative species abundances across the plots), we can perform a Monte Carlo permutation test, randomly assigning plot membership to individual rows of the inflated data table. This ensures that the number of individuals per plot is constant across permutation and corresponds to a permutation scheme for the original data table, in which the existing abundance values are randomly shifted through the original data table, with the restriction that the number of individuals in individual plots is kept at their original values.

But there is one important issue, concerning this permutation test when the occurrence-based inflation is used. Given the way the inflated table was generated (i.e. the total abundance of each species in a single plot generated one inflated table row), we are actually permuting the abundance values of each species, not the individuals. But the original logic behind Simpson's diversity index refers to a probability that two *individuals* randomly selected from a sample belong to the same species. From this perspective the individuals, not their collection within a plot, should be the permutable units.[16]

The need to permute individuals, not their sums in individual cases, is perhaps even more acute when decomposing the gamma diversity based on species richness (as described below): during the permutations, the number of rows belonging to individual plots cannot change (and hence the α richness does not change) and the fact that we get any variation at all in the β component across the permutations is purely due to the artefact of Canoco implementation, where the case weights do not travel with their original rows during the permutation test. In other words, the permutation test based on the occurrence-inflated (rather than individual-inflated) data table is not appropriate for the decomposition of richness and somewhat questionable for the other two indices.[17]

Canoco 5 is able to inflate compositional data table using both approaches and the last section of Chapter 12 illustrates how to correctly test the β diversity using individual-based inflation of the data table. In this case, the number of individuals per plot is kept constant in the null model and individuals are randomly selected (without replacement) from the pool of all individuals observed in the study. Using this permutation, our null model considers individuals to be independent units. This might be a reasonable

---

[16] This is, however, something we cannot achieve in our example table in Figure 10–4(a), where values like 2.8 or 0.5 cannot be interpreted as individual counts.

[17] But again, we cannot do any better when our data do not represent individual counts.

assumption when using e.g. *pairs* of birds (as we do in Chapter 12) and not bird individuals as independent permutable units.

But it will always be a questionable approach with plants or other groups of organisms, for which it is difficult to clearly define individuals. But even if we work with organisms with clearly separated individuals, we can expect some type of clumping, often due to dispersal. Thus, significant β diversity will be then very probably a consequence of the fact that the individuals of the same species are clumped in the real data, but completely randomly distributed in individual permutations under the null model.

To similarly decompose the species richness (number of present species) or Shannon–Wiener diversity index, we must modify the weighted linear ordination by adding explicit weights to individual species as well. These weights are calculated from the species relative frequencies (f_j) already introduced above: $1.0/f_j$ to decompose species richness or $\ln(1.0/f_j)/(1.0 - f_j)$ to decompose the Shannon–Wiener H. In either case, the total variation must be divided by the sum of row weights (17.3 in our example) and the resulting value is equal to the total number of species *minus one*[18] or to the H value, respectively.

The β component calculated for species richness represents the ratio of the total number of species in the data set (N) and the average species richness of plots (α_{avg}), i.e. the original β-diversity proposal by Whittaker (1960).[19] Alternatively, it can be seen as the ratio between an additive β component (calculated as $N - \alpha_{\text{avg}}$) and the α_{avg} value. Note, however, that the α component calculated by the constrained ordination (as the scaled residual variance, after accounting for plot identity) is not equal to α_{avg}.

We can alternatively decompose the species richness by applying the correspondence analysis (CA) to the inflated species composition table, with only the case weights (stored in the *Weight* variable in Figure 10–4(a)) applied. The total variation (inertia) reported by CA is then equal (without the need to divide by the total of the rows) to the total number of species minus one.

The additive decomposition of diversity indices was recently criticised (see, e.g. Jost 2007 or Chao et al. 2012) and the use of equivalent numbers of species (Hill 1973; Jost 2006) and of their subsequent multiplicative decomposition into α and β components was demonstrated to be a more appropriate approach, although the additive partitioning was also shown to have its own merits (Chao et al. 2012). The simple transformation of the variation decomposed by constrained ordination methods into diversity measures is nevertheless valuable e.g. for exploring the contribution of individual species to diversity components or for interpreting the diversity patterns in terms of available exploratory variables.

10.6 Predicting community composition

The method of constrained ordination relates the response data (e.g. community composition data table) to explanatory variables (e.g. the environmental properties) in a

---

[18] This is due to centring done in the linear ordination.
[19] This ratio is, however, decreased by one.

10.6 Predicting community composition

way similar to classical regression (see Section 4.8). As in a regression model, we can therefore also use known values of explanatory variables to predict the response data values, e.g. to predict the expected community composition under given environmental conditions. Canoco 5 offers this procedure in a form of creating a new 'data' table with predicted/fitted[20] values, based on a selected constrained ordination analysis.

Probably the most important use is the prediction for new environmental data and the following comparison of predicted community composition with that of already known community records. These techniques are illustrated and explained in more detail in the Canoco 5 manual, section 6.2.5.4. When using the predictions in the context of unimodal constrained ordination (i.e. CCA), one must be aware that there the ordination model predicts only the values relative to the case total and therefore, when predicting the composition of an unknown community, the total abundance is not known and the average total of recorded cases is used instead. In the context of linear ordination (i.e. RDA), on the other hand, the predicted values can be negative, unless we have chosen log transformation in the analysis set-up.

The calculation of fitted or predicted values is not available for constrained ordination analyses where covariates were present, standardisation or centring by cases was selected in a linear method, or where detrending procedure was used in a unimodal method.

---

[20] Following the terminology used in classical regression methods, the Canoco 5 manual distinguishes fitted values (calculated from explanatory variable values used to fit the constrained ordination model) from predicted values (calculated from additional data, not used in the analysis on which the predictions are based).

11 Visualising multivariate data

The primary device for presenting the results of an ordination model is the ordination diagram. The contents of an ordination diagram can be used to approximate the response data table, the matrix of distances between individual cases, or the matrix of correlations or dissimilarities between individual response variables. In ordination including predictor variables (either explanatory or supplementary variables), we can use the ordination diagram to approximate, among others, the relationship between the response and the predictor variables. The first two sections in this chapter summarise what we can deduce from ordination diagrams that result from linear and unimodal ordination methods.

Before we discuss the rules for interpreting ordination diagrams, we must stress that the absolute values of scores (i.e. coordinates of objects, such as cases, response and predictor variables) in ordination space do not have, in general, any meaning.[1] When interpreting ordination diagrams, we use relative distances of symbols, relative directions of arrows, or relative ordering of projection points. A detailed description of an ordination diagram created in Canoco 5 can be obtained by the *Graph | Describe contents* menu command or using the 'lifebuoy' button in the toolbar.

With only one type of scores plotted, we usually call the diagram a **scatter diagram**; with two types of scores combined the diagram represents a **biplot** (but only when a biplot rule can be used for its interpretation, see Table 11–2 for additional details on joint plots). A diagram combining symbols and/or arrows for three different types of entities is called a **triplot**.

For both linear and unimodal ordination methods, the following sections discuss the impact of the chosen score scaling type upon the interpretation of ordination diagrams. Unlike earlier versions, we do not specify in Canoco 5 the scaling type in the Analysis Setup Wizard. Instead, we can choose the scaling flexibly for any executed analysis before we create an ordination diagram.[2] To do so, it is easiest to choose the 'Edit

---

[1] This does not preclude a choice that helps interpretation. We mention two situations. The first is a scaling that allows the value of a linear correlation between a (response and/or explanatory) variable and the case scores on an ordination axis to be read off the diagram by projecting the tip of the arrow of that variable onto the ordination axes. The second is in unimodal analyses with Hill's scaling and the focus on case distances (see Figure 11–1). Distances between case points then measure their dissimilarity in composition turnover (or SD) units.

[2] There are some exceptions though, namely the unimodal analyses with detrending by segments and also specific analyses, where a particular scaling is required for correct interpretation of their diagrams (linear discriminant analysis or principal response curves).

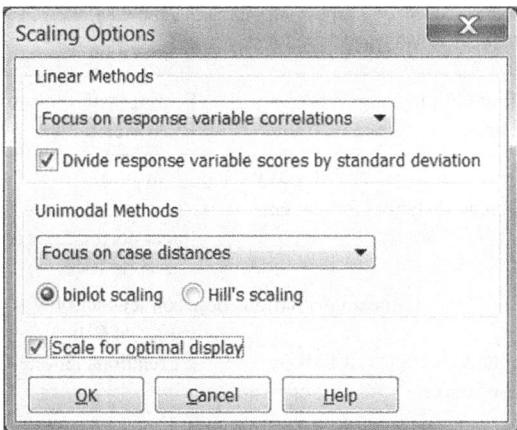

Figure 11-1 Scaling Options dialog.

scaling options' button in the main toolbar (sixth from the left, between QuickWizard and Print tools). This displays a dialog box illustrated in Figure 11-1.

Note that the scaling options are set separately for linear and unimodal methods (in the upper and lower part, respectively) and apply to all analyses/projects until you change them. The options are somewhat similar for linear and unimodal methods, as both are based primarily on the decision whether the viewer's attention is focused on the cases or on the response variables.

For linear methods, we must further decide whether the length of arrows should mirror the differences in the extent of variation of individual response variable values (with more varying variables generally having longer arrows) or whether the variation of individual response variables should be standardised to the same range.[3] In that case the length of each arrow only expresses how well the values of the response variable are approximated by ordination axes.

In the case of unimodal methods, we should also decide about the method of interpreting the ordination diagram. For data with very long composition gradients (i.e. with a large beta diversity across the cases), the distance rule is more appropriate and so is **Hill scaling**. Otherwise **biplot scaling** provides ordination diagrams that can be interpreted in a more quantitative way.

The *Scale for optimal display* option affects the scatter diagrams with only one type of entity. When it is checked and, for example, a scatter diagram with cases is plotted for a linear ordination, the scaling for this diagram is chosen as if the *Focus on case distances*

---

[3] This can be achieved in two, not fully equivalent ways: if you choose to *Center and standardize* the response variables in the *Ordination Options* page of setup wizard, the standardisation implies that all the response variables have the same weight in (same effect on) the resulting ordination. The option *Divide response variables scores by standard deviation* then has no effect on the look of ordination diagrams. If, on the other hand, you only *Center* the response variables, but you check the *Divide response variables scores...* option, the response variables enter the ordination algorithm non-standardised and hence with their effect proportional to variation of their values, but in the ordination diagram, the lengths of their arrows are adjusted so that they reflect only their fit to ordination axes.

Table 11-1 Relations between response variables, cases, and explanatory (or supplementary) variables that can be read from an ordination diagram for a linear method in two types of score scaling.

Compared entities	Scaling with focus on case distances	Scaling with focus on RV correlations
RVs vs cases	(fitted) RV values in cases	
cases vs cases	Euclidean distances among cases	n.a.
RVs vs RVs	n.a.	linear correlations (covariances) among RVs
RVs vs EVs	linear correlations between RVs and EVs	
cases vs EVs	n.a.	values of EVs
EVs vs EVs	independent effects of EVs on case scores	correlations among EVs
RVs vs factor EVs	mean RV values within case classes	
cases vs factor EVs	membership of cases in classes	
factor EVs vs factor EVs	Euclidean distances between case classes	n.a.
EVs vs factor EVs	n.a.	averages of EVs in case classes
Biplot with this scaling is called	distance biplot	correlation biplot
RV arrow length shows	contribution of RV to the ordination subspace definition	approximate std. deviation of RV values (if not post-standardised)
Length of the radius of equilibrium contribution circle	$\sqrt{d/p}$	$s_j \sqrt{d/p}$

$EV(s)$ means explanatory (or supplementary) variable(s), $RV(s)$ means response variable(s). In the definition of the equilibrium contribution circle, d expresses the dimensionality of the displayed ordination spaces (usually the value is 2), p means the dimensionality of the whole ordination space (usually equal to the number of response variables), and s_j is the standard deviation of the values of the j-th response variable.

was selected in the *Linear Methods* area, whatever the actual choice is. This reflects the fact (see Table 11–1, the *cases vs cases* row) that for other scaling choices, there is no optimum interpretation of the distances among case symbols. Similarly a separate scatter of response variable arrows (for linear method) or symbols (for unimodal method) is plotted with the best scaling for judging variable correlations or distances.

The choice among the scaling options (with focus on either cases or response variables) is not so important when the amount of variation explained by the displayed ordination axes is of similar magnitude. In particular, if the eigenvalues of the two axes are equal, a change in scaling focus does not change the diagram.

11.1 Reading ordination diagrams of linear methods

An ordination diagram based on a linear ordination (PCA or RDA) can display the scores of cases (represented by symbols), response variables (represented by arrows), quantitative predictor variables (represented by arrows), and factor predictors (represented by points – centroids – corresponding to individual levels of the factor variable). Table 11–1 (after Ter Braak 1994) summarises what can be deduced from ordination

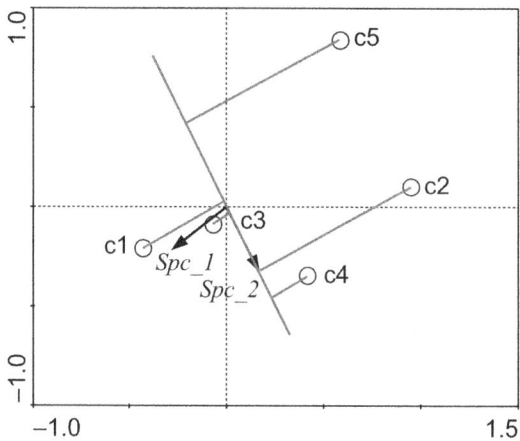

Figure 11-2 Projecting case points onto response (e.g. species) vector (of species *Spc2*) in a biplot from linear ordination. Here we predict the largest abundance of *Spc2* in cases *c4* and *c2*, then case *c3* is predicted to have a lower abundance near the average of *Spc2* values, and the expected abundance decreases even more for *c1* and *c5*.

diagrams based on these scores. Additionally, Table 11–1 explains the meaning of the length of response variable arrows in the two types of ordination scaling, and provides information about calculating the radius of the equilibrium contribution circle.

In linear ordination diagrams, the **equilibrium contribution (EC) circle** displays the expected positions of the heads of the response variable arrows, under the assumption that a variable contributes equally to the definition of all the ordination axes (see Figure 12–4 for an example). The EC circle has a common radius for all response variables only in a biplot diagram focusing on inter-case distances (see Legendre & Legendre 2012, section 9.1.3, for additional discussion). The EC circle can be shown with properly chosen scaling, using the options available from the *General* page of the dialog box shown by the *Analysis | Plot creation options* menu command.

Response variables are shown as arrows in the ordination diagrams of linear methods, corresponding to the assumed linear change of these variables along the ordination axes. There is a useful symbolism in the use of arrows: the arrow points in the direction of maximum increase in the value of the variable across the diagram, and its length is proportional to this maximum rate of change. A more precise interpretation is as follows. For a response variable, if we project the case points perpendicular to it, we obtain an approximate ordering of the values of this variable across the projected cases (see Figure 11–2). The perpendicular projection of points onto the vectors (arrows) is called the **biplot rule** in this context. If we use the case scores that are a linear combination of the explanatory variables[4] (*CaseE* scores, typically in RDA), we approximate the **fitted**, not

---

[4] Canoco uses the CaseE scores for cases in the ordination diagrams from constrained ordination (RDA, CCA). This default option can be changed, however, for a particular analysis in the dialog shown by the *Analysis | Plot creation options* menu command.

Visualising multivariate data

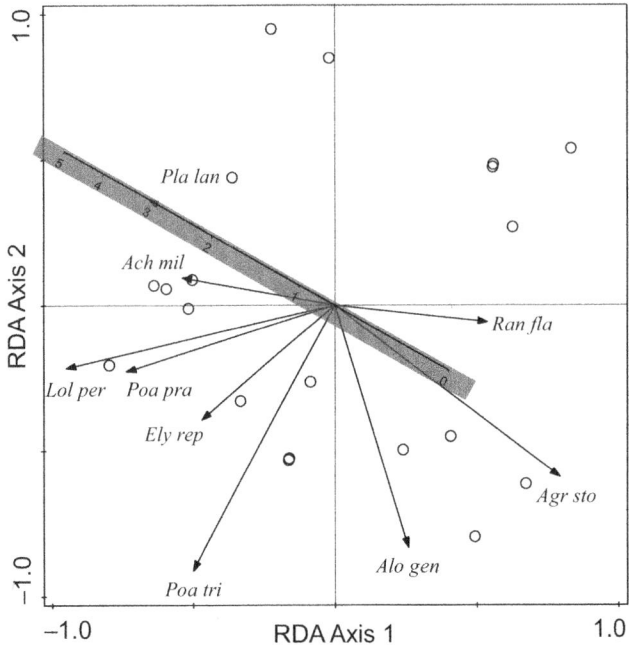

Figure 11–3 Variable calibration tool used with a response variable (*Pla lan* species) in an RDA ordination diagram. Predicted species abundances are shown on the original scale in the data table, yet their non-linear scaling reflects the log transformation applied to response data within the analysis.

the observed values of those abundances.[5] This interpretation is correct for both kinds of scaling. If centring by response variables was performed,[6] a case point projecting onto the beginning (origin) of the coordinate system (perpendicular to a response variable arrow) is predicted to have an average value of the corresponding variable. The cases projecting further from zero in the direction of the arrow are predicted to have above-average values, while the case points projecting in the opposite direction are predicted to have below-average values.

Canoco 5 offers a new tool (calibrate variable arrow) that allows you to interactively explore the biplot rule for response variables in linear ordination methods and for explanatory (or supplementary) variables in both linear and unimodal methods. To use it, select the calibration tool (by clicking the rightmost button in the graph toolbar) and then click either a variable name or corresponding arrowtip. If the calibration can be performed,[7] a 'measure tape' is overlaid upon the arrow, such as the plant species arrow in Figure 11–3. Its scale reflects also any transformation applied to the data during

---

[5] See Section 4.8 for a description of the linear ordination techniques in terms of fitting models of a response variable's linear change.
[6] This is the case in the majority of analyses using linear ordination methods.
[7] There exist circumstances when the calibration is not available, most of them being listed in Canoco 5 manual, p. 399.

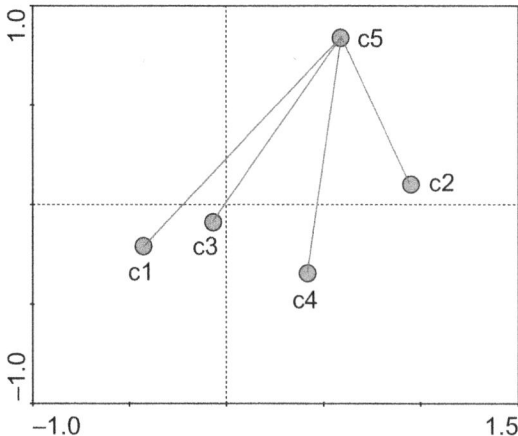

Figure 11-4 Distances between case points in an ordination diagram. If we measure the dissimilarity between case *c5* and the other cases using Euclidean distance, the distance between case *c5* and *c2* is predicted to be the shortest one, the distance to cases *c3* and *c4* the next shortest, and finally the distance to case *c1* the longest.

the analysis, such as the log transformation of the response data in our example. Each ordination diagram can contain at most one calibration axis, so clicking another arrow removes the original axis. To remove an axis, click with the tool active into an empty space of the diagram. Additional details about calibration axis construction can be found in Graffelman and Van Eeuwijk (2005).

The distance between case points approximates their dissimilarity, expressed using Euclidean distance in the scaling focused on case distances (see Figure 11–4).[8]

The relative directions of the response arrows approximate the (linear) correlation coefficients among the response variables in the scaling focused on response variable correlations. Arrows pointing in the same direction correspond to response variables that are predicted to have a large positive correlation, whereas variables with a large negative correlation are predicted to have arrows pointing in the opposite direction (Figure 11–5).

A more precise interpretation rule is as follows. If the response variables were centred and standardised or their scores were post-transformed (divided by their standard deviations), we can estimate the correlations by perpendicularly projecting the arrow tips of (other) response variables onto the arrow of a particular response variable (see Figure 11–6), analogously to the way we infer about its values in individual cases. Variables at right angles to the one considered project into the origin of the coordinate system. The origin thus indicates the value of zero correlation along the arrow. In most cases,

---

[8] Note that the actual distance measure approximated in ordination diagrams may be different from the raw Euclidean distance. For example, if you specify standardisation by case norm in Canoco, the distances between case points approximate the so-called Chord distance (see Legendre & Gallagher 2001 and Section 6.2). Canoco 5 adds the possibility of applying the Hellinger transformation that turns Euclidean distances into Hellinger distances, an approach recommended by Legendre and Gallagher (2001) – see Section 6.2.

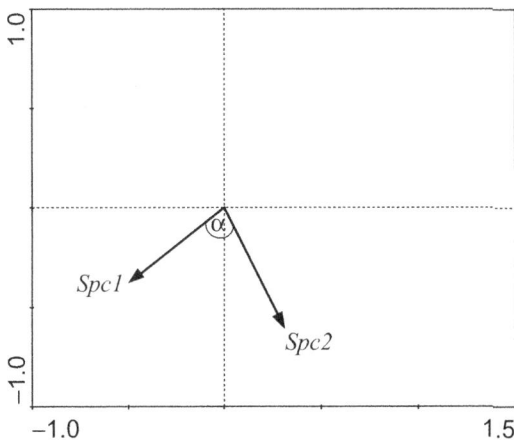

Figure 11-5 Angles between the response variable arrows in a diagram from a linear ordination. As the arrows for the two variables *Spc1* and *Spc2* meet nearly at a right angle, the variables are predicted to have a low (near-to-zero) correlation. A more precise approximation in the default scaling options (with response scores being post-transformed) is achieved by the biplot projection rule (see Figure 11-6).

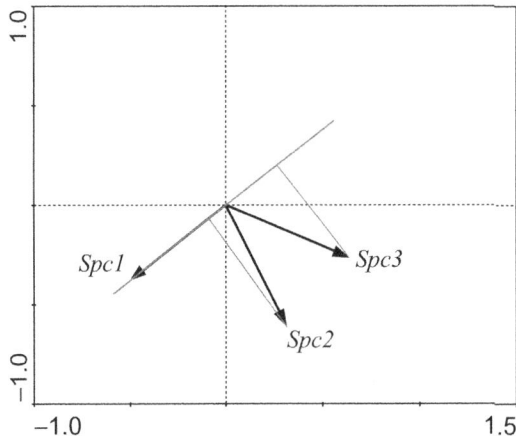

Figure 11-6 Perpendicular projection of arrow tips of other response variables on the selected response variable arrow.

we obtain very similar conclusions with both alternative interpretation rules (i.e. angle comparison and perpendicular projection).

In case of covariance-based biplots (where response variable scores are not post-transformed) the approximated correlation between two variables is equal to the cosine of the angle between the corresponding arrows (Figure 11-5).

We can apply a similar approximation when comparing a response variable with a (quantitative) explanatory variable (see Figure 11-7). For example, if the arrow for an explanatory variable (e.g. A in our example) points in a similar direction to a response

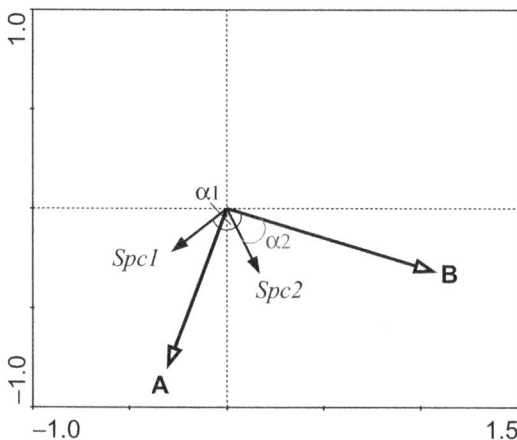

Figure 11-7 Angles between arrows of response and explanatory variables in an ordination diagram from a linear ordination. When comparing the correlation of explanatory variable B with the two response variables, we predict from the angles that B has an intermediate positive correlation with the response variable $Spc2$ and has an intermediate negative correlation with variable $Spc1$ (the higher the value of B, the lower the expected value of $Spc1$). The biplot projection of response arrow tips onto the arrow of an explanatory variable provides a more precise approximation.

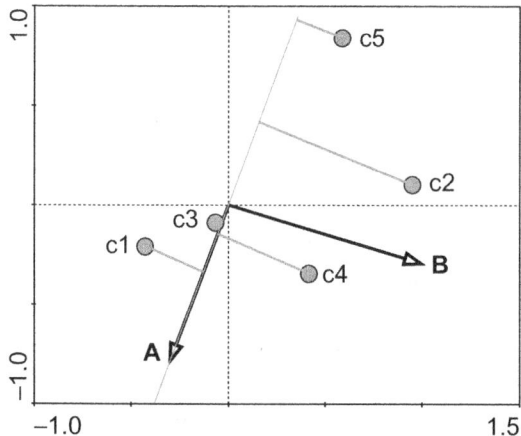

Figure 11-8 Projecting case points onto the arrow of a quantitative explanatory variable. Variable A is predicted to have similar values (near the variable average) for the cases $c3$ and $c4$, even though they are located at different distances from the explanatory variable line.

arrow (e.g. $Spc1$), the values of that response variable are predicted to be positively correlated with the values of the explanatory variable. This interpretation can be used with both types of scaling of ordination scores.

The case points can also be projected perpendicularly to the arrows of explanatory variables (see Figure 11-8). This gives us an approximate ordering of the cases by increasing values of the explanatory variable (if we proceed towards the arrow tip

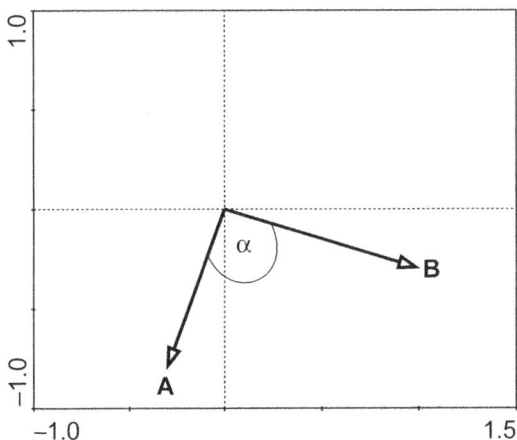

Figure 11-9 Measuring angles between arrows of quantitative explanatory variables. The angle between the two variables suggests that they have independent effects upon the community composition and the correlation between A and B is low.

and beyond it). The explanatory (and supplementary) variables (and covariates) are always centred (and standardised) before the ordination model is fitted. Thus, similar to projecting the case points on the response variable arrows, a projection point near zero (the coordinate system origin) corresponds to the average value of that particular explanatory variable in that observation (case).[9] This biplot rule can be also explored with the *Calibrate values of variable arrow* tool, discussed earlier in this section.

The angle between the arrows of explanatory variables can be used to approximate the correlations among those variables in the scaling focused on response variable correlations (see Figure 11-9). Note, however, that this approximation is not as good as the one we would achieve if analysing the explanatory data table as the primary data in a PCA.[10] If the scaling is focused on inter-case distances, we can interpret each arrow independently as pointing in the direction in which the case scores would shift with an increase of that explanatory variable's value. The length of the arrow allows us to compare the size of such an effect across the explanatory variables (remember that all the explanatory variables enter the analysis with a zero average and a unit variance).

Canoco ordination diagrams employ a different (and usually a more useful) visualisation for the factor variables used as predictors (explanatory or supplementary variables). Such variables are represented by multiple symbols (as many as there are levels of that factor) which are placed at the centroids of the scores for cases that have the particular factor level. We can view the original factor variable as representing a classification

---

[9] Note, however, that the constrained ordination model aims to approximate the relation between response data and the explanatory variables, rather than to provide an optimal approximation of the explanatory variable values for individual cases.

[10] This is again due to the predictor role played by these variables, as noted in the footnote for the preceding paragraph. For example, two arrows of explanatory variables pointing in a nearly identical direction imply similar effects of the change of values for the two variables upon the response data values (e.g. community composition), but not necessarily a positive correlation between the two explanatory variables.

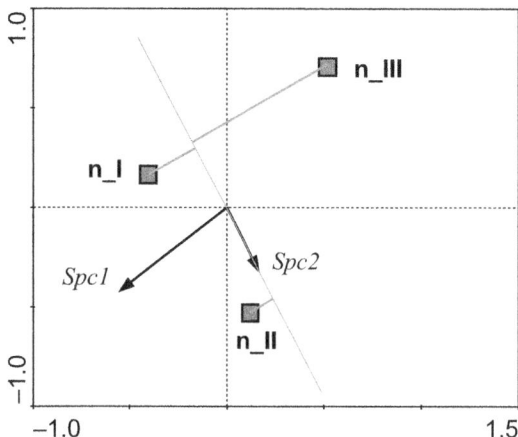

Figure 11-10 Projecting centroids of a factor predictor onto response variable arrow in an ordination diagram from a linear method. Here we predict the largest average abundance of response variable (species) *Spc2* for the cases belonging to the class *n_II*, with cases in the *n_I* and *n_III* classes having a lower (and similar) average predicted abundance.

and then the individual centroid symbols correspond to individual case classes. The positions of those symbols are called the centroid scores (referred to as *CenE* or *CenS* in the Canoco 5 manual – for explanatory and supplementary variables, respectively).

Variables representing fuzzy coding of a categorical (qualitative) variable (see the end of Section 2.2) have numerical contents and so they are represented by arrows in the ordination diagram. But given their nature, it might be of interest to use the centroids for their presentation. To do so for an analysis, you must select the *Analysis | Plot creation options* menu command and check one of the two *Customize* <variable-term> options (corresponding, respectively to explanatory and supplementary variables, if present in the analysis). When you then create an ordination diagram, Canoco displays the *Customize predictor presentation* dialog box, where you can choose which variables are shown as centroids and which as arrows.[11]

If we project centroids of a qualitative predictor variable onto a response variable arrow, we can approximate the average values of this response variable in the individual classes (Figure 11-10). Similarly, the distance between centroids of a predictor variable approximates (in the scaling focused on inter-case distances) the dissimilarity of the response variable values among the classes (typically the dissimilarity of species composition), expressed using Euclidean distance (i.e. how different is the species composition of the classes).[12]

---

[11] Not all explanatory/supplementary variables have the centroid scores computed, however: Canoco 5 does not compute the scores for variables with negative values (before centring).

[12] Note, however, that if we do compare the positions of explanatory variable centroids in a constrained ordination model (RDA or, with the unimodal model, CCA), the underlying ordination model is 'optimised' to show differences among the classes. An unbiased portrait of the difference among the classes can be obtained from an unconstrained ordination with the centroids of a supplementary factor variable post hoc projected into the ordination space.

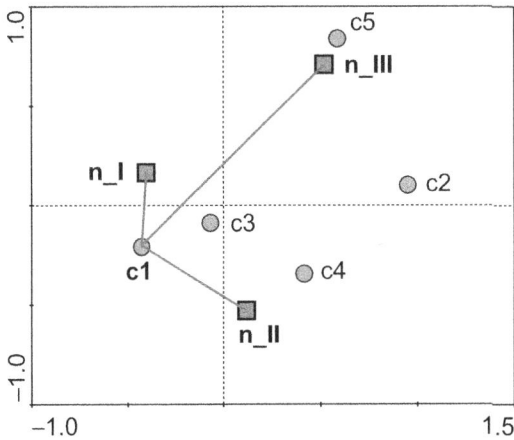

Figure 11-11 Measuring the distance between case points and centroids of an explanatory factor variable. We can predict here that case *c1* has the highest probability of belonging to class *n_I* and has the lowest probability of belonging to class *n_III*.

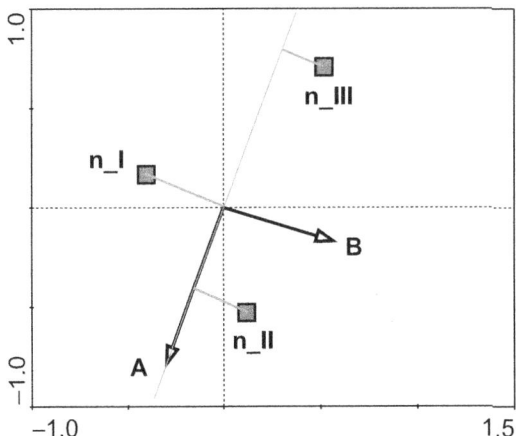

Figure 11-12 Projecting centroids of explanatory variables onto an arrow of a quantitative explanatory variable. The cases from the class *n_II* are predicted to have the largest average value of variable *A*, followed by cases from class *n_I* (near the average of variable A); the cases from *n_III* have the lowest average value of the variable *A*.

In both types of scaling, the distance between the centroids of individual case classes and a particular case point allows us to predict class membership of that case (Figure 11–11). A case has the highest probability of belonging to the class with its centroid closest to that case point. We must note that in a constrained analysis, where a factor is used as the only explanatory variable, the constrained case scores (*CaseE* scores) have identical coordinates to the *CenE* scores of the classes to which the cases belong.

If we project the centroids of explanatory factor variables onto an arrow of a quantitative explanatory variable, we can deduce the approximate ordering of that variable's average values in the individual case classes (see Figure 11–12).

Table 11-2 Relation between response variables, cases, and explanatory variables (here this term encompasses also the supplementary variables) that can be read from an ordination diagram of the weighted-averaging (unimodal) ordination for two types of scaling of ordination scores.

Compared entities	Scaling with focus on case distances and Hill's scaling	Scaling with focus on RV distances and biplot scaling
RVs vs cases	(fitted) relative RV values (e.g. abundances) in cases	
cases vs cases	turnover distances among cases	χ^2 distances among cases (if λs are comparable)
RVs vs RVs	n.a.	χ^2 distances among RV distributions across cases
RVs vs EVs	weighted averages – RV optima with respect to particular EV	
cases vs EVs	n.a.	values of EVs in cases
EVs vs EVs	independent effects of EVs	correlations among EVs
RVs vs factor EVs	relative total values (e.g. abundances) in case classes	
cases vs factor EVs	membership of cases in the classes	
factor EVs vs factor EVs	turnover distances between case classes	χ^2 distances (if λs comparable) between case classes
EVs vs factor EVs	n.a.	averages of EVs in case classes
RV – cases diagram with this scaling is called	joint plot	biplot

EV(s) means explanatory (or supplementary) variable(s), *RV(s)* means response variable(s), λ represents eigenvalue.

11.2 Reading ordination diagrams of unimodal methods

The interpretation of ordination diagrams based on unimodal ordination is summarised in Table 11-2 (following Ter Braak & Verdonschot 1995). It has many similarities with the interpretation we discussed in detail for the linear ordination model in the preceding section and we will point the reader to the preceding section when required.

The main difference in interpreting ordination diagrams from linear and unimodal ordination methods lies in the different model of the response variable change along the constructed gradients (ordination axes) – i.e. the response model. While a linear (monotonic) change was assumed in the preceding section, here (many of) the response variables are assumed to have an optimum position along each of the ordination axes with their abundance (or probability of occurrence for presence–absence data) decreasing in all directions from that point.[13] The estimated position of the response variable (species)

---

[13] The question about how many response variables (most often species) in one's data should show a linear or a unimodal response with respect to the explanatory variables frequently arises among Canoco users in trying to decide between linear and unimodal ordination. The key idea for answering this is that the ordination model is simply a **model**. We must pay for the generalisation it provides us with some (sometimes quite crude) simplification of the true community patterns. We should try to select the type of ordination model (linear or unimodal), possibly improved by an appropriate transformation of the response data, which fits better than the alternative one, not the one which fits 'perfectly'. While it might sound like a highly disputable view, we do not think that data sets where the choice of one or the other ordination type would be entirely inappropriate occur very often.

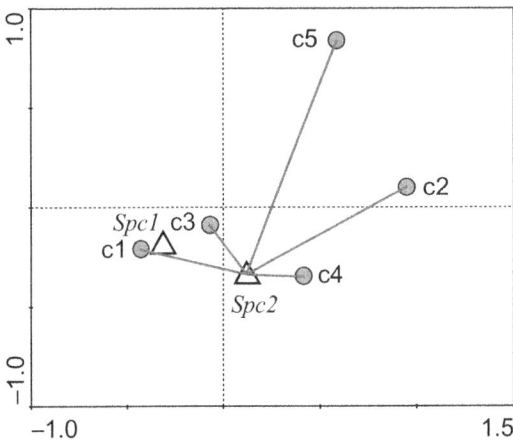

Figure 11-13 The distance between a response variable point and the case points. The response variable (species) *Spc2* is predicted to have the highest relative abundance in cases *c4* and *c3* and the lowest abundance in case *c5* (it is probably absent from it).

optimum is displayed as its score, i.e. as a point (symbol) – not as an arrow used in linear methods. These scores are calculated as the weighted averages of case positions with weights related to the response (species) values in respective cases, see Section 4.5.

Another important difference of unimodal ordination diagrams is that the dissimilarity between cases is based on the chi-square metric, implying that any two cases with identical **relative** values (say two plots with three species present and their values being 1 2 1 and 10 20 10, respectively) are judged to be identical by a unimodal ordination. The dissimilarity of the distribution of different response variables is judged using the same kind of metric, being applied to a transposed (90-degrees rotated) data matrix.

The mutual position of case and response variable points allows us to approximate the relative values in the response data table. The response variable scores are near the points for cases in which they have the highest relative values (e.g. species abundance) and, similarly, the case points are scattered near the positions of response variables that tend to have highest relative proportions in those cases (see Figure 11–13). This kind of interpretation is called the **centroid principle**. Its more quantitative form works directly with the distances between points. If we order the cases based on their distance to a point for a particular response variable, this ordering approximates the ordering based on the decreasing relative values of that response variable in the respective cases.

For shorter gradient lengths (less than 2 SD units, approximately) we can also interpret the positions of response variables and cases in the ordination plot using the **biplot rule** (see Figure 11–14). This is similar to the interpretation used in a diagram based on linear ordination methods. We simply connect a response variable point to the origin of the coordinate system and project the case points perpendicular to this line.[14]

---

[14] Even here the predicted response variable values (species abundances) are the relative proportions, not the original values.

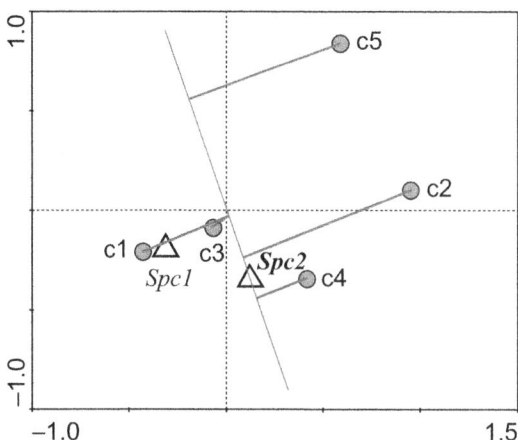

Figure 11–14 Biplot rule applied to response variable and case points in an ordination diagram from a unimodal ordination. The response variable (species) *Spc2* is predicted to have the highest relative frequency in cases *c4* and *c2* and the lowest one in case *c5*. This species is predicted to occur in cases *c3* and *c1* with its average relative frequency.

The distance between case points approximates the chi-square distances between cases in the biplot scaling with the focus on response variables, but only if the ordination axes used in the ordination diagram explain a similar amount of variability (if they have comparable eigenvalues).

If we use Hill's scaling with focus on inter-case distances, then the distance between the cases is scaled in 'turnover' units[15] also shown by the tick mark values of the ordination axes. The cases that are at least four units apart have a very low probability of sharing any positive values of response variables (i.e. sharing any taxa for community data), because a 'half change' distance in the case values (e.g. plot composition) is predicted to occur along one SD unit.

The distance between response variable points in the biplot scaling with focus on the response variable distances approximates the chi-square distance between the response variable distributions (see Figure 11–15).

If we project response variable (namely species) points onto an arrow of a quantitative explanatory (or supplementary) variable, we get an approximate ordering of those species' optima with respect to the chosen explanatory variable (see Figure 11–16). Similarly, we can project case points onto the arrow for a quantitative explanatory variable to approximate values in the explanatory data table, but only if biplot scaling is used (and with a low reliability, see the discussion in the preceding section). Canoco 5 supports these biplot rule interpretations with an interactive *Calibrate values of variable arrow* tool, discussed in the preceding section.

---

[15] In the context of community ecology called **species** turnover units. These units are also called **SD units**, a term derived from the 'standard deviation' (i.e. a width measure) of a unimodal response curve.

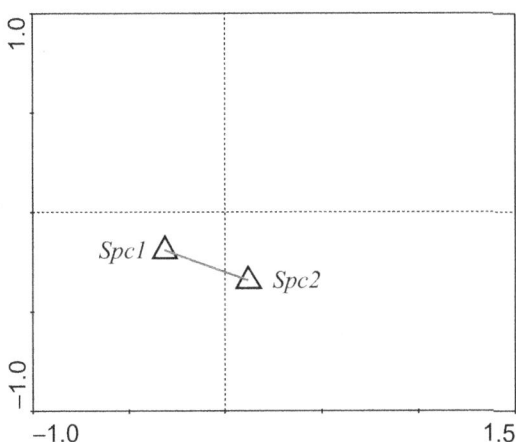

Figure 11-15 Measuring the distance between response variable points.

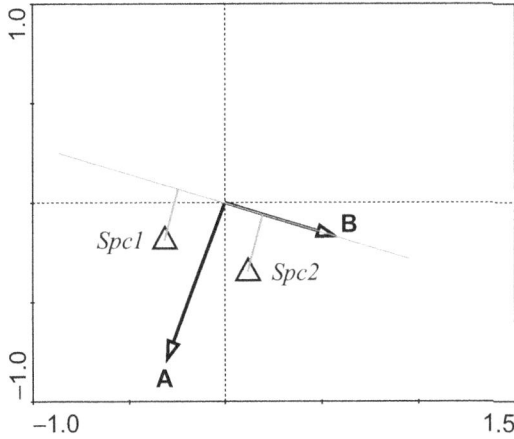

Figure 11-16 Projecting response variable points onto an arrow of a quantitative explanatory variable. We can interpret the graph by saying that the response variable (species) *Spc2* is predicted to have its optimum with respect to explanatory variable *B* at higher values of that variable than species *Spc1*.

Interpretation of the relationships among explanatory (or supplementary) variable arrows (either using the angle between the arrows or comparing the relative directions and size of the impact) is similar to the interpretation used in linear ordination, and described in the preceding section (see also Figure 11–9).

For centroids of explanatory (or supplementary) factor variables (see previous section for an explanation of their meaning), we can use the distance between the response variable (species) points and those centroids to approximate the relative sum (sum of relative abundances) of the response variable (species) in the cases belonging to the considered class – see Figure 11–17.

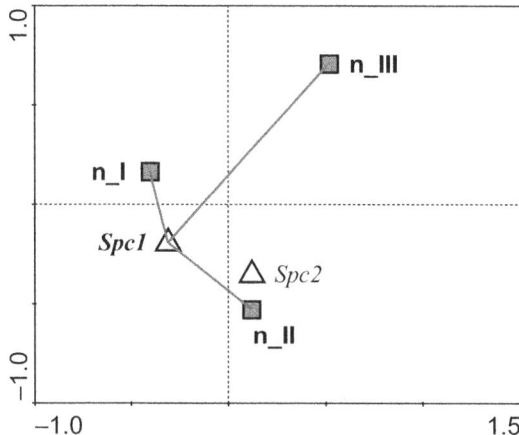

Figure 11-17 Measuring the distance between response variable points and centroids of a factor variable. The average relative frequency of response variable (species) *Spc1* is predicted to be highest in class *n_I*, followed by *n_II* and then by class *n_III*.

Comparison between the case points and centroids of factor variables and between the centroids and the arrows for the quantitative explanatory variables proceeds as in the diagrams of linear ordination models, and it was described in the previous section.

The distance between the centroids of a factor explanatory variable is interpreted similarly to the distance (dissimilarity) between the case points. In this case the distance refers to the average of the chi-square distances (or turnover, depending on the scaling) between the cases of the corresponding two case classes.

11.3 Attribute plots

The results of ordination methods can be summarised in one or a few ordination diagrams; in addition, the determined compositional gradients (ordination axes) can be used as a framework in which we can study the variation of individual variables, various case characteristics (such as species richness or diversity for community data), or even the relationship between various variables.

The relationships between variables (including independent variables, as well as the ordination scores of cases or response variables) can be abstracted ('formalised') with the help of regression models and their use is illustrated in the tutorial in Section 8.8. Canoco 5 also contains other visualisation methods, which we illustrate here. These more specialised diagrams, plotting a particular type of information (attribute) into an ordination plot are called **attribute plots**.

We can create, for example, a plot displaying the values of a particular explanatory (or supplementary) variable in ordination space. The actual value of the selected variable in a case is displayed at the case position using a symbol with its size proportional to the value (Figure 11-18(a)). If we expect variable values to change monotonically across

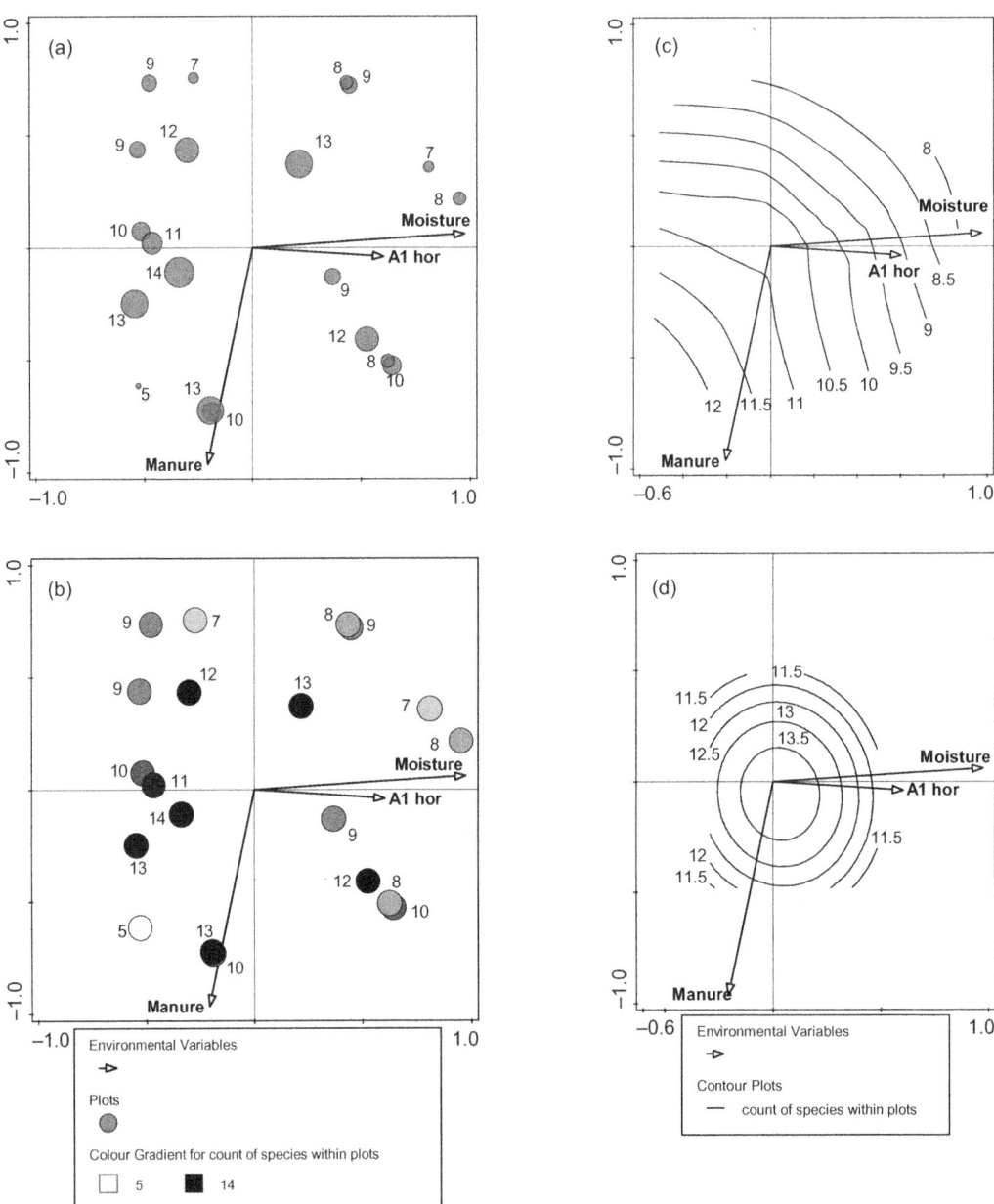

Figure 11–18 Various types of attribute plots: (a) symbol plot encoding an attribute with symbol size; (b) symbol plot encoding an attribute with fill colour; (c) contour plot based on a fitted loess model; (d) contour plot based on a fitted generalized linear model (second-order polynomial). All attribute plots present the species richness (number of species found in each plot) within the ordination space of the first two axes of redundancy analysis (RDA).

11.3 Attribute plots

the ordination (sub)space (or even linearly, as in the constrained ordination model), we might recognise such a pattern from this **symbol plot**. Alternatively, an attribute can be encoded not in the symbol size but in its fill colour, changing along a predefined colour gradient (see Figure 11–18(b)). But plotting individual case values for even a medium-sized data set often does not allow for an efficient comprehension of the pattern. Under such circumstances, some kind of regression model (GLM, GAM, or loess smoother, see Chapter 8) should be used. Such fitted models can be then presented with a contour plot, as illustrated in Figure 11–18(c)(d).

Ordination diagrams and attribute plots can also be used to check how well the data analysed with an ordination fulfil the assumptions of the underlying model. The first obvious check concerns our assumption of the shape of the response variable change along the ordination axes, which represent (for the case of community data) the 'recovered' gradients of the community composition change.[16] We can fit regression models describing the change of response variable (species) values along the ordination axes with the *Graph | Attribute plots | <Response variable> response curves* command. In the case of both linear and unimodal ordinations, we should probably let Canoco select the regression model complexity, whether we use generalized linear or generalized additive models. You can find the necessary steps, as well as additional discussion, in Section 8.8.

Another assumption of constrained ordination methods (both RDA and CCA) is that the gradients of response data variation (e.g. community composition change) depend on explanatory variables (e.g. environmental descriptors) in a linear way, similarly to the linear predictor of (generalized) linear models (see Section 8.2). This is enforced by the constrained ordination methods, where the constrained ordination axes are defined as **linear** combinations of the submitted explanatory variables (see Section 4.6).

Sometimes the scale at which our predictors were measured, does not correspond well to this assumption.[17] Nevertheless, often a simple monotone transformation of a predictor can lead to its substantially higher predictive power. This happens, for example, if we study the successional development of a community during secondary succession (e.g. after destruction of the original habitat). The pace of community change is usually highest at the beginning and slows down continually. If we use the time since the start of succession as a predictor, the community change is usually better explained by the log-transformed time. This corresponds to a non-uniform response of community composition along the successional time gradient.

Canoco also computes for each ordination so-called **ordination diagnostics**, which are statistics describing how well the properties of individual cases and of individual response variables are characterised by the fitted ordination model. The most useful

---

[16] With an underlying interpretation (e.g. in terms of environmental differences) in the case of constrained ordination.

[17] To detect such a situation, you might use an unconstrained ordination where the considered predictors are used as supplementary variables, and to create a separate XY plot for each variable, plotting its values against *CaseR.n* scores (where *n* is the axis number) and possibly supplementing it by a loess model curve.

Visualising multivariate data

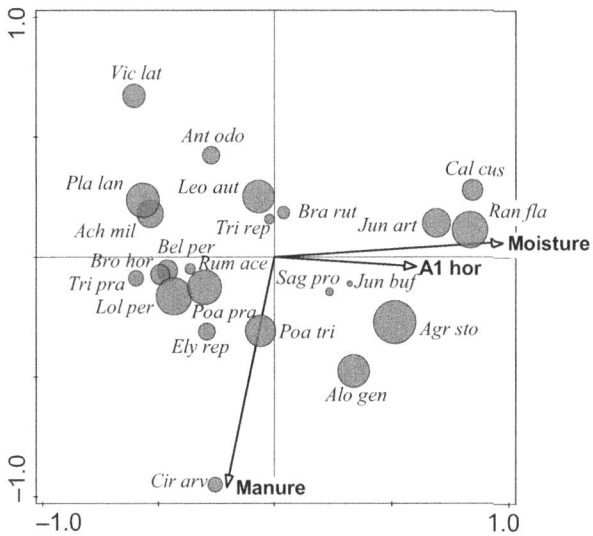

Figure 11-19 Attribute plot showing the values of the fit of response variables (the *CFit* statistics in Canoco output) into the ordination space of the first two CCA axes.

statistic for response variables is their fit – the percentage of variation of their values accounted for by individual axes. This statistic is given in a cumulative form (for the first, first two, first three, . . . ordination axes) and is named *CFit* (see Canoco 5 manual, section 5.6.5.2). We can plot its values (in cumulative or in differenced form) as in Figure 11–19, but perhaps more importantly, it can be used to select which response variables are shown in the ordination diagram (see Section 12.1, step 6, for an example).

11.4 Visualising classification, groups, and sequences

Data items present in a Canoco 5 project can often be distinguished into multiple classes representing e.g. the type of management (for cases) or functional groups (for taxa). Canoco 5 allows us to define multiple classifications for any data item type available in the project, using the menu commands present in the *Project | Classifications* submenu. We can use the selected ('active') classification to create a scatter of the items visually differentiated based on their class membership (see Figure 11–20).

Such scatters can either be shown as separate ordination diagrams (as in Figure 11–20) or be combined into biplots or triplots. If the symbols representing a particular class tend to be assembled in a specific part of the diagram, it might be useful to present their grouping by drawing envelopes, enclosing all symbols of particular class. Canoco 5 supports minimum enclosing polygonal envelopes the same way the CanoDraw program in CANOCO 4.5 did (see Figure 11–21(a)).

11.4 Visualising classification, groups, and sequences

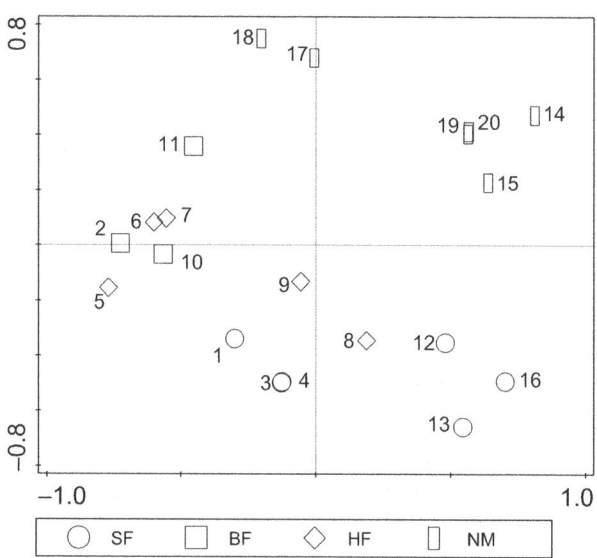

Figure 11-20 Scatter diagram of plots in dune meadow data (see Jongman et al. 1987 for the detailed description of this data set), with individual plots (cases) classified by the type of grassland owner. Axes represent the first two axes of redundancy analysis (RDA).

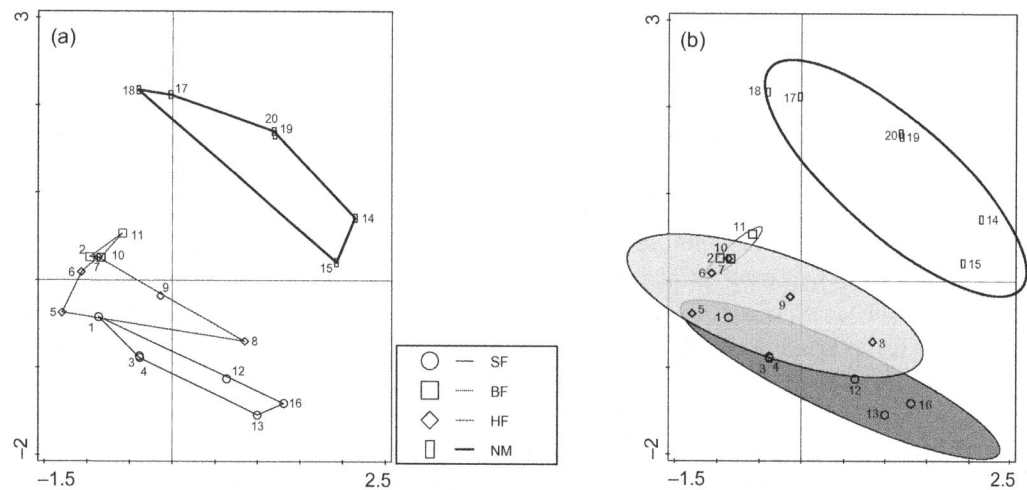

Figure 11-21 Scatter of classified case symbols (for dune meadow data) with members of individual classes (representing the type of grassland owner) enclosed in polygons (a) or in ellipses (b). The polygons/ellipses are plotted in the space of the first two PCA axes.

As a new feature, Canoco 5 supports ellipses as alternative forms of enclosing envelopes (Figure 11–21(b)), but we believe the polygons provide a more informative (despite aesthetically less pleasing) variant. For both polygons and ellipses, you can set their area fill colour and in doing so, we recommend to set the colour partially transparent (as in Figure 11–21(b)).

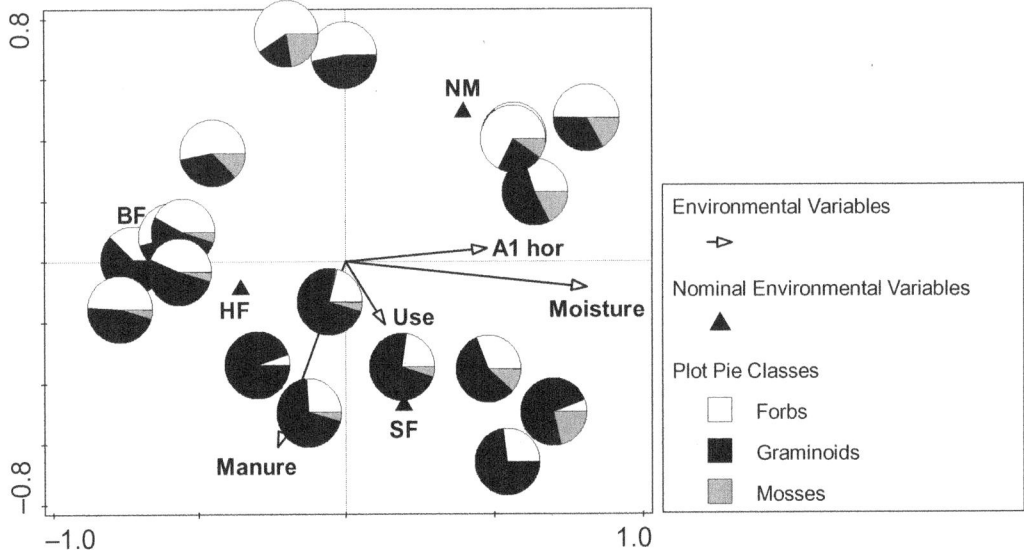

Figure 11-22 Pie plot for cases of the dune meadow data set, displaying relative abundance of three groups of plant species within an ordination space of the first two CCA axes.

The existing classification of cases or response variables can also be used to define pie plots for the other type (i.e. response variables or cases) of data items (see Figure 11–22).

The diagram in Figure 11–22 is based on a classification of plant species and displays, for each case, the wedges representing the fraction of the total abundance taken by the members of a particular species group (forbs, graminoids, and mosses). A similar plot can also be created for a case classification. In our example data set, a scatter of plant species pies could display the relative share of each species cover within the cases of individual classes. To create a pie plot for species or cases, the other data item type (cases or species) must have a classification active and, in addition, you must check one of the boxes shown in the *Use pies instead of symbols* area in the *General* page of the dialog displayed by the *Analysis | Plot creation options* menu command.

Beside classifying data items in a project, we can define a subset, called *group*, using the commands in the *Project | Groups* submenu. With a group defined, we can select in the dialog boxes creating attribute plots (either ordination-space based or XY diagrams) to plot the number of group members, the sum of their values, the proportion of group members out of the total number of non-zero values, or a fraction of the values taken by group members out of the total (as shown in Figure 11–23). Additional information can be found in the Canoco 5 manual, p. 457.

Canoco further supports definition of series collections that allow us to define a temporal or spatial sequence of data items (typically cases) and visualise it in ordination diagrams using polylines connecting item symbols (Figure 11–24). You can define series collections using dialog boxes available from the *Project | Series* submenu.

11.5 T-value biplot 205

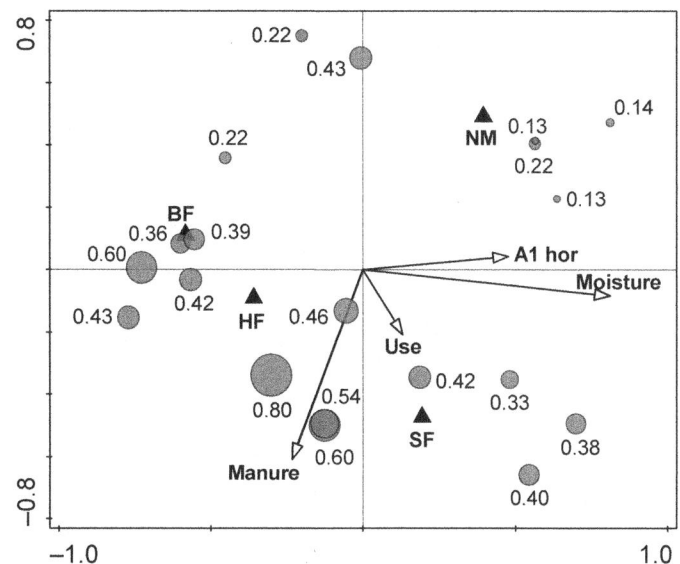

Figure 11-23 Proportion of grasses (on a 0–1 scale) in the total plot abundance of a grassland community, plotted in the ordination space of the first two CCA axes.

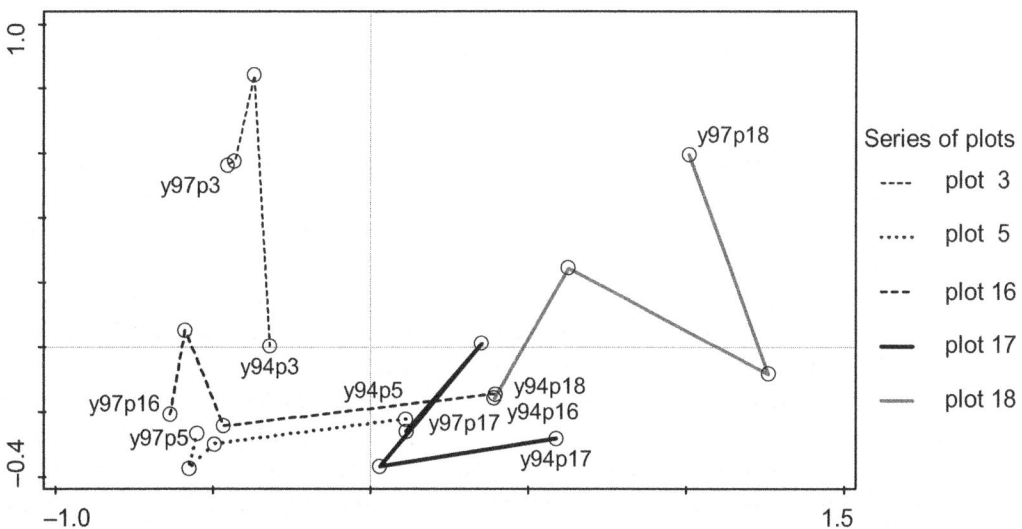

Figure 11-24 Series line plot for five permanent plots measured through four years, in the space of the first two CCA axes.

11.5 T-value biplot

This section covers more advanced material, as it discusses visualisation of the significance of individual regression coefficients in a multivariate regression. A t–value biplot is a diagram containing arrows for the response variables and arrows or symbols for

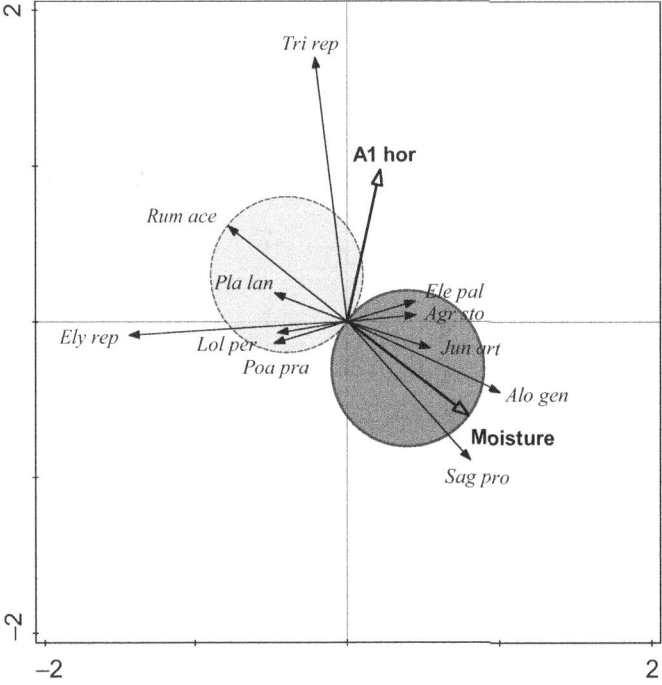

Figure 11-25 An example of a t-value biplot, displaying its first two axes with van Dobben circles drawn for *Moisture* variable.

the explanatory variables. It is primarily used to reveal statistically significant pair-wise relationships between response and explanatory variables (i.e. which response variables depend on which explanatory variables). We interpret the t–value biplot using the **biplot rule** (see Section 11.1).

Using biplot projections, we approximate the t-values of the regression coefficients, which we would get for a multiple regression with the particular response variable and all the explanatory variables as predictors. The choice made in a t-value biplot is to put the arrow head of a response variable at a t-value of 2.0, as this is the critical value for a significant positive (negative) relation at P = 0.05. The advantage of this choice is elaborated in the next paragraph, but note also a disadvantage: variables with the shorter arrows are more related to the explanatory variables than the ones with longer arrows, which is kind of counter-intuitive.

To interpret a t-value biplot (see Figure 11–25 for an example), we project the tips of explanatory variable arrows (or their symbols) onto a line overlaying the arrow of a particular response variable. If the response variable projects on the line further from the origin than the response variable arrowhead, we deduce that the t-value of the corresponding regression coefficient is larger than 2.0. This also holds true for a projection onto the part of the line going in the opposite direction, with the t-value being less than −2.0. Projection points between these limits (including the coordinate system origin) correspond to t-values between −2 and +2, with the predicted value of

11.5 T-value biplot

a T statistic being 0.0 for the coordinates' origin. The importance of a T statistic of value +2 (or −2) is in the fact that for a sufficiently large data set, this is the critical value for a significant positive (negative) relation at $P = 0.05$.

When we need to find which response variables have a significant **positive** relationship to a particular explanatory variable, we can draw a circle with its centre midway between the arrowhead (or symbol) for that variable and the origin of the coordinate system. The circle's diameter is equal to the length of that variable's arrow. Response variables with arrows that end in that circle have a positive regression coefficient for the explanatory variable with corresponding t-value larger than 2.0. We can draw a similar circle in the opposite direction, corresponding to significant negative regression coefficients. These two circles are so-called **Van Dobben circles** (see Canoco 5 manual, section 5.6.8) and they are illustrated in Figure 11–25 for the variable *Moisture*. We can deduce (under the assumption of sufficient number of cases – data points) that increasing moisture positively affects species like *Jun art* or *Agr sto*, and negatively affects the abundance of species such as *Pla lan*, *Poa pra*, or *Rum ace*. There are as many independent pairs of Van Dobben circles as there are explanatory variables.

When assembling the list of response variables with significant relation with the chosen explanatory variable, we must notice that this also represents the case of many individual tests done in parallel and inflating overall Type I error (see Section 5.8). As a minimal protection, the t-value biplot diagrams should not be interpreted when the permutation test of the constrained ordination does not suggest a significant relation between response and explanatory variables.

Additional information on t-value biplots can be found in the Canoco 5 manual, sections 5.6.8 and 6.2.11. Canoco 5 further supports plotting of so-called regression biplots that visualise the regression (or canonical) coefficients of explanatory variables (see Canoco 5 manual, sections 5.6.6.1 and 5.6.8 for additional details).

12 Case study 1: Variation in forest bird assemblages

The primary goal of the analyses demonstrated in this case study is to describe the variability of bird communities and to relate it to the differences in their habitat.

This data set originates from a field study by Mirek E. Šálek et al. (unpublished data) in the Velká Fatra Mts. (Slovak Republic) where bird assemblages were studied using a grid of equidistant points placed over a selected area of montane forest, representing a mix of spruce-dominated and beech-dominated stands. There was a varying cover of deforested area (primarily pastures) and individual quadrats differed in their altitude, slope, forest density, cover and nature of shrub layer, and other characteristics – see below for a description of recorded variables. The response data table contains the numbers of nesting pairs of individual bird species, estimated by listening to singing males at each point. Each value represents an average of four censuses (performed twice in each of two consecutive seasons).

The data in the Canoco 5 example project *Birds.c5p* originate from the Excel spreadsheet file *Birds.xlsx*, sheet *birds*. This single sheet contains both the response data table (average abundance of individual bird species) and the explanatory data table (description of habitat characteristics for individual quadrats centred at the grid points) and these two tables must be imported into separate tables of the Canoco 5 project. Make sure you check Section 2.2 first, for a brief description of the Excel import wizard and note the following two points:

- the first data table is contained in the columns *A* to *AL* (column *A* contains plot labels and is followed by data for 37 bird species) and the explanatory data table in the columns *AN* to *AY*
- while the plot (row) labels are present in just a single column for each table (and they can be imported either as full or as brief labels – see Section 2.2), the labels of variables are presented in both full form (in the first sheet row) and brief form (in the second row). This fact must be properly reflected in the options set for each data table in the Excel import wizard (in the lower part of the sheet preview wizard pages)

There is only one quantitative variable, *Altitude* (*Altit*), specifying the average altitude of the sampled quadrat in metres above sea level. The next eight columns represent a mixture of semi-quantitative and strictly ordinal variables,[1] describing: forest cover

---

[1] Ordinal variables are treated in Canoco in the same way as quantitative variables: it is only your (more cautious) interpretation that distinguishes their nature.

in the whole quadrat (*Forest Cover*), average density of forest stands (*Forest Density*), relative frequency of broad-leaved tree species in the tree layer (*Broadleaf Trees*, with 0 value meaning spruce forest, and 4 only broad-leaved trees), total shrub layer cover (*E2 Layer*), percentage of coniferous species (spruce) in the shrub layer (*E2 Conifers*), cover of the herb layer (*E1 Layer*), and its average height (*E1Height*). The variable *Slope* has a more quantitative character; it corresponds to slope inclination in degrees, divided by five. The last two variables represent two factors: presence of larger rocks in the quadrat (*Rocks* with levels *yes* and *no*) and position of the quadrat on the sunny (southeast, south, and southwest) slopes (*Exposition* factor with levels *warm* and *cold*).

12.1 Unconstrained ordination: portraying variation in bird community

In this first analysis, you will summarise the variation in bird community and relate this summary to measured habitat characteristics. Even though an unconstrained ordination does not need the variables present in the second data table (*Landscape*), you will keep them in the analysis. They will be passively projected into the resulting ordination space and can suggest interpretations for the computed ordination axes. To create the analysis, you should follow the steps described below. Resulting analysis can also be found in the example project as *Unconstrained-suppl-vars* and you can use it as a reference, namely if you get lost at some stage.[2]

1. You can create this new Canoco 5 analysis most efficiently with the help of the Canoco Adviser. Click the *New* button in the lower left corner of the Canoco 5 workspace below the list of *Analyses*.[3]
2. On the first two pages of the wizard titled *Canoco Adviser: Create New Analysis* keep the default choices (both data tables used and *Birds* is the focal table for the analysis), progressing through the pages with the *Next* button. On the third page choose the *Unconstrained-suppl-vars* template in the first section (named *Standard Analyses*) – this choice might be already pre-selected. You can see that the assignment of your data to the two roles available in this analysis template is described in the parenthesis following the template name (*bird species ~ [landscape parameters]|**).
3. After you click the *Finish* button, the first page of the Analysis Setup Wizard appears, where you can specify options for the selected analysis. The Analysis Setup Wizard can be shown in two modes, differing in the number of details shown on its pages and also in the total count of pages displayed. After installation, Canoco 5 works in so-called QuickWizard mode (see Section 2.3), in which only a subset of wizard pages is shown, and we will assume here that you have the QuickWizard mode set. If so, the first page you will see is named *Selection of supplementary variables* and allows you to specify which of the landscape parameters will be projected into the ordination

---

[2] You can find all the analyses described in this and other case studies in example projects that can be downloaded from the book web page (see Appendix B).
[3] The list is presumably empty, if you have created the Canoco 5 project yourself by importing data from the Excel file.

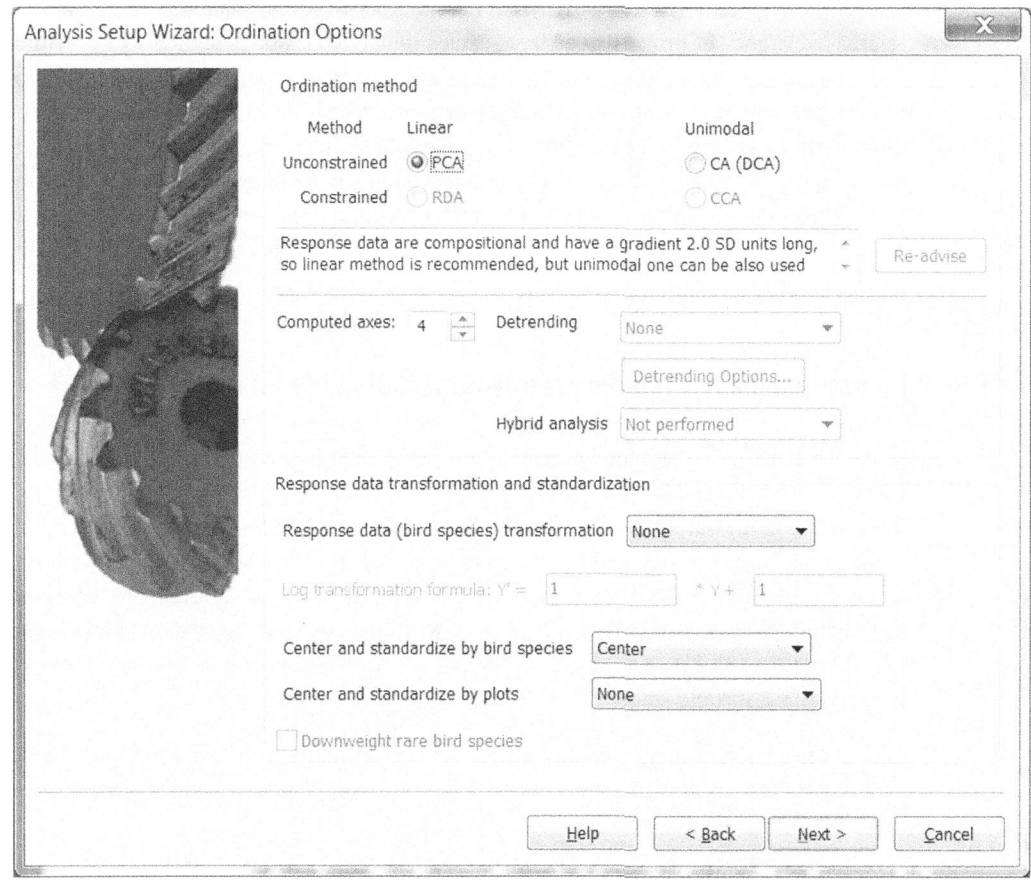

Figure 12-1 Ordination Options page of the Analysis Setup Wizard for unconstrained ordination of bird data.

space of the unconstrained ordination. We suggest you leave all the variables selected (in the right-hand list).

4. After you click the *Next* button again, the *Ordination Options* page is shown – see Figure 12-1.

In contrast to previous versions of CANOCO, Canoco 5 sets the default choices in this page based on the properties of the analysed data. First, you can see that the linear ordination (*PCA*) is pre-selected and this choice is further justified in the text field below the method choices: the Canoco Adviser performed a test analysis of the data using DCA and recommends the use of linear method based on the length of the gradients in SD units (see Section 2.3 for an explanation of this approach).

Second, the *Response data (bird species) transformation* field says *None*. The log transformation was considered, but rejected by the expert system, based on the range of response data values.

Finally, the *Center and standardize by bird species* option is set to *Center*. The centring is appropriate in almost any linear analysis, while the standardisation (usually

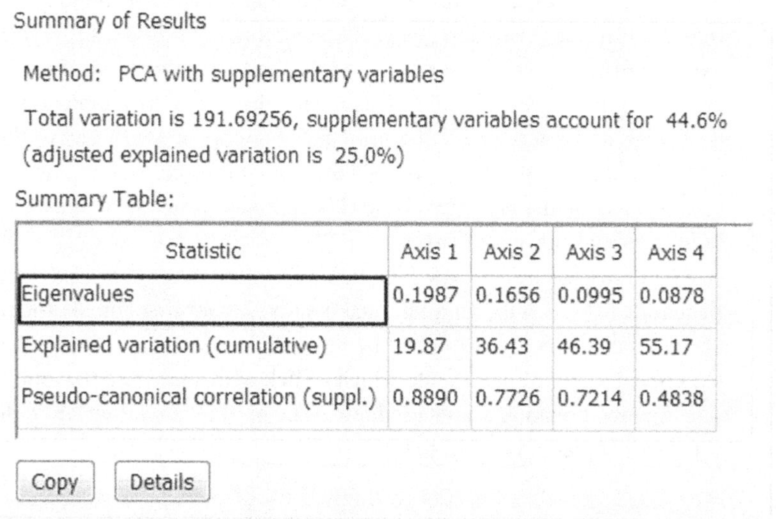

12.1 Unconstrained ordination

Summary of Results

Method: PCA with supplementary variables

Total variation is 191.69256, supplementary variables account for 44.6% (adjusted explained variation is 25.0%)

Summary Table:

Statistic	Axis 1	Axis 2	Axis 3	Axis 4
Eigenvalues	0.1987	0.1656	0.0995	0.0878
Explained variation (cumulative)	19.87	36.43	46.39	55.17
Pseudo-canonical correlation (suppl.)	0.8890	0.7726	0.7214	0.4838

Copy   Details

Figure 12–2 *Summary of Results* area of the Summary page of the PCA of bird data.

performed in addition to centring) is not such an easy choice (see Section 2.3). If you stress the fact that the abundances of all species are measured in the same units (counts of nesting pairs over identical area) and are therefore comparable, you might prefer not to standardise. If, on the other hand, you take the view that different bird species with different territory size, aggregation, and feeding habits differ in their potential frequency of occurrence in the landscape, you might prefer to make the counts of different species comparable by their standardisation. We suggest that in this tutorial you will follow the 'centre-only' path.

5. After you click the *Next* button again, the *Finish* page is shown. There is only one option present at this page, namely *Execute this analysis after Finish*. Make sure it is checked so that after the *Finish* button is clicked, the analysis is executed immediately. After execution, Canoco 5 displays the Graph Wizard window, where ordination diagrams considered appropriate for the analysis type are offered (see Figure 2–15 in Section 2.4 for an illustration). While we recommend that you normally start with the graphs suggested by Graph Wizard, in this tutorial we want you to focus first on analysis summary, so click the *Cancel* button in the lower right corner. Canoco 5 then displays the analysis notebook with its *Summary* page in the foreground. What is most important for you is – at this moment – its lower area labelled *Summary of Results* and shown in Figure 12–2.

You can see that the first two PCA axes (principal components) explain 36.4% (0.1987 + 0.1656) of the variance in bird species data. The amount of variability explained by individual axes gradually decreases, so you must face the difficult problem of how many ordination axes to present and interpret. One approach is to select all the ordination axes explaining more than the average variability explained per axis. This threshold value can be obtained by dividing the total variability (which is set to 1.0^4 in linear ordination methods in Canoco) by the total number of axes. In our linear unconstrained analysis, the total number of axes is equal to the smaller value of the number of response variables and of the number of cases decreased by one, i.e. min (37, 43–1) = 37. This means that using this approach we would need to display and interpret all the principal components with eigenvalues larger than 0.0270 (=1.0/37); therefore not only the first four principal components, but possibly more. The conclusions drawn from this approach probably overestimate the number of interpretable ordination axes.

Alternatively, the so-called broken-stick model can be used (see Legendre & Legendre 2012, p. 449, for additional details). There we compare the relative amount of the total variability explained by the individual axes, with the relative lengths of the same number of pieces into which a stick with a unit length would separate when selecting the breaking points randomly. The predicted relative length for the j-th axis (j-th longest piece) is equal to

$$E(length_j) = \frac{1}{N} \sum_{x=j}^{N} \frac{1}{x}$$

where N is the total number of axes (pieces). For the number of axes equal to 37, the broken-stick model predicts the values 0.1136, 0.0865, 0.0730, and 0.0640 for the first four principal components. Again, this implies that the fractions of variability explained by the four axes exceed values predicted by the null model and that all the axes describe non-random, interpretable variation in the response data.[5] See Jackson (1993) for additional discussion of selecting the number of interpretable axes in indirect ordination methods. For simplicity, we will limit our attention here to just the first two (most important) principal components.

The decision about the number of interpretable ordination axes is somewhat simpler for constrained ordinations, where we can test the significance of all individual constrained (canonical) axes.[6] The distinction between the uncertainty in selecting an interpretable number of unconstrained axes and the much more clear-cut solution for the constrained axes stems from the semantic limitations of the word *interpretable*. In an unconstrained analysis, we do not provide a specific criterion to measure 'interpretability', while in the constrained ordination we look for interpretability **in relation to the values of used explanatory variables**.

6. The results of this unconstrained ordination (PCA) are best presented by a biplot diagram with response variables (bird species) and supplementary variables (landscape parameters). To create the biplot shown in Figure 12–3, you must select the *Graph | Biplots | Bird species + Landscape parameters* menu command. There are

---

[4] This refers to the *Eigenvalues* row: individual eigenvalues express the fraction of total variability explained by a particular axis and they sum (over all axes, not just the first four shown in the table) to 1.

[5] The application of the broken-stick model to the variation explained by unconstrained ordination axes is graphically implemented in Canoco 5's analysis template *Compare-constrained-unconstrained* – see Section 13.2 for an example.

[6] See Section 5.3 for additional information on how to perform this test with Canoco 5.

12.1 Unconstrained ordination

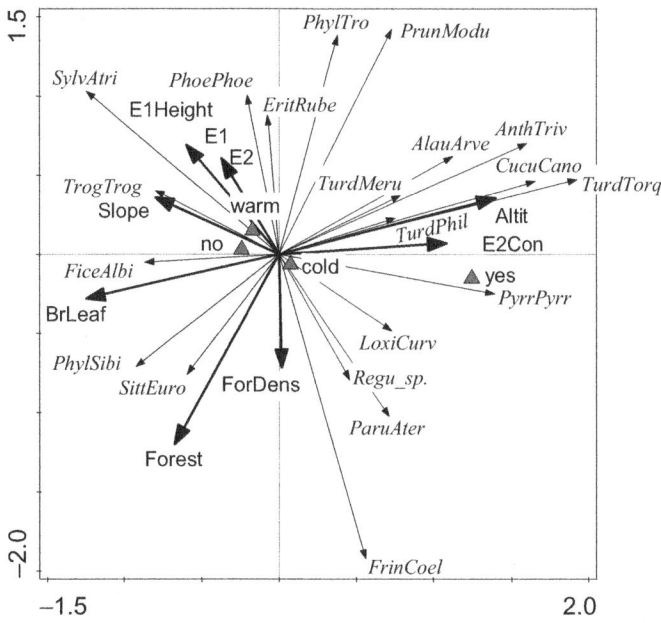

Figure 12-3 Species-environment biplot diagram from PCA (the first two PCA axes are plotted) with landscape parameters passively projected into the ordination space. In default set-up, the species arrows and landscape parameter labels are blue, while the arrows and triangles for supplementary variables have grey colour.

too many species arrows and labels in the resulting diagram, so you should display only the bird species, which are well characterised (fitted) by the first two principal components. To do so, use the *Analysis | Plot creation options* command and in the *Bird species Selection* page check the box before the *show at most* text, specify value 20 in the same line, and close the dialog with the *OK* button. This rule will limit the set of displayed bird species to 20 species with the highest fit to shown ordination axes. Then apply the new rule to your graph using the *Graph | Recreate graph* menu command (or the green double-arrow button in Canoco 5 toolbar).

The biplot diagram shown in Figure 12-3 is an improved version of the original one. Most importantly, the positions of some labels were adjusted (by dragging them with left mouse button pressed) and the background of labels interfering with diagram lines was set to opaque white by selecting them and using the *Fill* page of the Attribute Editor (shown using the *Graph | Show Attribute Editor* menu command; see section 7.7.6 of Canoco 5 manual for more details). In addition, the coloured arrows and labels were set to black for use in this book and the arrows of supplementary variables were made more prominent by modifying line width in the *Outline* page of the Attribute Editor. Check the box at the end of Section 2.4 for information about labelling ordination axes with the method name and axis number (here e.g. *PCA Axis 1*), if you prefer such a style of ordination diagrams.

Variation in forest bird assemblages

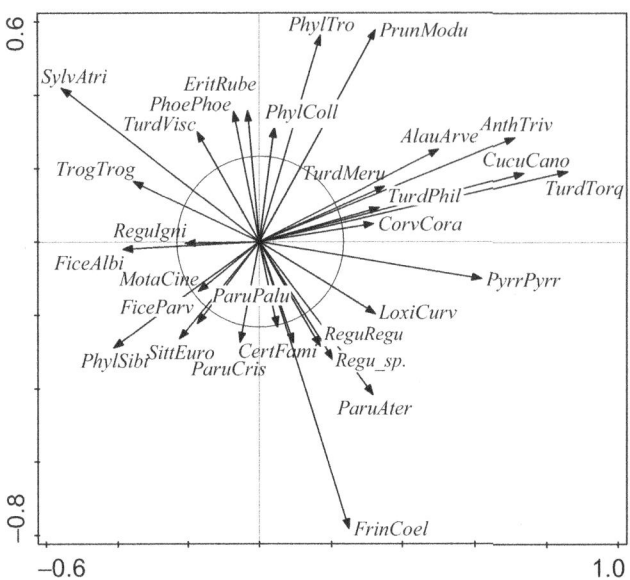

Figure 12–4 Scatter of 30 best-fitting bird species from PCA (with the first two axes plotted) with the equilibrium contribution circle drawn.

The two factor variables are displayed as four centroids (because each of the factors has two levels) and you are advised to consult Section 11.1 for more information about interpreting the centroids of categorical variables. The first principal component is correlated mainly with the altitude (*Altit*, increasing from the left to the right side of the diagram) and with the relative importance of broad-leaved trees in the tree layer (*BrLeaf*, increasing in the opposite direction). The species like <u>*Turdus torquatus*</u> or <u>*Cuculus canorus*</u> or <u>*Anthus trivialis*</u> tend to have larger abundance (or higher probability of occurrence) at higher altitudes (their arrows point in similar directions as the arrow for *Altit*), and also in the spruce forest (as the species arrows point in an opposite direction to the *BrLeaf* arrow). The second ordination axis is more correlated with the cover of herb layer (*E1*), height of the herb layer (*E1Height*), forest density (*ForDens*), and the cover of the shrub layer (*E2*). The *ForDens* versus the other descriptors have their values negatively correlated (because their arrows point in opposite directions).

The pairs of triangle symbols of the related dummy variables (*yes* and *no* for the *Rocks* factor or *cold* and *warm* for *Exposition* factor) in Figure 12–3 are not distributed symmetrically with respect to the coordinate origin. This is because their frequency in data is not balanced: for example the cases that have rocks present are less frequent, so the corresponding symbol for *yes* lies further from the origin.

Instead of selecting the best-fitting bird-species, as described above, you can alternatively separate the species more correlated with the plotted ordination axes from the less correlated ones using the equilibrium contribution (EC) circle, discussed in Section 11.1. To plot it (see Figure 12–4), you must create an ordination diagram (with plotted bird species arrows) with a focus on correlation among response variables in an analysis with standardised response data or post-standardised response variable scores (see the introductory part of Chapter 11) and you must also select the *EC circle* option in the *General* page of the dialog displayed by the *Analysis | Plot creation options* menu command.

The information provided by the length of response variable arrows is strictly related to the fit of these variables, because it reflects (due to the (post–)standardisation of response variable scores) their correlation with the ordination axes. Consequently, the best-fitting species are the ones most likely to exceed the central area delimited by the EC circle, as you can see in Figure 12–4 where the majority of arrows end outside the circle.

12.2 Simple constrained ordination: the effect of altitude on bird community

There are several questions you can address with data sets concerning the relationship between environmental variables and community composition. We restrict this tutorial only to detection of the extent of differences in composition of bird assemblages explainable by the quadrat average altitude (this Section 12.2), and to studying the effects that the other landscape descriptors have in addition to the altitudinal gradient (in the next Section 12.3). To quantify the effect of altitude upon the bird community, you will perform a constrained ordination, using *Altit* as the only explanatory variable (see also *Constrained* analysis in the example project *Birds.c5p*).

1. Click on the *New* button to invoke the New Analysis Wizard: keep the default suggestions on the first two pages and then select the constrained analysis template (*Constrained (bird species ~ landscape parameters)*) on the third page and click the *Finish* button.
2. In the Analysis Setup Wizard, the first page[7] offers the selection of explanatory variables, so to test the effect of altitude, you must select all the variables in the right-hand list except the altitude and then click the << button. After you click the *Next* button, the *Ordination Options* page appears, with the same options as described in the preceding section, except the *PCA* was replaced by its constrained form *RDA* (redundancy analysis).
3. After you click the *Next* button again, instead of the *Finish* page you see a page titled *Testing and Stepwise selection*. This page refers to Monte Carlo permutation testing and allows you to specify whether the test is performed at all and what kind of test to use. The default selection (*All constrained axes test*) is for this analysis identical with the following choice (*First constrained axis test*), because with only one explanatory variable just one constrained axis will be produced. Keep the *Unrestricted permutations* option selected in the lower part of the page, as well as the choice of 499 permutations. When you click the *Next* button this time, the *Finish* page is shown, so keep the checkbox selected there and close the setup wizard with *Finish*.

---

[7] We assume you run Canoco 5 with QuickWizard mode switched on, see Section 2.3.

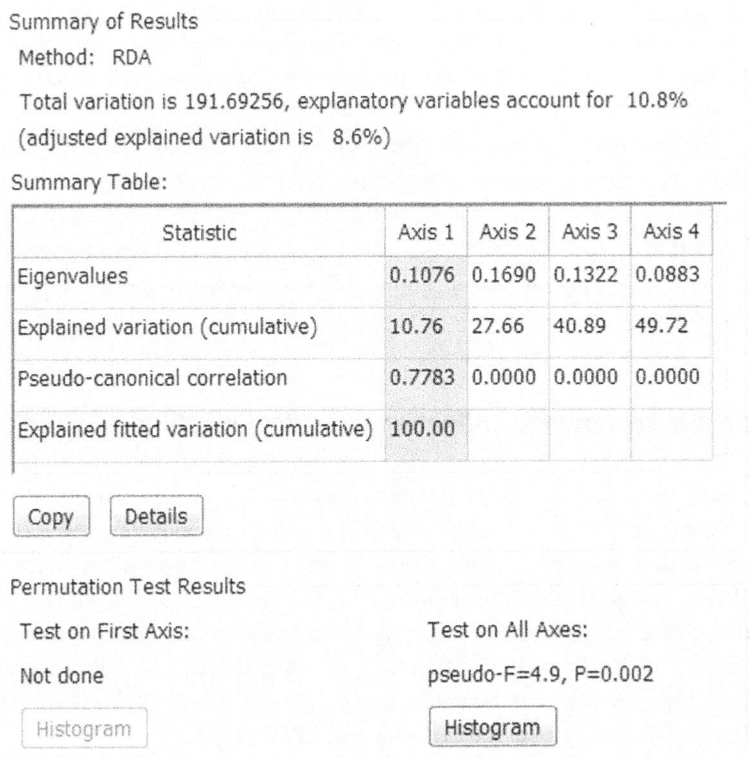

Summary of Results
Method: RDA

Total variation is 191.69256, explanatory variables account for 10.8%
(adjusted explained variation is 8.6%)

Summary Table:

Statistic	Axis 1	Axis 2	Axis 3	Axis 4
Eigenvalues	0.1076	0.1690	0.1322	0.0883
Explained variation (cumulative)	10.76	27.66	40.89	49.72
Pseudo-canonical correlation	0.7783	0.0000	0.0000	0.0000
Explained fitted variation (cumulative)	100.00			

Permutation Test Results

Test on First Axis:

Not done

Test on All Axes:

pseudo-F=4.9, P=0.002

Figure 12–5 Part of the Summary page of the notebook for *Constrained* analysis with altitude as the only explanatory variable.

4. You will again ignore the Graph Wizard for reasons discussed below. So click the *Cancel* button and look at the *Summary* page of the notebook shown for this new analysis (Figure 12–5).

The first thing you will note in the *Summary Table* is that the column corresponding to the first ordination axis has a light blue background, distinguishing the constrained axis from the other, unconstrained axes. The value in the *Explained variation (cumulative)* row shows that the altitude explains about 11% (10.76) of the total variance in bird species data. This is, in fact, less than what the two following unconstrained axes explain individually (16.9 and 13.2, respectively, for the second and third axis[8]). But there is nothing wrong with it, this pattern merely suggests there might be environmental factors explaining the bird community better than the altitude does.

Yet the effect of altitude is significant, as confirmed by the reported result of the Monte Carlo permutation test in the lower right area of Figure 12–5. The reported

---

[8] These values can be worked-out either by multiplying the values in *Eigenvalues* row by 100 or by taking the difference of the consecutive values in the *Explained variation (cumulative)* row.

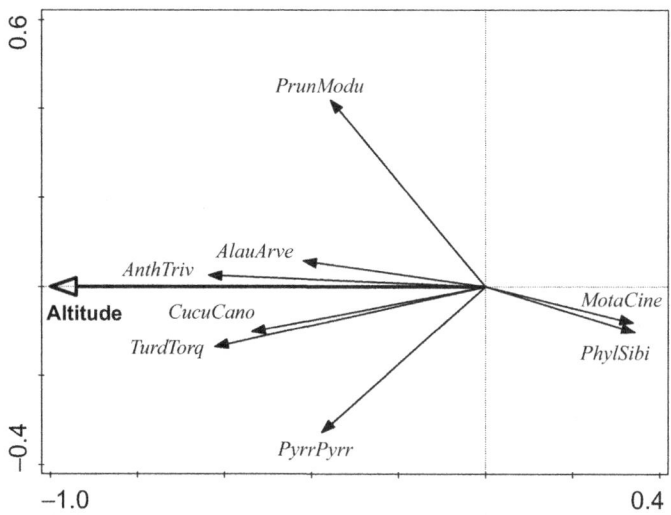

Figure 12–6 Biplot diagram from RDA summarising the differences in bird assemblages along the altitudinal gradient (represented by the first – horizontal – RDA axis). Eight bird species, best fitted by altitude, are shown.

significance level estimate (P = 0.002) is the lowest achievable value given the chosen number of permutations: (0 + 1)/(499 + 1). This means that none of the 499 ordinations based on the permuted data set achieved as good a result (as high a pseudo-F statistic) as the 'true' one (see Section 5.2). You can check the distribution of actual pseudo-F values generated during the permutation test by clicking the *Histogram* button. This will show you a window where the pseudo-F value for real data (i.e. value 4.9) is shown by a red vertical line, while the other 499 values of this statistic, computed during the permutation test, are summarised with a blue frequency histogram. You can see nicely how the real pseudo-F value does not match the distribution generated under the null hypothesis (i.e. the hypothesis of no relation between bird assemblages and the altitude).

5. You can again summarise the results of this constrained analysis using a biplot with bird species and altitude, as shown in Figure 12–6. Before doing that, however, please note that the diagram has 'mixed contents': the first (horizontal) axis is constrained (canonical), representing the variation in bird abundances explainable by the quadrat altitude, while the second (vertical) ordination axis is already unconstrained, representing a part of the residual variation that is not explained by the *Altit* variable. It is, therefore, advisable to display in this biplot diagram only the species that have their abundances well explained by the first ordination axis, i.e. by the altitude, and ignore their fit on the second (vertical) axis.

Such species can be selected as the ones well fitted by the case scores on the first ordination axis (see Section 4.8 for additional details about interpreting PCA/RDA axes as regression predictors). Because the first RDA axis explains approximately 11% of the variability in species data, you can say that a species with an average 'explainability' by the first ordination axis will have at least 11% of the variability in its values explained by that axis. You can set this threshold in Canoco 5 using a dialog box invoked by the *Analysis | Plot creation options* menu command, after selecting the *Bird species Selection* page. You must use the field located below the *On horizontal axis:* prompt, and set its value to 11 – only eight bird species pass this criterion. You can create the biplot diagram afterwards (see Figure 12–6).[9]

The ordination diagram (again mildly improved including the use of the full label for *Altitude*) shows six bird species increasing their abundance with the altitude (those with arrows pointing to the left side, i.e. in the same direction as the Altitude arrow), as well as two species obviously preferring lower altitudes.

> If the look of this or other ordination diagrams shown in the case studies is different from what you see on screen (namely the relative lengths of the individual arrows and/or the ratio of their length to the positions of symbols), this might be due to different scaling options active in your Canoco 5 installation. Check the introductory part of Chapter 11 for additional information about the Scaling Options dialog. Note also that the different scaling choices have different efficiency when interpreting particular content of the diagrams, as explained in Tables 11–1 and 11–2.

Look again at the ordination diagram in Figure 12–6. Do you find anything inconvenient about it? Some people do: probably due to the left-to-right writing system in many regions, some researchers prefer to have the arrow for *Altitude* (playing the central role in this analysis) pointing in an opposite direction than it presently does, i.e. to the right from the coordinate origin. If you check Chapter 11 about interpreting ordination diagram contents, you will find that none of the rules are affected by flipping axis direction. To do it, you can use the dialog shown by the *Analysis | Plot creation options* menu command. In its *General* page, you can check a box in *Flip axes* row to achieve axis mirroring. After you close the dialog, you must re-create the graph. Flipping (only) the horizontal axis for Figure 12–6 will naturally change the horizontal direction not only for *Altitude* but also for the displayed bird species, but will not change the vertical coordinate of any of the arrows.

12.3 Partial constrained ordination: additional effect of other habitat characteristics

You will continue this case study with a more advanced partial constrained ordination, which helps to address the following question:

---

[9] Note that the Graph Wizard (automatically displayed after analysis execution, but also available from Canoco 5 menu and toolbar) does not support selection of response variables based on their fit on a single ordination axis. You must create such plots using the *Scatterplots*, *Biplots*, or *Triplots* submenu of the *Graph* menu.

12.3 Partial constrained ordination

Can we detect any significant effects of the other measured habitat descriptors upon the bird community composition, when we have already removed the compositional variability explained by the average quadrat altitude?

This question can be addressed using a partial redundancy analysis where the *Altit* variable is used as a covariate, while the other descriptors are used as explanatory variables (example analysis in the *Birds.c5p* project is named *Constrained-partial*).

1. Create your analysis using the New Analysis Wizard (starting with the *New* button), but this time select the *Constrained-partial* analysis template from the *Advanced Constrained Analyses* folder.[10]
2. The first Analysis Setup Wizard page that appears (at least in the QuickWizard mode) after you close the New Analysis Wizard is named *Definition of Groups* and there you must distinguish the covariates (*Altit[ude]* in your analysis) from the explanatory variables (the remaining landscape parameters). Initially both groups (shown in the top list) are empty, so start with the *Explanatory Variables* group selected and select in the lower-left list all variables except *Altit* and make them members of the group by clicking the >> button. Then click the name of *Covariates* group, select the remaining *Altit* variable, and click the >> button again.
3. Do not change anything on the next wizard page, as the ordination method and related options are suggested correctly. On the following *Testing and Stepwise selection* page, make sure that the *All constrained axes test* option is selected, because you cannot a priori expect that one constrained axis is sufficient to describe the relation of the 10 explanatory variables to the composition of the bird community. Close the setup wizard on the next page using the *Finish* button.
4. In the Graph Wizard first page, keep only the first offered diagram (biplot) chosen and on the second wizard page check the *limit to* box for the bird species scores and decrease the count of shown species to 15, before closing the Graph Wizard at its third page with the *Finish* button. This creates a biplot diagram summarising the additional effects of the other habitat characteristics, when the altitude effects are already accounted for – see Figure 12–8, discussed below.
5. After the analysis is executed, the *Summary* page of the analysis notebook is displayed. You can see that the amount of variability explained by the other habitat descriptors (in **addition to** the information provided by the quadrat altitude) is quite high (38.0%), but this value is a proportion relating just to the variation left after the variation explained by altitude was taken out. You can see it more clearly by clicking on the *Details* button below the Summary Table, displaying dialog illustrated in Figure 12–7.

The total variance in linear ordination methods is always set to 1.0 (through the rescaling of original data, by dividing them by the value of *Tau* as shown in the dialog). But the *All eigenvalues* entry shows, for a partial ordination, the variance part left after excluding the effect of covariates (here 0.892; compare it to the 10.8% of the variance explained by altitude in the earlier analysis: 1.0−0.892 = 0.108). The

---

[10] The *Advanced Constrained Ordination* folder is initially closed and to unfold it, you must double-click its title.

Variation in forest bird assemblages

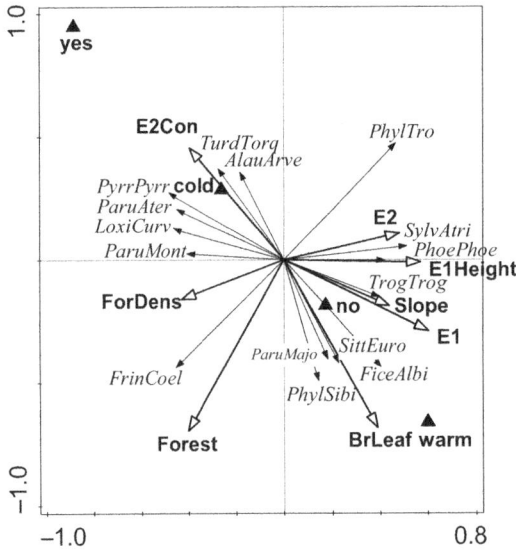

Figure 12-7 Upper part of the Additional Results dialog displayed by the Details button in the Summary page.

Figure 12-8 Biplot diagram (using the first two RDA axes) summarising the effects of habitat descriptors upon bird communities, after removing the effect of altitude changes. Fifteen best-fitting species are shown.

Canonical eigenvalues entry shows the amount of variance explained by the explanatory variables (here 0.339), so the value reported in the *Summary* page itself (38.0) is 100*0.339/0.892 (but without the rounding, of course). When comparing the variation explained by altitude (ca 11%) with that explained by the other variables (38%), you should realise that the current ordination model is much more complex (with $DF = 10$, see the *Explanatory vars* line near the top of the *Summary* page) and so it is quite natural that more variation is explained here, even when each of the predictors explains less than the altitude does.

6. The Summary Table for this analysis is followed by a report on the Monte Carlo permutation test. From it, you can see that the additional contribution of the 10 landscape descriptors is highly significant (P = 0.002).
7. You can go now back to the biplot diagram created in step 4 and illustrated in Figure 12–8. Similarly to the graph in the preceding section (Figure 12–6), only the 'well-fitting' species were included. But here the species fit refers to both ordination axes of the diagram, as both are constrained and summarise therefore the effect of the landscape descriptors.

You can see in the diagram that the first axis (running in the horizontal direction) partly reflects the changes in forest density (increasing from right to left) and shrub layer (E2) cover (increasing in the opposite direction), as well as the differences in the development of the herb layer (more developed in the sites on the right side of the ordination diagram, see *E1* and *E1Height* arrows). The second ordination axis correlates (negatively) with the total cover of the forest in quadrats and with the difference between the beech-dominated stands (at bottom, see *BrLeaf* arrow) and the spruce forest (near the diagram top).

12.4 Separating and testing alpha and beta diversity

This section describes an advanced analysis task, illustrating topics presented in Section 10.5, using the birds data. Two example projects are stored in separate files, named *birds-counts.c5p* and *birds-inflated.c5p*.

The *birds-counts.c5p* project originates from the same data as the *birds.c5p* project, but the compositional data table *Birds* represents here the counts of bird pairs recorded over the four visits of each plot (spread over two years), rather than their averages used in the original project. Consequently, all values are integers and represent the nesting pairs.[11]

To separate the α and β diversity components and to test for the existence of real β diversity (with null hypothesis $\beta = 0$), you must first create an inflated data table. To do so, open the *birds-counts.c5p* project and select the *Project | Create derived project | Expand occurrences* menu command. In the dialog box select the *Birds* table and then choose the *individuals* option for the inflation method. This option is enabled because the contents of the *Birds* data table can be interpreted as individual counts (all the values are integers). Based on your choice, the inflated data table will contain a separate row for each individual (here pair) recorded in the data. This is needed to correctly test the β-diversity (see Section 10.5). The resulting state of the dialog before pressing the *OK* button is illustrated in Figure 12–9.

The new project (already available in the *birds-inflated.c5p* file) is created, but it is not automatically opened. To do so, save the current project and select the newly created project from the *File | Recent files* submenu.

---

[11] Summing up the pairs recorded at four different times is partly controversial, as it is likely that many of the pairs were recorded repeatedly. This might be a problem particularly in the permutation test, where the individual pairs are considered as independent observations.

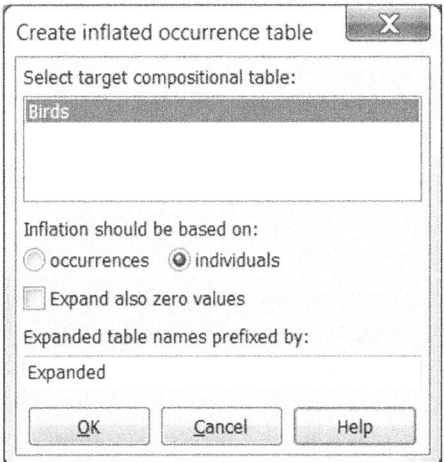

Figure 12–9 Dialog specifying options for the creation of a project with inflated data tables.

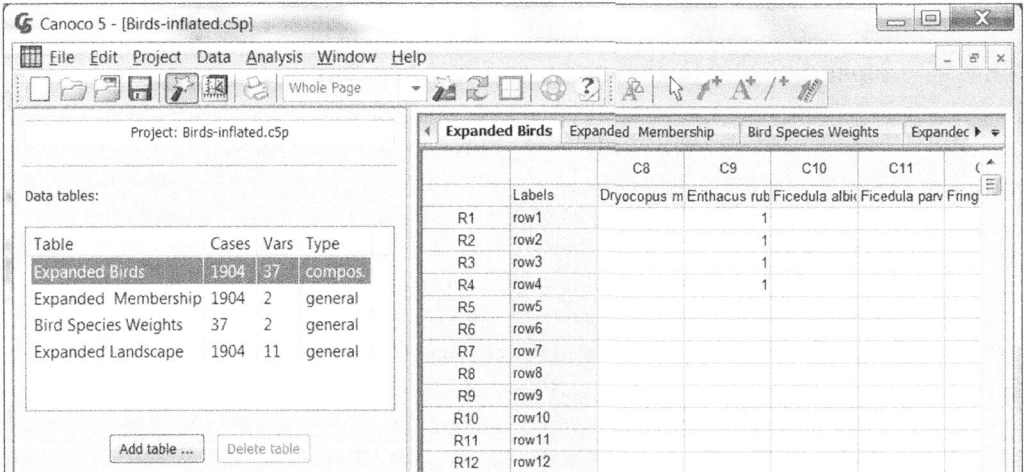

Figure 12–10 Data tables created in the *birds-inflated.c5p* project, which is the result of expanding data tables of the *birds-counts.c5p* project.

You can see (Figure 12–10) in the *Expanded Birds* data table (as well as in two other tables) that the expanded table has 1904 rows, so there were 1904 bird pairs recorded in the original data (i.e. this is the sum of all values in the *Birds* table of the *birds-counts.c5p* project). Consequently, the *Expanded Birds* table contains just zeros (probably shown as empty cells an your display) and ones. For example, you can see in Figure 12–10 that in the original first quadrat four pairs of robin (*Erithacus rubeculla*) were recorded. Due to the expansion (by individuals), each row of this table contains only one 1 value, the other cells are zero.

The data table *Expanded Landscape* represents an inflated version of the *Landscape* table in the parent project, obtained by repeating the landscape parameter values for the corresponding plot. The *Expanded Membership* data table codes, using two factor variables, the essential information linking the rows of the three inflated data tables to the rows (cases, here plots or quadrats) and columns (response variables, bird species) of the parental table.[12] You will use the *Row Membership (RowMemb)* variable in the following analysis, testing for non-random β-diversity.

You will start with an analysis decomposing the γ diversity into α and β components based on the Simpson diversity index, and testing the β diversity (see the *Simpson-beta-test* analysis in the example *birds-inflated.c5p* project). When using the Simpson diversity index, you can use a standard linear ordination (with no weighting of response variables), so this analysis can be set up even in the QuickWizard mode. Click the *New* button below the list of analyses to open the New Analysis Wizard, uncheck the last two of the offered data tables (keeping only *Expanded Birds* and *Expanded Membership* selected), keep the *Expanded Birds* table as focal on the second wizard page, and finally select the *Constrained* analysis template on the third page (within the first, *Standard Analyses* folder).

In the Analysis Setup Wizard (here we assume that you run it in QuickWizard mode, so if you do not, simply progress through the additional wizard pages not mentioned here), move the *Bird Species Membership* variable from the right- to the left-hand list in the *Selection of explanatory variables* page, keeping *Plot Membership* as the only explanatory (constraining) variable. In the *Ordination Options* page, unimodal method (CCA) is suggested by the Canoco Adviser, but change the method to RDA. Keep the other choices with suggested values (no data transformation and centring by bird species). In the *Testing and Stepwise selection* page, you can increase the *Number of permutations* to a higher number (we have used 1999 in the example analysis), but leave the choice of *Unrestricted permutations*. After finishing the analysis set-up and executing the analysis, close the Graph Wizard with *Cancel* button. The *Summary* page of the analysis notebook (illustrated in Figure 12–11) provides essential information for decomposing the γ diversity.

To calculate the total (γ) value of the Simpson diversity index for the analysed data set, the total variation (1748.756) must be divided by the sum of the analysed data table values. Luckily, as there is exactly a single value 1.0 in each row, with the rest being zeros, you know this total is equal to 1904, i.e. the number of rows in the inflated data table (see the *Cases:* row in the *Summary* page). Hence, the Simpson index value you would obtain from the data table with all rows summed, is $1748.756/1904 = 0.918464$. The other variant of the Simpson index (calculated as $1/D$ instead of $1 - D$, see Section 10.5) can be therefore calculated easily as $1.0/(1.0-0.918464) = 12.26$, which can be seen as the number of 'effective' species in the data set (see the discussion of the N2 coefficient in the Canoco 5 manual, p. 208).

To obtain the part of γ diversity index representing the variation among the plots (i.e. to quantify the β diversity), you should note that the plot membership explains 4.4% of

---

[12] Note also that with the chosen inflation type (by individuals), the weights for inflated table rows (as shown in Figure 10–4(a), and discussed there) are not generated, as they all have (an implicit) value 1.0. The inflated tables were generated here as in Figure 10–4(b).

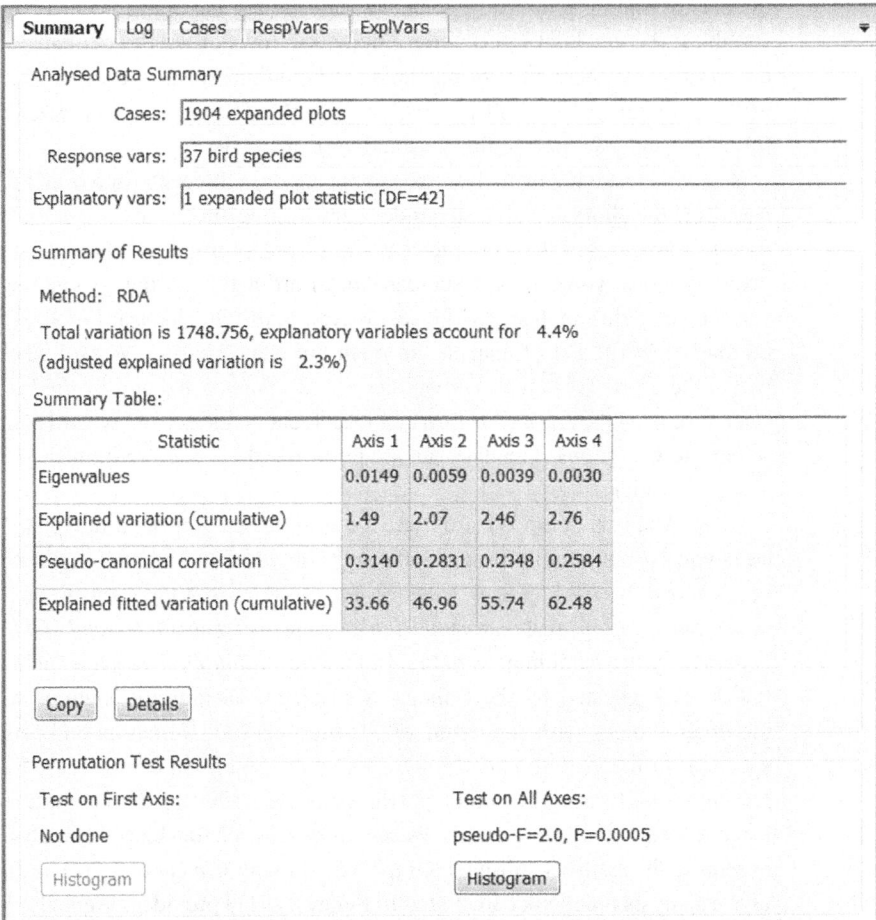

Figure 12-11 Summary page of the *Simpson-beta-test* analysis.

the total variation. Hence the β component estimate is $1748.756*0.044/1904 = 0.04041$, leaving 0.878052 for the α diversity component.

The example project *birds-inflated.c5p* contains one similar analysis named *Richness-beta-test*, where the decomposition of γ-diversity is based on bird species richness, rather than on the Simpson index. You can create such an analysis using the following steps: (1) switch off the QuickWizard mode using the toolbar button (see Figure 2-9 in Section 2.3); (2) select the analysis described in the preceding paragraphs, performing the decomposition of the Simpson diversity index (or use the existing *Simpson-beta-test* analysis in the example project), click the *Modify* button below the list of analyses, choose the *Clone the analysis...* option, and possibly adjust the analysis name; (3) at the second page of the Analysis Setup Wizard check the *Set weight for response variables* option in the upper part of the page and in the following page click the *Import weights* button, selecting the *Richness Weight* from the *Bird Species Weights* data table[13];

---

[13] You would use the *Entropy Weight* when decomposing diversity based on Shannon–Wiener index H.

(4) keep the other choices as set in the parental analysis and calculate the total (γ) and α and β components as before. The resulting value of the γ richness (36, calculated as 68544.001/1904) is smaller by 1 than the actual number of species (37). The β component does not represent an average 'turnover' of species among plots, however, but rather a ratio between the total richness and average plot richness, minus 1 (see Section 10.5).

The decomposition of γ diversity based on bird species richness can also, and perhaps more easily, be carried out via CCA. This is illustrated in the example analysis *Richness-beta-test-using-CCA* (no response variable weights are needed here). The resulting total inertia is 36 and the constrained inertia is 1.59274 (click the *Details* button in the *Summary* page and use the displayed values of *Total variation* and *Canonical eigenvalues* statistic), giving a γ diversity of $36 + 1 = 37$ and a β diversity of $1.59274 + 1 = 2.59274$. It is perhaps illuminating to note that the constrained inertia (*Canonical eigenvalues* statistic) of this CCA is equal to the total inertia of the *Correspondence analysis* performed on a non-inflated data table (see the example project *birds-counts.c5p*). This suggests that the CA (or CCA) methods can be therefore seen as related to the β diversity measure based on the species richness (in other words, they attribute the same importance to scarce and dominant species, see Pélissier et al. 2003).

Finally, the *Shannon-beta-test* analysis demonstrates the decomposition of the Shannon–Wiener H index of diversity, but except using a different set of species weights, this analysis is analogical to the *Richness-beta-test* analysis.

13 Case study 2: Search for community composition patterns and their environmental correlates: vegetation of spring meadows

In this case study, we will demonstrate a common application of multivariate analysis: search for a pattern in a set of vegetation records with available measurements of environmental conditions. The primary data (courtesy of Michal Hájek – see Hájek et al. 2002) comprise classical vegetation relevés (records of all vascular plants and bryophytes, with estimates of their cover using the Braun-Blanquet scale) in spring fen meadows in the mountain ranges of the westernmost Carpathians. The data represent an ordinal transformation of the Braun-Blanquet scale, i.e. the r, $+$, and 1 to 5 values of the Braun-Blanquet scale are replaced by numeric values 1 to 7.[1] The data presented here were extracted from a database using the TURBOVEG program (Hennekens & Schaminee 2001).[2]

The relevés are complemented with environmental data – chemical analyses of the spring water, soil organic carbon content, and slope of the locality. The ion concentrations represent their molarity: remember that predictors are standardised in all Canoco procedures and, consequently, the choice of units does not play any role, as long as the various units are linearly related. The distributions of ion concentrations are highly skewed, i.e. there are usually few highly positively outlying values, which would have a strong effect on the results. Moreover, we can expect that the vegetation composition will be better related to the logarithm of concentrations (i.e. the size of the response will be similar when we double the concentration rather than when we add a constant amount). Consequently, we have first log-transformed ion concentrations,[3] but not the organic carbon (*Corg*), pH, conductivity, and slope. The transformation was done directly in the Excel file before import.[4] We will use the log-transformed values in the analyses. The aim of this case study is to describe basic vegetation patterns and their relationship with available environmental data, mainly with the chemical properties of the spring water.

---

[1] Note that the scale is not linear with respect to species cover and roughly corresponds to logarithmic transformation of the cover, see Section 1.3.
[2] Information about TURBOVEG is available at www.synbiosys.alterra.nl/turboveg/
[3] Using the base-10 logarithm, but whatever log is used, the values will be the same after centring and standardisation implicit for explanatory variables in Canoco.
[4] The same effect could be achieved by importing only the non-transformed variables into Canoco 5 project and selecting their transformation using the *Data | Default transformation and standardization* menu command. The Canoco Adviser then suggests log-transformation for the same variables we have transformed explicitly (but also for *Corg* and *conduc*tivity variables). See also Section 13.5.

The data are stored in the *meadows.xlsx* file, where one worksheet (*Species*) represents the species composition data and the other one the environmental data (*EnvVar*; in comparison with the original ones, the data were slightly simplified). You will first create the *meadows.cp5* project, by importing the data to Canoco using the *File | Import project | From Excel* command (or you can use the existing example project). Two tables must be created in the project, *Species* and *Environment*. In the Excel file, the species data are transposed (i.e. species as rows, relevés as columns). This will often be the case in similar broad-scale surveys. In fact, the analysed data set (70 cases, 285 species) is rather small in the context of recent phytosociological surveys.

13.1 Unconstrained ordination

In the first step (as recommended by the Canoco Adviser), you will calculate an unconstrained ordination – the detrended correspondence analysis (DCA) – see the example analysis *Unconstrained-suppl-vars*. No transformation of response data is needed, because the ordinal transformation (done in Excel) has a logarithmic nature with respect to cover and provides reasonable weighting of species dominance. In DCA with detrending by segments and Hill's scaling, the length of the longest axis provides an estimate of the beta diversity in the data set: value 3.9, computed by the Canoco Adviser for our data set (see Section 2.3), suggests that the use of unimodal ordination is quite appropriate here. The unconstrained ordination will provide the basic overview of compositional gradients present in the data.

It is also useful to include the environmental data in the analysis. In this unconstrained ordination, they will not influence the species and relevé ordination, but they will be projected afterwards to the ordination diagram. For an unconstrained analysis, you will specify them as *Supplementary variables*. If you imported to the environmental data table both the original and log-transformed ion concentrations (as in our example project), do not forget to omit the non-transformed ones here. Because there are many species, some of them with rather low frequency, the downweighting of rare species is a good choice.[5] You will further carry out various analyses, including some constrained ones, and also a comparison of constrained with unconstrained analyses. Please keep in mind that the results are directly comparable only when you do not change the downweighting option in subsequent analyses.

You will first inspect the Summary Table shown in the *Summary* page of the analysis notebook and illustrated in Figure 13–1.

You can see that the first gradient (DCA axis) is by far the longest one (3.48[6]), explaining about 16% of the total variability in species composition (which is a lot, given the number of species in this data set), whereas the second and higher axes explain much less. The results of fitting the environmental variables show that there is very

---

[5] See Section 2.3 for more details about the downweighting option.
[6] Note that this value is smaller than 3.9 reported by Canoco Adviser: this is because we have chosen the downweighting option that generally 'shortens' computed ordination axes.

Summary of Results

Method: DCA with supplementary variables

Total variation is 3.09630, supplementary variables account for 35.5%
(adjusted explained variation is 17.6%)

Summary Table:

Statistic	Axis 1	Axis 2	Axis 3	Axis 4
Eigenvalues	0.4950	0.1431	0.0876	0.0650
Explained variation (cumulative)	15.99	20.61	23.44	25.54
Gradient length	3.48	2.13	1.41	1.58
Pseudo-canonical correlation (suppl.)	0.9700	0.4914	0.6080	0.4388

Figure 13–1 Lower part of the *Summary* page of the analysis notebook for DCA of spring meadow data.

strong correlation of the species-based axis with environmental data (Pseudo-canonical correlation = 0.970 on the first axis) – of this, the correlation with log(Ca) itself is −0.928.[7] The negative value of this correlation is important for the ordination diagrams, but the orientation of ordination axes is arbitrary, so you need not consider it now.

The two above correlation values are much higher than if you were to use the non-transformed ion concentrations (then, they would be 0.893 and 0.772, respectively). The second (and higher) axes not only explain much less, but also their correlation with measured environmental variables is low.

Now you will create ordination diagrams. We suggest that you present the results in the form shown in Figure 13–2, i.e. as separate scatterplots for species, relevés, and environmental variables.[8] It is not possible to include all the 285 species in an ordination diagram without total loss of clarity. When reducing the set of displayed species in DCA, the only possibility is to select them according to their weight, i.e. according to their sum over all the cases.[9] Only one tenth of the total number of species is shown in Figure 13–2(a), based on their weight. Although this is probably the most feasible way

---

[7] To check the correlation of individual variables with the first DCA axis, look at the *CorS.1* column in the *SupplVars* page of the analysis notebook. This page is however visible only when the notebook is displayed in its non-brief mode (see the *Brief and non-brief view of analysis notebooks* box near the end of Section 2.3).

[8] This choice of separating all three item types into individual scatter plots, rather than combining them in, say, two biplots or a single triplot, cannot be taken as a general recommendation. Rather, such a choice depends on many factors, including how you want to interpret the graph contents (see Chapter 11) and the balance between the number of items shown and overall graph readability.

[9] The fit of response variables (usually species), which is another statistic useful for their selection into ordination diagram, is not available for analyses with detrending by segments. You can, therefore, consider detrending by polynomials or a non-detrended analysis, and eventually define a group of well-fitting species from such analysis and use it for species selection in this one.

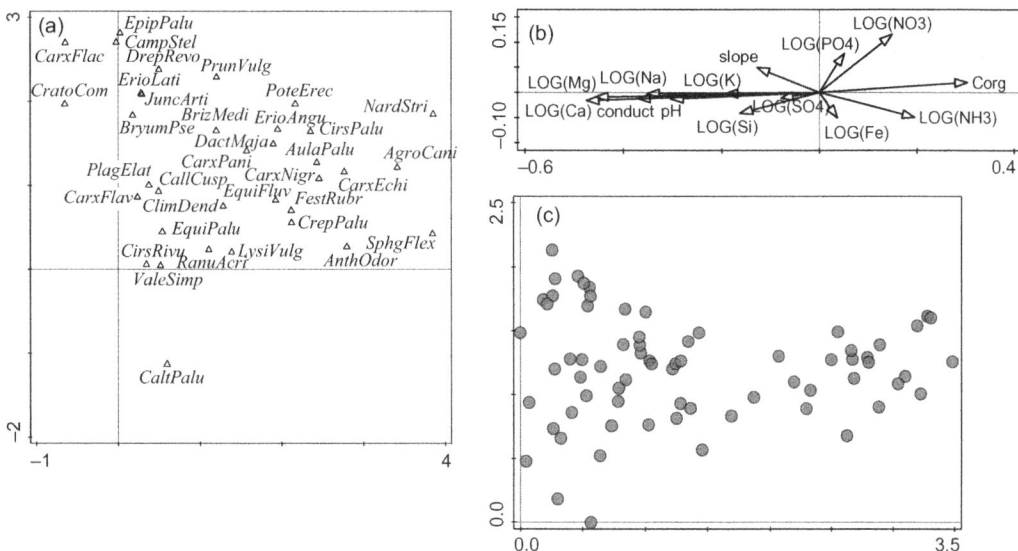

Figure 13–2 The results of DCA of the whole data set: species scatterplot (a) (only the species with the highest weight are shown), post hoc projection of environmental variables (b), and positions of individual relevés (c). The first two DCA axes are used in all three subplots.

to show the diagram in a publication, you should be aware that there is a substantial loss of information. When you create such a diagram, it is therefore necessary to pass to the reader the information about the subsetting of displayed items – how many are shown and on which criterion is the selection based.

You can create several scatterplots of species of varying frequency: the *Analysis | Plot creation options | Species selection* page enables you to first select the species with weight higher than 30%, then species with weight in the 25% to 30% interval, etc. Journal editors will probably not allow you to include all the graphs in your paper, but you can learn about the species preferences from their contents. Another possibility for this data set is to put the bryophyte species into one plot and the vascular plants into another one. Positions of individual relevés in ordination diagrams are usually not very informative for readers of your papers (and so you will probably not present the scatterplot of relevés there), but it might be useful for you to inspect this scatterplot, if you are familiar with the relevés, and also, to check whether there are some distinct groups formed by your relevés.

The relevés are shown in Figure 13–2(c) without labels. Inspection of their general distribution suggests that there is a continuous variation of species composition in the whole data set and that we are not, therefore, able to find distinct vegetation types in the data set. The projection of environmental variables reveals that the first axis is negatively correlated with the *pH* gradient, with conductivity (*conduct*), and with the increasing concentration of cations (*Ca*, *Mg*, and also *Na*), and positively correlated with the soil organic carbon (*Corg*). The position of individual species supports this interpretation – with *Carex flacca*, or *Cratoneuron commutatum* being typical for the calcium-rich spring fens, and *Aulacomnium palustre*, *Carex echinata*, and *Agrostis canina* for the acidic ones. The relationships of the species to the pH gradient are generally well known, and probably any field botanist with basic knowledge of local flora would identify the first axis with the pH gradient, even without any measured chemical characteristics.[10] The second axis is more difficult to interpret, as there are several variables weakly correlated with it.

The positions of arrows for environmental variables suggest that there is a group of variables that are mutually highly positively correlated (*pH, Ca, Mg, Na, conductivity*), and which are negatively correlated with organic carbon (*Corg*). But you should recall that the diagram with environmental variables is based purely on their effect on species composition. A closer inspection of the correlation matrix (present in the *Log* page of the analysis notebook, visible only in its non-brief form) shows that the variables are indeed correlated, but in some cases the correlation is not large (particularly the negative correlation with organic carbon).[11] The correlation matrix also confirms that the correlation of all the measured variables with the second axis is rather weak.

13.2 Constrained ordination

Now, you can continue with the constrained ordinations. Whereas in DCA, you first extract the axes of maximum variation in species composition and only then you fit the environmental variables, now you will directly extract the variation that is explainable by the measured environmental variables. You will start with canonical correspondence analysis (CCA) using the same environmental variables that were passively projected in the DCA (and keeping the option of downweighting of rare species chosen) – this is the *Constrained* analysis in the example project.

In the *Testing and Stepwise selection* page of the Analysis Setup Wizard, specify that both types of permutation tests are performed. Both the test (see Figure 13–3) on the first axis and the test on all axes (on the trace) are highly significant (P = 0.001, which is the maximum under the given number of 999 permutations). However, the pseudo-F value is much higher for the test on first axis (F = 9.5) than for the test on the trace (F = 2.0). The same pattern also appears in the Summary Table, where the first axis explains more than the second, third, and fourth axes do together.

You can compare this summary table with that from DCA you have seen earlier. You will notice that the percentage variance explained by the first axis is very close to that explained by the first axis in the unconstrained DCA (15.0 in comparison with 16.0), and also that the pseudo-canonical correlation (i.e. the correlation between species-based and environmental-variables based axes) is only slightly higher. This suggests that the measured environmental variables are those responsible for species composition

---

[10] The situation might be rather different with other types of organisms, or in less known areas, where such intimate knowledge of individual species ecology is not available.
[11] To obtain a better graphical summary of correlation among the environmental variables, you would have to use them as response variables in a principal component analysis (PCA), with necessary centring and standardisation of these variables.

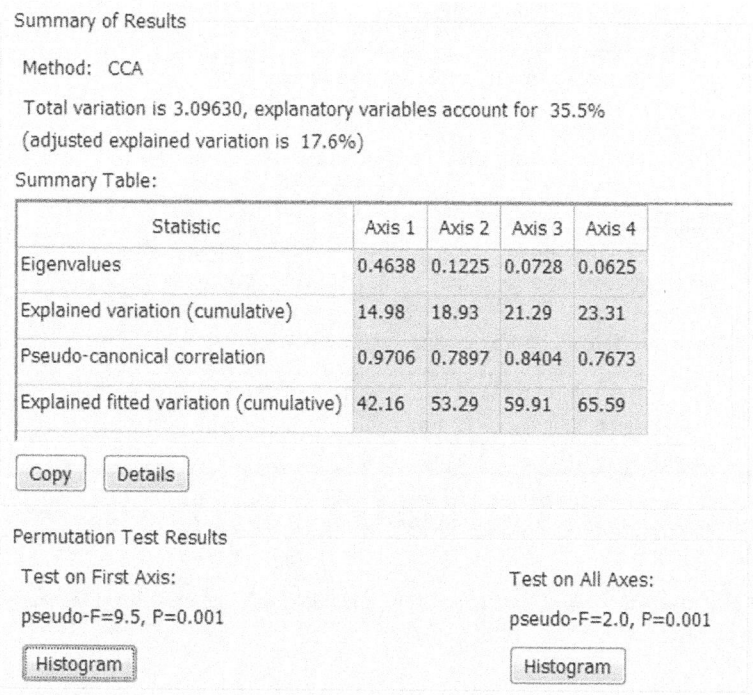

Figure 13–3 Lower part of the Summary page of analysis notebook for the *Constrained* analysis (CCA).

variation. And indeed, in the ordination diagrams of DCA (Figure 13–2) and CCA (not shown here), the first axis of CCA is very similar (both for the species and for the case scores) to the first axis of DCA. But the second axes differ: the CCA shows a remarkable arch effect – the quadratic dependence of the second axis on the first one. Rather than trying the detrended form of CCA (i.e. DCCA), we suggest that the arch effect is caused by a redundancy in the set of explanatory variables (many highly correlated variables), and that you should try to reduce this set using forward selection.

You can make the same comparison of unconstrained and constrained ordinations more directly using a predefined template of the Canoco Adviser, named *compare-constrained-unconstrained*, which will show you the explanatory efficiency of the predictor variables (as a group, not individually, see Figure 13–4). This analysis is illustrated in the example project as the *Compare-constrained-unconstrained* analysis.

The *Comparison* page also shows that the first eigenvalue is higher, and the other eigenvalues are lower than expected under the broken-stick model, demonstrating again that there is just one strong explanatory gradient. To obtain a more fitting comparison, you could opt for having both analyses detrended, or both without detrending, but the results would be very similar.

Before using the forward selection, it might be interesting to know whether the second and higher axes are worth interpreting, i.e. whether they are statistically significant.

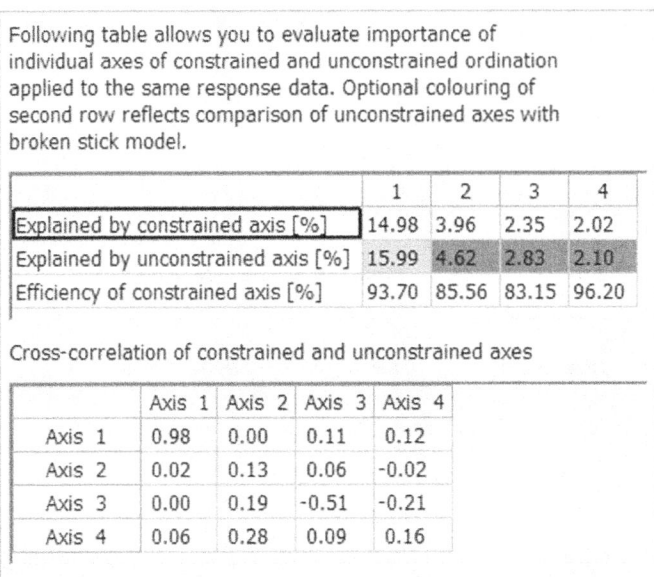

Figure 13-4 The *Comparison* page of the notebook for *Compare-constrained-unconstrained* analysis.

The test of the second axis can be performed as the test of the first axis in an analysis, where the case scores on the first axis are taken as a covariate. Similarly, we can continue with testing the third and higher axes, taking all the lower axes scores as covariates (see Section 5.3). The Canoco Adviser provides a predefined analysis template *Test-constr-axes* (and the example analysis uses the same name), which tests in this way the significance of the first four axes (or how many you ask for in the setup wizard). In the specialised page of the *Test-constr-axes* analysis notebook you can see that the second axis, although much weaker than the first one, is still significant, whereas the third and fourth axes are not.

You now know that there is a close correlation between environmental variables and species composition. You will use the forward selection to build a simpler model (with fewer explanatory variables), but one that still sufficiently explains the species composition patterns. First, it is useful to inspect the simple effects (i.e. the independent effects of individual variables, called in CANOCO 4.x the marginal effects) of all explanatory variables. You will get these results by asking for *Summarize effects of expl. variables* on the *Testing and Stepwise selection* page of the setup wizard. To be on the safe side, you can also consider some P-value correction for multiple testing – the false discovery rate approach is a reasonable option (see e.g. Verhoeven et al. 2005). Because all the correction methods imply some kind of multiplication of the original P-values, it might be a good idea to increase the number of permutations in the permutation test to enable low P-values in cases where the relationship is really strong. We will use 9999 permutations – but be prepared to wait for a while, because many permutation tests

Table 13-1 Simple (marginal) and conditional effects of explanatory variables.

Name	Explains%	pseudo-F	P	P(adj)
Simple term effects:				
LOG(Ca)	13.7	10.8	0.0001	0.00025
LOG(Mg)	12.8	10.0	0.0001	0.00025
LOG(Na)	9.0	6.7	0.0001	0.00025
conduct	8.7	6.5	0.0001	0.00025
Corg	6.6	4.8	0.0001	0.00025
pH	6.4	4.7	0.0001	0.00025
LOG(NH3)	3.9	2.8	0.0005	0.00107
LOG(Si)	3.6	2.5	0.0012	0.00195
LOG(K)	3.6	2.5	0.0013	0.00195
LOG(NO3)	3.6	2.5	0.0011	0.00195
LOG(SO4)	2.2	1.5	0.0431	0.05387
LOG(Fe)	2.2	1.5	0.0403	0.05387
slope	2.1	1.5	0.0548	0.06323
LOG(PO4)	1.5	1.0	0.3412	0.36557
LOG(Cl)	1.2	0.9	0.6982	0.6982
Conditional term effects:				
LOG(Ca)	13.7	10.8	0.0001	0.0015
LOG(NO3)	2.3	1.8	0.0008	0.003
LOG(Na)	2.2	1.8	0.0007	0.003
LOG(Si)	2.2	1.8	0.0005	0.003
Corg	1.9	1.6	0.004	0.012
LOG(NH3)	1.7	1.4	0.0215	0.05375
LOG(Fe)	1.6	1.4	0.0322	0.069
pH	1.5	1.2	0.0943	0.15717
conduct	1.5	1.2	0.0827	0.15506
LOG(SO4)	1.3	1.1	0.291	0.4365
LOG(Mg)	1.3	1.1	0.3218	0.43882
LOG(K)	1.2	1.0	0.4939	0.61738
LOG(PO4)	1.1	0.9	0.6846	0.73371
slope	1.1	0.9	0.6848	0.73371
LOG(Cl)	1.0	0.9	0.7728	0.7728

are executed during the analysis (example analysis is named *Constrained-Simple-And-Conditional-effects*). When finished, you will find the tables of simple and conditional effects in the *Summary* page and they are reproduced in Table 13–1.

From the table with simple effects, you can see that the calcium concentration (*LOG(Ca)*) is the most important factor for species composition, followed by the log-transformed values of *Mg* and *Na*, and by *conduct*ivity. All these variables are closely correlated – conductivity is in fact a measure of dissolved ions. It is, consequently, not surprising that, after the *LOG(Ca)* variable is selected, the conditional effects of the correlated variables decrease dramatically. In fact, if you apply the false discovery rate (FDR) correction, only five variables qualify for the final model with the 0.05 threshold for adjusted P-value: *LOG(Ca)*, *LOG(NO3)*, *LOG(Na)*, *LOG(Si)*, and *Corg*. *LOG(Si)* and *LOG(NO3)* have both relatively small (but significant) simple effects, but as they are independent of the other variables (note their low inflation factor values seen in the *VIFE* column on the *ExplVars* page of the non-brief analysis notebook), they add a significant explanatory power to the previously selected variables.

You should note that the final selection tells you, in fact, that this is a sufficient set of predictors, and that further addition of variables does not significantly improve the fit. You should be extremely careful, however, when trying to interpret the identified effects as the real causal relationships, e.g. by concluding that the effect of Mg is negligible, because it is not included in the final selection. If variables in a group are closely correlated, then only a limited number of them are selected. It is often simply a matter of chance which of them is best correlated with species composition.[12] After the best variable is selected, the conditional effects of variables correlated with it will drop, sometimes dramatically. Even if they are functionally linked with the response (i.e. community composition), the test is rather weak. Also, the results of selection sometimes differ after even a relatively minor change in the set of predictors, from which you select the final model. But this is a general problem of all observational studies with many correlated predictors.[13]

Two graphs are suggested by the Canoco Adviser: the species–environmental variables biplot (Figure 13–5) and the diversity biplot, where the positions of relevés and arrows for environmental variables are shown, with species richness visualised by the size of relevé symbols (Figure 13–6).[14]

The ordination diagram (Figure 13–5) reveals the same dominant gradient that you found in the previous analyses. Also, the effects of individual selected predictors can be better distinguished here. When plotting the diagram, you can select the displayed species either according to their weight or according to their fit to axes. But you should usually combine both methods – the species with highest weights are useful for comparison with the DCA, whereas the species with the highest fit are often ecologically more interesting.[15]

The comparison of the two diagrams in Figure 13–5 shows the possible danger in selecting species according to the highest weight.[16] *Potentilla erecta*, *Festuca rubra*, *Carex nigra*, *Anthoxanthum odoratum*, and *Carex echinata* are suggested as most typical species for stands with a high soil carbon content; in fact, they are not very typical of such habitats. Because the acidic habitats were rare in comparison with the calcium-rich stands, their typical species were present in few relevés, and consequently

---

[12] You are not obliged to always pick the 'best' predictor: if there is another one that you find easier to interpret and it has similarly strong (and significant) effect, you can choose it instead.

[13] In the first edition of this book, we have not log-transformed the ion concentrations and the resulting selection of explanatory variables in the final model was quite different, which again shows how unstable the selection could be when you have many correlated predictors.

[14] Note that both figures demonstrate two alternatively created graphs: the Graph Wizard creates just one of these at any time.

[15] Use the *Analysis | Plot creation options – Species selection* in the menu bar, to change the default selection settings.

[16] But similar problems can occur even when you combine fit and weight as selection criteria, because the species have to pass both limits to be included in the graph.

13.2 Constrained ordination

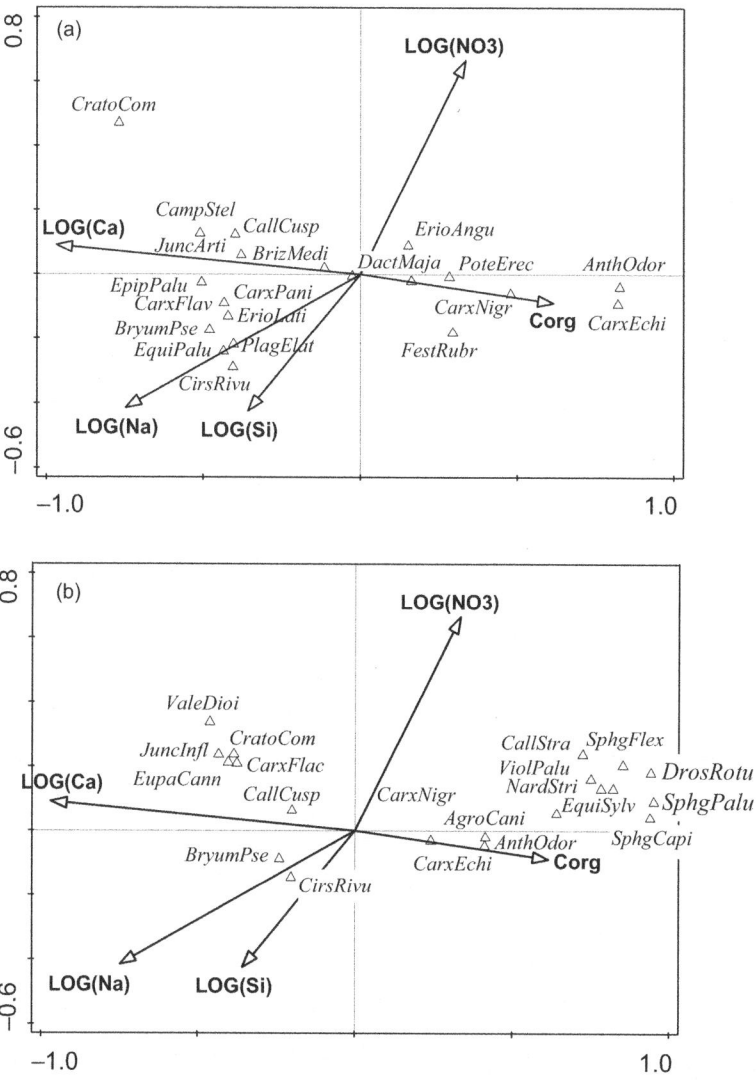

Figure 13–5 The species–environmental variables biplot of CCA (the first two axes) with environmental variables selected by the forward selection procedure. The 20 species with the highest weight are shown in the upper diagram (a), whereas the 20 species with the highest fit are presented in the lower diagram (b). Diagram was improved by adjusting label positions.

they do not have sufficient weight in the analysis. The selection according to the fit displays *Sphagnum* species to be typical for acidic bog habitats, which is ecologically more meaningful.

Now, you can also inspect some community characteristics, using the second graph suggested by the Canoco Adviser, the species diversity diagram. When created by the adviser (through the Graph Wizard), the CaseR scores are used in the diagram (see Section 4.6 for the explanation of differences between CaseR and CaseE scores). If you

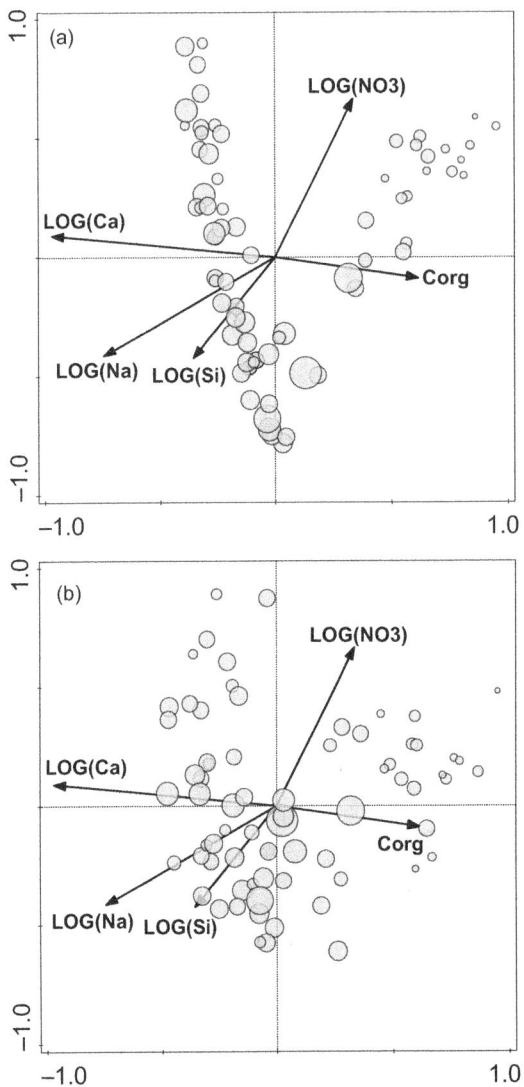

Figure 13-6 CCA biplot (the first two axes) with symbol size corresponding to the number of species in the relevé. Relevé positions represent CaseR scores (a) or CaseE scores (b).

want to use the CaseE scores, you must create the graph with the *Graph | Attribute plots | In ordination space* menu command. In both cases (shown in Figure 13-6), the size of the symbols corresponds to the species richness in individual relevés.

You can see that the use of CaseR scores results in a very pronounced arch effect, but no such effect appears when you use the CaseE scores. When working with the number of species, one should be sure that the size of the plots was constant. In our case (as in other similar phytosociological surveys), plot size varies slightly among cases. Nevertheless, the noise that can be caused by the variability in plot size is small relative to the richness differences in the data.

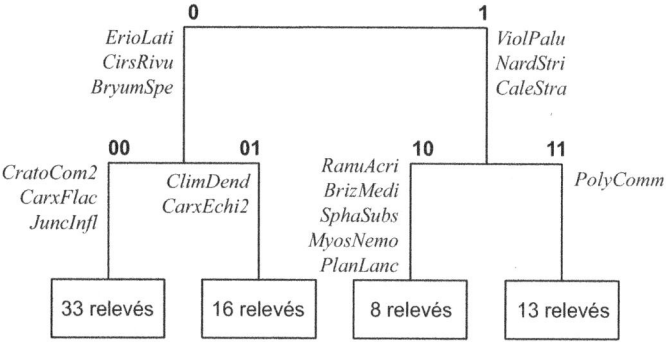

Figure 13-7 The results of the TWINSPAN classification. Each division is accompanied by indicator species. More explanations are given in the text.

13.3 Classification

Another insight for this data set can be obtained from its classification. The TWINSPAN results are presented here as an example (Figure 13–7). The TWINSPAN for Windows program was run with the default options (see Section 7.4), and the pseudo-species cut levels were 0, 2, 4 and 6. We will present the results only up to the second division (i.e. as a classification with four groups), although there were more divisions described in the program output.

There are several possibilities as to how to use the environmental variables when interpreting classification (e.g. TWINSPAN) results. You can run, for each division, a stepwise discriminant analysis for the two groups defined by the division; in this way, you will get a selection of environmental variables and a discriminant function, constructed as a linear combination of the selected environmental variables. The other possibility is to inspect, for each division, which environmental variables show the greatest differences between the groups (e.g. based on ordinary t-tests of individual variables; in such a heuristic procedure, you need not be too scrupulous about the normality). The environmental variables selected by the stepwise procedure will often not be those with the largest differences between the groups, because the stepwise selection is based on the largest conditional effects in each step, which means that the predictors correlated with variables already selected are excluded. Both approaches, however, provide information that is potentially interesting.

You can also compare the environmental values among the final groups only (as in Figure 13–8) or you can try to predict class membership from environmental data using a classification tree method (see Section 8.7). But in both cases, you will not obtain information about the importance of individual variables for individual TWINSPAN divisions. Because these procedures involve data exploration rather than hypothesis testing, we are not concerned here about multiple testing and P-value correction.

The search for a pattern is an iterative process, and is (at least in plant community studies) usually based on the intimate biological knowledge of the species. Unlike hypothesis testing, at this stage the researcher uses his/her extensive field experience (i.e. the information that is external to the data set analysed). We have demonstrated

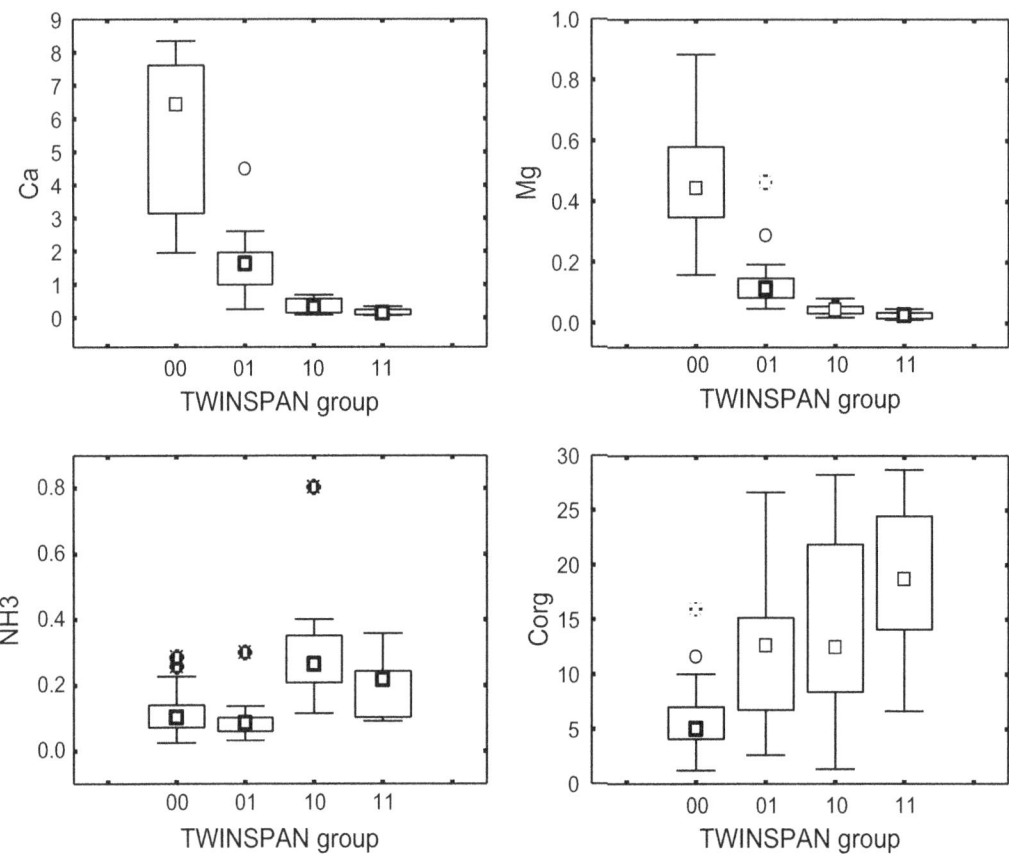

Figure 13-8 Box-and-whisker plots of molarity values of calcium (Ca), magnesium (Mg), NH_3 in the spring water, and of the weight percentage of soil organic carbon, categorised according to the four groups defined by the TWINSPAN classification. Whiskers reach to the non-outlier extremes; outliers are shown as points.

here what can be inferred from multivariate data analyses. Further interpretations would be based on our knowledge of the biology of the species.

13.4 Suggestions for additional analyses

There are many other analyses that one could do during data exploration. Here we suggest just four interesting approaches:

1. The Ca and Mg ions are usually highly correlated in nature. You can be interested in the effect of one of them when the other one is kept constant. This can be achieved with two partial analyses: each of the variables will be a (single) explanatory variable in one of the analyses and a (single) covariate in the other one.

2. You might be interested in how well correlated the species composition gradients are in vascular plants and in bryophytes. You can calculate the unconstrained ordination (DCA) first based on vascular plants only and then using the bryophyte taxa only, and finally you can compare the case scores on the ordination axes. Alternatively, the community variation within these two taxonomic groups can be compared more directly using the co-correspondence analysis. We illustrate both approaches in the following section.
3. You might like to find which of the environmental variables are the most important ones for vascular plants, and which for bryophytes. You can calculate separate constrained ordinations, and compare the importance of individual environmental variables (for example, by comparing the models selected by the forward selection).
4. Further analyses are also possible with community characteristics,[17] such as species richness, diversity and their relation to the environmental variables.

We have stressed several times that we are concentrating on a search for patterns in this case study, and not on hypotheses testing. If you are interested in testing the hypotheses and you have a large enough data set, you should consider using the method suggested by Hallgren et al. (1999). They split the data set into two parts. The first one is used for data diving, i.e. for a search for patterns, without any restrictions, i.e. you can try as many tests as you want, any 'statistical fishing' is permitted. By this procedure, you will generate a limited number of hypotheses that are supported by the first part of the data set. Those hypotheses are then formally tested using the second part of the data set, which was not used in hypothesis generation. See Hallgren et al. (1999) for more details.

13.5 Comparing two communities

In this section, we will briefly demonstrate two approaches available in Canoco 5 and allowing you to compare two kinds of biotic communities, focusing on their mutual links. Analyses discussed in this section are located in a separate example project, *meadows-compare.c5p*. It differs from the original *meadows.c5p* project mainly by storing the moss and vascular plant data sets as two separate tables, but both projects were imported from the same Excel file (*meadows.xlsx*). Please note well that the two community tables (*Mosses* and *Vascular*) are both flagged as compositional, which is important for successful set-up of the analyses illustrated below.

Another difference from the original project is that here we have imported only the original ion concentrations, not their log-transformed values. This allows us to show how the transformations can be specified in Canoco 5 projects. To do so with newly imported data, select first the corresponding data table (here named *Environment*) by clicking its name in the *Data tables* list and then choose the *Data | Default transformation and standardization* menu command. When you do it for the first time for the *Environment* data table, Canoco 5 displays a warning, pointing out that the log transformation is

---

[17] See Section 15.4 for a description of how to compute such characteristics in Canoco 5.

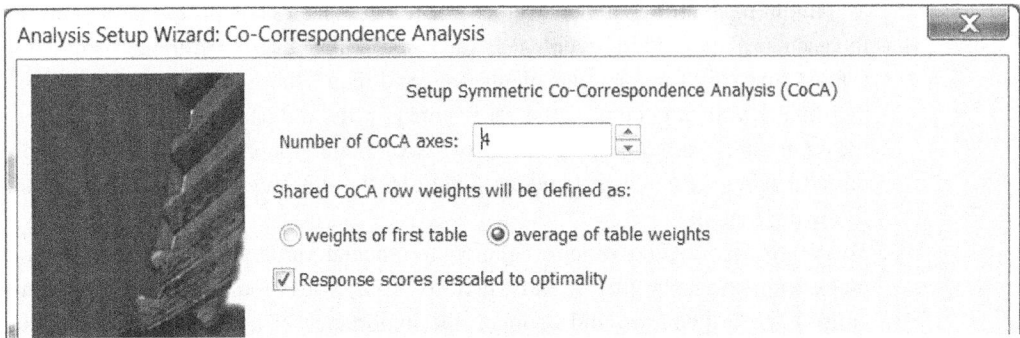

Figure 13–9 Co-correspondence analysis settings in the Analysis Setup Wizard.

suggested for 13 out of the 15 variables (i.e. for all variables except *pH* and *slope*). This suggestion is made by the Canoco Adviser and it also suggests different *A* and *B* constants for the generalised log transformation (see Section 1.3.1), depending on the presence of zero values and on the total range of values in the original variables. By closing the dialog box with *OK*, the suggested transformations are stored and will be applied to the variables each time they are used in an analysis.

You will start with the co-correspondence analysis (example analysis is named *CoCA-suppl-vars*), which finds ordination axes that maximise (weighted) covariance between the case scores computed for the two data tables (representing the two compared types of communities). The rows present in the two data tables must come from the same set of sampling locations and must be stored in identical order. In addition, you might provide another data table whose variables are treated as supplementary variables and therefore can be used to interpret the resulting ordination axes. Click the *New* button to start the New Analysis Wizard, keep all three project tables selected, and on the last page double-click the *Specialized Analyses* folder and choose the *CoCA-suppl-vars* template.[18]

In the Analysis Setup Wizard that appears next, the first two pages are quite usual, concerning the choice of supplementary variables (keep all selected) and the settings for the Monte Carlo permutation test. All the restricted types of permutations are supported also for the CoCA test, so you might perform it even with data sets coming from non-trivial sampling designs. For this data set, however, keep the *Unrestricted permutations* chosen. The next page is specialised for CoCA and illustrated in Figure 13–9.

You can specify in the first field how many axes you want to calculate. The next choice concerns the case weights used in the CoCA. Similarly to a correspondence analysis (CA or DCA), the case weights represent the sum of data values for the particular case. Yet here you have two data tables acting in mutually symmetric roles (Canoco 5 implements the symmetric type of CoCA[19]). The default choice (displayed in Figure 13–9) averages the weights coming from the two tables, but if you want to use as case weights the

---

[18] If you cannot find the template in this folder, make sure that both tables with vegetation data are marked as compositional in the project. This setting can be changed after the import using the *Data | Change table kind...* menu command.

[19] See Ter Braak and Schaffers (2004) for a description of the predictive CoCA method.

13.5 Comparing two communities

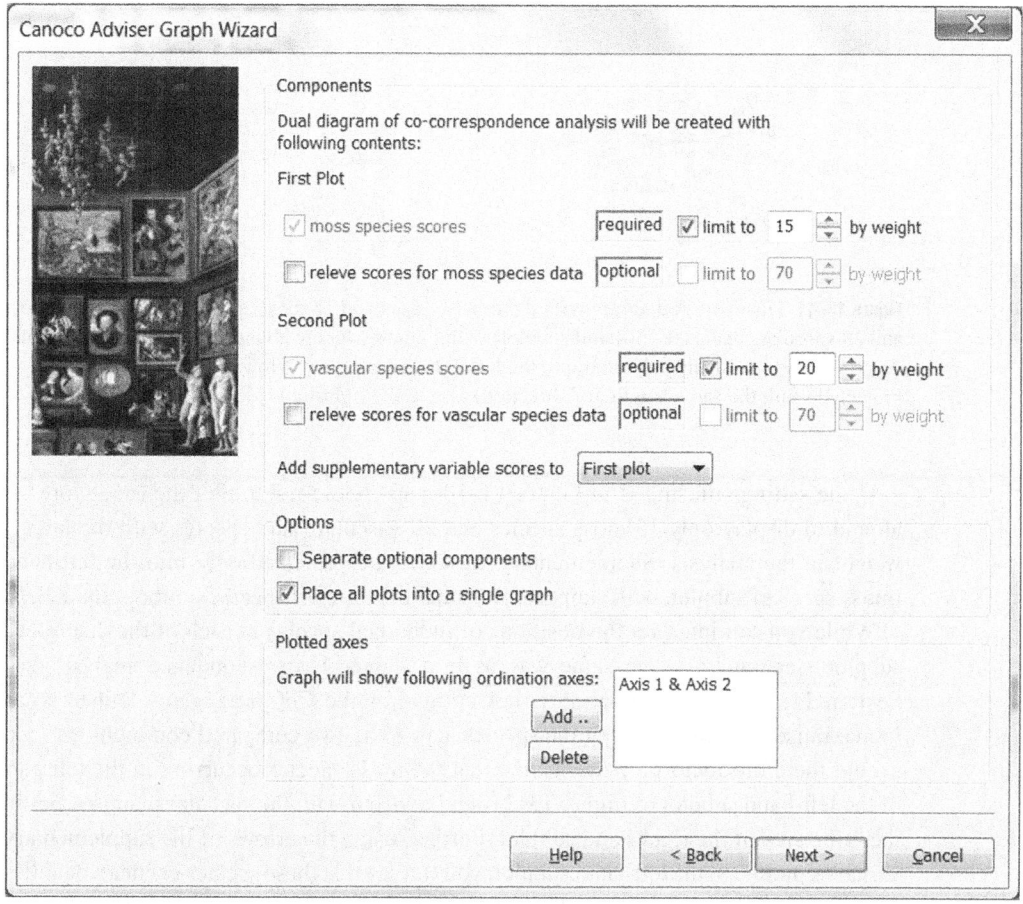

Figure 13-10 Graph Wizard page for setting up the contents of a dual CoCA diagram.

abundances from the table chosen as focal in the New Analysis Wizard, choose the *weights of first table* option.

After you close the setup wizard with the *Finish* button and the analysis is executed, the Graph Wizard window is shown as usual. But unlike the standard ordination methods, use of the Graph Wizard is essential here, because only through it can you create the specialised dual diagram of CoCA axes: two ordination plots displaying side-by-side the scores of response variables (usually species) and optionally the case scores, separately for each of the two compared communities. Additionally, when you have chosen the CoCA template with supplementary variables, these variables can be projected into one or both of these plots in the dual diagram.[20] The content of the Graph Wizard page specifying dual CoCA diagram is illustrated in Figure 13–10, and the suggested choices result in the diagram shown in Figure 13–11.

---

[20] Like for any other composite graph created by the Graph Wizard, you can also choose to place each subplot in a separate graph window, which might be a useful choice when you want to post-process and re-assemble the subplots using different software.

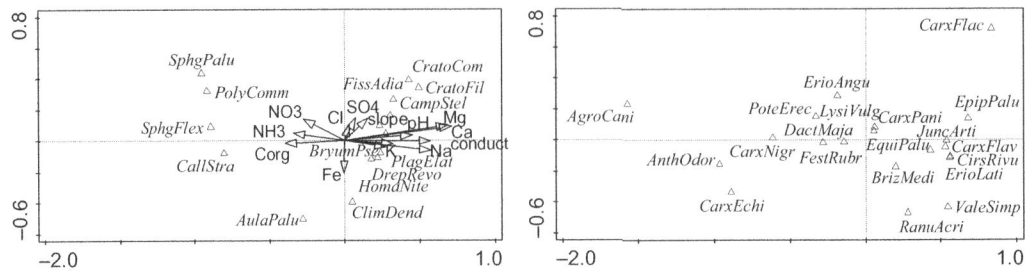

Figure 13–11 Dual CoCA diagram with the first two axes and 15 moss species (left-hand subplot) and 20 vascular species (right-hand subplot) with highest average abundance. The environmental descriptors are passively projected into the left-hand subplot, but arrows could have been projected (with the same length and direction) also to the right-hand plot.

In the settings illustrated in Figure 13–10, you choose not to plot the case scores at all and to display only 15 moss species and 20 vascular plant species with the largest weight in the analysis. Supplementary variables are projected only into the left-hand (moss species) subplot, with supposedly a smaller clutter of species symbols and labels.

While you can interpret the positions of individual species in each of the dual plot's subplots separately, in the same way as in a standard correspondence analysis (see Section 11.2), this interpretation is not optimal as the CoCA axes are defined so as to maximise the links between the species across the two compared communities. You should therefore focus on patterns like that the moss species occurring in the left part of the left-hand subplot of Figure 13–11 tend to occur with the vascular plant species in the same area of the right-hand subplot. Further, using the arrows of the supplementary variables plotted in the left-hand subplot, you can say that these species occur at localities with lower pH values and higher concentration of N ions and larger amount of organic carbon (so the first CoCA axis displays a similar gradient as the earlier CCA analysis, but with a mirrored direction).

The relation between the two types of biotic communities is significant, as documented on the specialised *Co-Correspondence Analysis* page of the analysis notebook with the results of permutation tests. The extent of cross-correlation between the case scores for the two community types can be seen in a table on the same notebook page. You can see that particularly the first CoCA axis is strong, with the cross-correlation value 0.951. But this (almost) perfectly correlated part of the variation of the two communities, reported by the CoCA axes, represents only a relatively small part of the total variation in these two communities: their total inertias, reported in the *Co-Correspondence Analysis* page (5.43 and 5.29 for, respectively, the mosses and vascular plants) are much larger than the total variation 0.716 captured by CoCA (this value is reported on the *Summary* page of the analysis notebook).

Finally, we will briefly discuss another approach to comparing two biotic communities, namely applying Procrustes rotation (Schönemann 1966; Legendre & Legendre 2012, p. 703) to separate ordinations of the two communities. The Procrustes rotation takes two tables with the same number of rows (representing a shared set of cases) and their

13.5 Comparing two communities

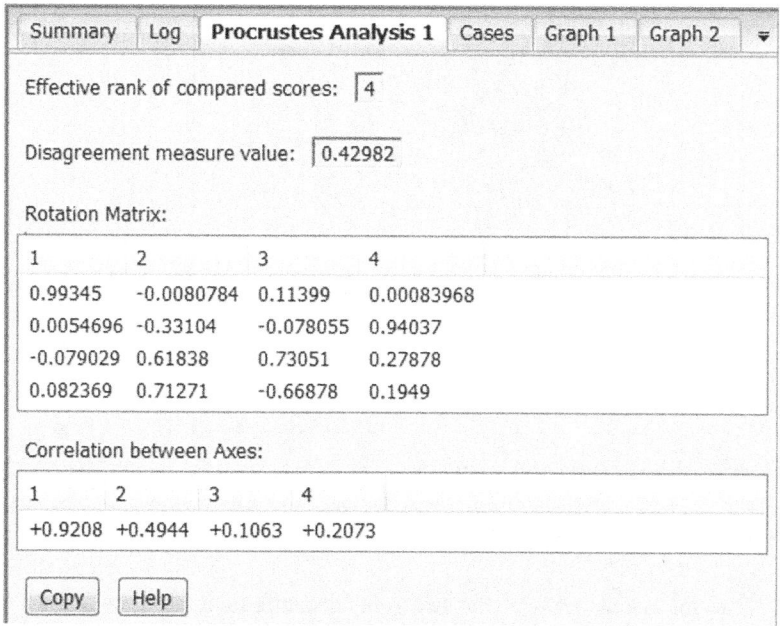

Figure 13-12 Summary of Procrustes analysis comparing DCA ordinations of the moss and vascular plant communities.

columns interpreted as coordinates of points (cases) in two multidimensional spaces, and attempts to rotate these two point configurations so that they match one another as closely as possible, based on the least-squares criterion (minimising the sum of squared distances between the corresponding points).[21] You must therefore start by creating two separate analyses, summarising the two compared data tables. In the example project, we have used detrended correspondence analysis (*DCA-mosses* and *DCA-vasculars*), but other methods can be used as well, including constrained ordination, likely with the same set of explanatory variables.

Once you have the two analyses ready, you can create the Procrustes analysis. This is, however, done in a very different way than for the other analyses, as this analysis will not work with project data tables. Instead of using New Analysis Wizard, select the *Analysis | Add new analysis | Compare ordinations* menu command and in the dialog box select the two ordinations you want to compare in the left and right list, respectively. When choosing an analysis with constrained ordination, you can choose between *CaseR* and *CaseE* scores (calculated, respectively, from the response variable scores and from the values of explanatory variables, see Section 4.6). The summary of the resulting analysis (called *Procrustes-mosses-vs-vasculars* in the example project) is shown in Figure 13-12.

---

[21] Procrustes rotation can be generally applied to any data matrix, including original community composition data and this might represent an alternative to the use of co-correspondence analysis when the linear model seems to better fit the analysed data. Canoco 5, however, offers an easy access to Procrustes rotation only when comparing case scores calculated by ordination methods. You can use the Procrustes rotation helper in customised analyses, however (see Canoco 5 manual, section 4.4.4).

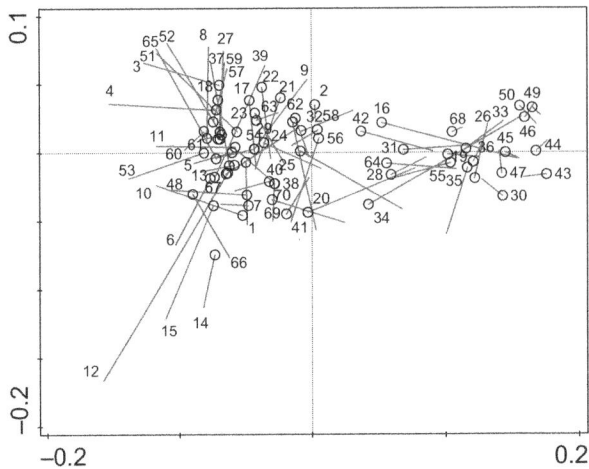

Figure 13–13 Scatter-diagram with cases from the Procrustes rotation analysis. The first two axes of the rotated space are plotted.

The mismatch between the two rotated configurations is about 43% (seen as the disagreement value 0.42982 at 0–1 scale) and you can see from the last table (*Correlation between Axes*) that the gradients recovered by the first ordination axis are almost identical, with correlation 0.9208, but the agreement quickly diminishes for higher axes.

Beside overall agreement measures, you can also compare the agreement for individual plots by visualising the difference in their positions. This can be achieved (in a surprisingly easy way) by plotting the scatter of cases (*Graph | Scatterplots | Releves*). The resulting diagram is present in *Graph 1* of the example analysis notebook and reproduced in Figure 13–13. Each case is represented by a symbol as usual, but additionally by a line segment starting from the symbol position, representing the case position in the ordination space of one of the two compared analyses, and ending at the case position in the other ordination space. The length of the segment therefore represents the extent of discrepancy in the community composition between the two communities for a particular site. You can clearly see that the discrepancy is generally lower for the cases at the right side of the diagram.

The easy way to create such a diagram was enabled by the fact that the Canoco Adviser pre-selected one special option for this analysis, as you can check by selecting the *Analysis | Plot creation options* menu command. In the *General* page of the dialog, you can see the option *Use case-shift plots* checked. This option is used, in a more general context, to graphically relate the *CaseR* and *CaseE* scores in the same ordination diagram, but in the case of Procrustes rotation, the positions of cases in one of the compared ordination spaces are stored as *CaseR* scores, while the positions from the other ordination space as *CaseE* scores.[22]

---

[22] You can find out which community is represented by which type of scores in the Analysis Setup Wizard, where the two source analyses are assigned to response and explanatory data (for our example analysis, the case positions for moss community data are represented by *CaseR* scores – circles in Figure 13–13).

When you first create the above diagram, you will not see all the line segments of Figure 13–13. This is because the range of the diagram axes is initially based only on the positions of symbols and it must be extended. To do so, use the *Graph | Range of axes* menu command or the corresponding toolbar button.

This alternative approach to comparing two biotic communities is often less effective than the use of co-correspondence analysis. This comes about due to independent summarisation of each data table by two separate analyses. The CoCA, on the other hand, constructs one shared set of axes that maximise the relation between the two communities. But Procrustes rotation is applicable to a wider range of problems than merely comparing the compositional variation of two communities.

14 Case study 3: Separating the effects of explanatory variables

14.1 Introduction

In many cases, the effects of several explanatory variables need to be separated, even when the explanatory variables are correlated. The example below comes from a field fertilisation experiment (Pyšek & Lepš 1991). A barley field was fertilised with three types of nitrogen fertiliser (ammonium sulphate, calcium-ammonium nitrate, and liquid urea) and two different total nitrogen doses. For practical reasons, the experiment was not established in a correct experimental design, the plots are pseudoreplicates, which limits correct statistical inference (see Section 3.4). The experiment was designed by hydrologists to assess nutrient runoff and, consequently, smaller plots were not practical.[1] In 122 plots, the species composition of weed community was recorded as classical Braun-Blanquet relevés (for the calculations, the ordinal transformation was used, i.e. numbers 1–7 were used for grades of the Braun-Blanquet scale: r, +, 1, ..., 5). The percentage cover of barley was estimated in all relevés.

The authors expected the weed community to be influenced both directly by fertilisers and indirectly through the effect of crop competition. Based on the experimental manipulations, the overall fertiliser effect can be assessed. However, barley cover is highly correlated with fertiliser dose. As the cover of barley was not manipulated, there is no direct evidence of the effect of barley cover on the weed assemblages. But the data enable us to partially separate the direct effects of fertilisation from the indirect effects of barley competition. This is done in a way similar to the separation of the effects of correlated predictors on the univariate response in multiple regression. The separation can be done using the variable of interest as an explanatory variable and the other one as a covariate.

Our example data set provides the simplest possible set-up for variation partitioning – with two groups, each containing only one predictor. But the illustrated approach can be readily extended in Canoco 5 to variation partitioning with three groups and to a more common set-up with multiple variables in each group. See Section 5.7 for additional discussion of issues concerning the selection of variables used in each group.

---

[1] We ignore this drawback in the following text.

14.2 Data

In this case study, you will work with simplified data, ignoring the fertiliser type and taking into account only the total fertiliser dose. Data are in the Excel file *fertil.xlsx* – the FERTSPE worksheet contains vegetation composition data, i.e. species composition of individual relevés, the FERTENV worksheet contains the explanatory variables (predictors), i.e. the dose of the fertiliser and cover of the barley. The dose is *0* for unfertilised plots, *1* for 70 kg of N/ha, and *2* for 140 kg N/ha. All the predictor variables used in Canoco analyses are automatically centred and standardised to z-scores, i.e. they become variables with zero mean and variance equal to 1,[2] so that the results would be exactly the same if you use the dose in its original units (i.e. as 0, 70, and 140 kg N per hectare).

If you want to import data into Canoco project yourself (although there is an example project *Fertil.c5p*, including the analyses discussed in this chapter), you need to create two data tables from the two sheets (remember that in both sheets, the first row is just a description, so that the data themselves start in the second row). The *FERTENV* worksheet also contains the number of species in each relevé. You will use it as a response variable in a univariate analysis (multiple regression of species numbers on the two predictors), but there is no need to import this variable into Canoco project (so do not import the D column into the second data table). RDA is in fact a generalisation of multiple regression (see Section 4.8), so you will see the match between similar analyses when the response is univariate (in multiple regression) and when the response is multivariate (in RDA).

14.3 Changes in species richness and composition

We will demonstrate the univariate analysis first. You have two predictors (explanatory, or 'independent' variables), *cover* and *dose*, and one response variable, the number of species *NSP*.[3] You will estimate the multiple regression model in R, using data frame *fertil*, which contains the data imported from the *FERTENV* sheet.

```
fertil.lm <- lm(NSP ~ dose+cover,data=fertil)
summary(fertil.lm)
...
```

---

[2] More precisely, for unimodal methods (not used in this case study), they are standardised to zero weighted average and unit weighted variance, using the same case weights as used in Step 2 of the weighted averaging algorithm (Section 4.5).

[3] This is also a simplification: one of the independent variables, *cover*, is dependent on the *dose* and consequently the use of path analysis might be a feasible solution here (see Legendre & Legendre 2012, p. 592). Further, the weed community might affect barley cover; but if that is so, it would be the total amount of weeds rather than the number of species in a weed community. Nevertheless, from the point of view of the weed community, *dose* and *cover* can be considered as two correlated predictors.

```
Coefficients:
              Estimate Std. Error t value Pr(>|t|)
(Intercept)   9.423662   0.388684  24.245  < 2e-16 ***
dose         -0.085011   0.361781  -0.235    0.815
cover        -0.061743   0.008999  -6.861 3.28e-10 ***
---
Signif. codes:  0 '***' 0.001 '**' 0.01 '*' 0.05 '.' 0.1 ' ' 1

Residual standard error: 1.821 on 119 degrees of freedom
Multiple R-squared:  0.4925,    Adjusted R-squared:  0.4839
F-statistic: 57.74 on 2 and 119 DF,  p-value: < 2.2e-16
```

In the multiple regression summarised above, two types of test are carried out. The complete model is tested by an analysis of variance, with the corresponding null hypothesis stating: **the response is independent of all the predictors**. Test results are shown in the last line of the regression summary ($F_{2,119} = 57.74$, $p < 2.2 \cdot 10^{-16}$).

The null hypothesis was clearly rejected and you can see that the regression sum of squares is roughly half of the total sum of squares (multiple $R^2 = 0.4925$), which means that this model explains half of the variability in the species counts. But you still do not know which of the two explanatory variables is more important. This can be seen from the t-value tests shown in the *Coefficients* section of the output.

The results show (in the *Estimate* column, with the estimates of regression coefficients) that the number of weed species decreases with both *cover* and *dose*, that the direct effect of *cover* is much more important (based on the magnitude of its value in the *t value* column), and also that the effect of *cover* is significant, while that of *dose* is not (see the $Pr(>|t|)$ column). If you calculate the univariate regressions on each of the predictors separately (i.e. their independent effects), you will see that both of them are highly significant. This means that *dose* itself might be a good predictor for the number of species, but does not significantly improve the fit when added to the predictor describing the cover of barley. On the other hand, *cover* is a good predictor for the number of species and significantly improves the fit when added to *dose*. We conclude that within the context of the two available predictors, *cover* is sufficient to explain the number of present weed species.

You will now proceed with the multivariate analysis.[4] Similarly as in the univariate analysis, you will be interested in the common effect of both variables, as well as in simple (marginal) and partial effects of individual variables. The following steps are recommended:

First, when you set up an analysis with the Canoco 5 setup wizard, the Canoco Adviser calculates for you in the background a DCA, to obtain the length of the longest DCA axis in turnover units. The resulting value (3.8) is in a 'grey zone' where both linear

---

[4] Note the difference: the species composition might change even when the number of species does not; in contrast, the test with the number of species might, in some cases, be stronger than the test with species composition. Consequently, the results might differ.

and unimodal methods should perform reasonably well.[5] But you can also calculate an unconstrained ordination first, to inspect the total variation in the species composition and to project passively the explanatory variables – this is done in the *Unconstrained-suppl-vars* analysis in the example *Fertil.c5p* project.

Then, calculate a constrained ordination with both explanatory variables (this is the *Constrained* analysis in the *Fertil.c5p* project file). The authors of the published paper (Pyšek & Lepš 1991) used CCA, but in this case study, you will use RDA. RDA enables you to use both the standardised and non-standardised analyses. Standardisation by cases allows you to differentiate between the effect upon the total cover and the effect upon the species composition. Use RDA (with the fertiliser *dose* and the barley *cover* as explanatory variables) first on data not standardised by the case norm. Its results will reflect **both** the differences in total cover and the differences in the relative species composition (if they exist, of course). But use also RDA on data standardised by the case norm where the results will reflect **only** the differences in the relative proportions of particular species.[6] In this study, however, the effects of explanatory variables and also the variation partitioning results with response data standardised by case norm or non-standardised were rather similar (and were also similar to the output of CCA[7]). Only the results of non-standardised RDA will be therefore shown in this tutorial.

In your analysis, test the significance by a Monte Carlo permutation test using unconstrained permutations. This is a test of the following null hypothesis: *there is no effect of the explanatory variables on species abundances*, i.e. the effect of both variables is zero. Because the analysis does not standardise the species values, even a proportional increase of all the species is considered a change in species composition. The rejection of the null hypothesis means that at least one of the two variables has some effect on the species composition of the vegetation. The meaning of this test is analogous to an overall ANOVA on the multivariate regression model.

Next, inspect the analysis results. The effects of explanatory variables are highly significant ($P = 0.002$ with 499 permutations – the lowest possible P-value with this number of permutations). The Canoco Adviser suggests the species–explanatory variable biplot, which clearly shows the relationships of the weed species to the explanatory variables (Figure 14–1). The species with the highest fit on the axes are selected (20 best fitted species in Figure 14–1). The diagram shows that the two explanatory variables are positively correlated and that most of the weed species are negatively correlated with both the fertiliser dose and cover of barley. Only *Galium aparine*, a species able to climb the stems of barley, is strongly positively correlated.

The diversity plot (in Figure 14–2) is the second graph suggested by the Canoco Adviser. The case scores based on the response (species) scores (*CaseR* scores) are displayed, with the size of each symbol corresponding to the number of species in the relevé. Both explanatory variables are also projected onto the diagram. The diagram

---

[5] See Section 2.3 for further details.
[6] You should note that the interpretation of standardised data based on the species cover values estimated on an ordinal scale may be problematic.
[7] This relates to the fact that the length of the longest DCA axis in turnover (SD) units indicates that both linear and unimodal methods would perform well.

250   **Separating the effects of explanatory variables**

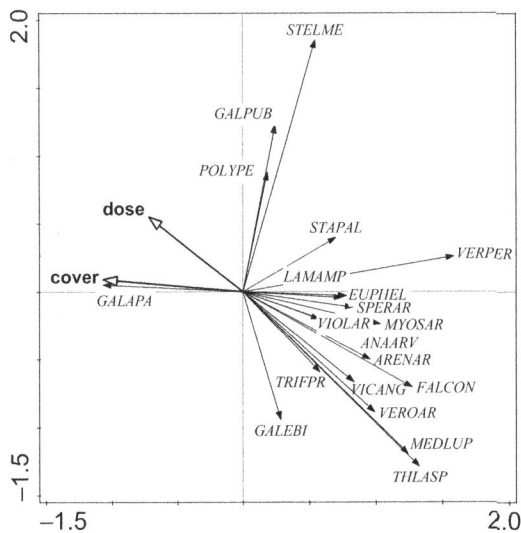

Figure 14-1 Ordination diagram of RDA (the first two axes) on non-standardised data, with two explanatory variables, *cover* and *dose*. Twenty best fitted weed species are plotted.

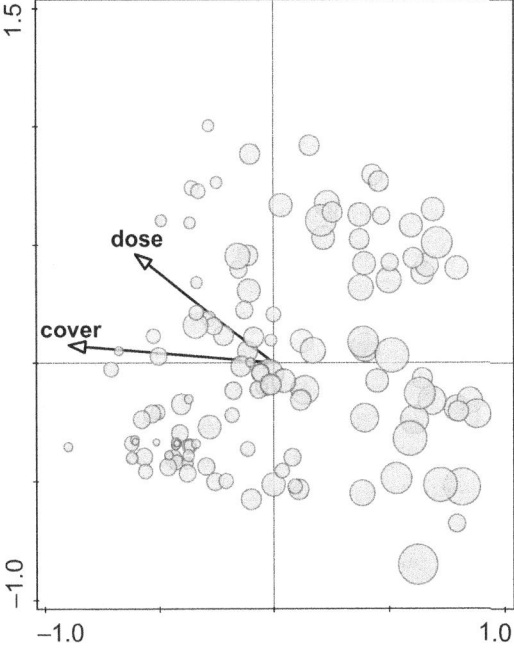

Figure 14-2 Biplot diagram (the first two RDA axes) with explanatory variables and cases (relevés), with the size of case symbols corresponding to species richness (number of species in each relevé).

14.3 Changes in species richness and composition 251

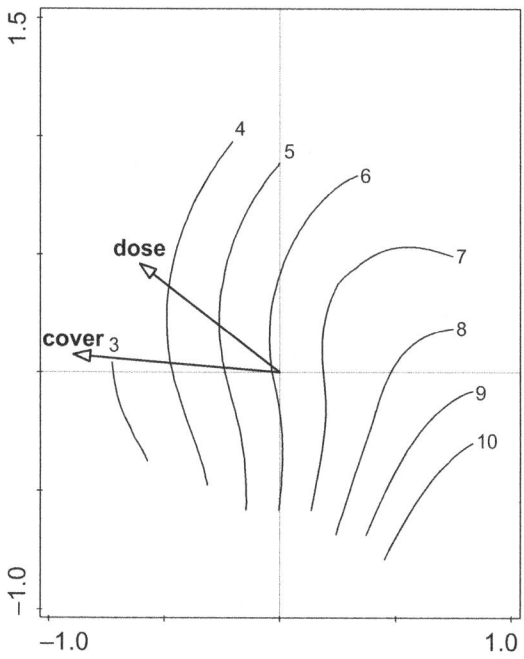

Figure 14-3 The isolines of species richness of relevés, plotted into the RDA ordination diagram (with the first two RDA axes).

displays clearly how the species richness decreases with both dose and cover. To get the diagram with positions of individual cases, you should ask for a symbol plot in the Graph Wizard.

The same information can also be presented by smoothed data (particularly when you have many cases) – to do so, ask for a *contour plot* in the Graph Wizard. The plot in Figure 14-3 was obtained by using the default values for the loess method.

As you can see in the *Summary* page of the analysis notebook for this RDA (Figure 14-4), the first two axes explain only 14.3 percent of the variance of species data. This seems to be rather a low amount, but you should take into account that this includes also the reduction of dimensionality (from the original 44 species to two constrained axes). Consequently, the explanatory power of the two variables can be reasonably judged only by a comparison of the constrained ordination with corresponding unconstrained ordination (PCA).

The Canoco Adviser provides a pre-defined analysis template to compare constrained and unconstrained analyses using the same pair of data tables. Figure 14-5 shows the upper part of the *Comparison* page of the *Compare-constrained-unconstrained* analysis' notebook.

The third row of the table in Figure 14-5 shows that the efficiency of the first constrained axis is 40.56% of the first unconstrained axis, which looks much better than the 9.7% explained by the first axis. Nevertheless, it is not a really high number and also the original results (see Figure 14-4) show that in the constrained analysis, the third axis

Summary of Results

Method: RDA

Total variation is 2581.180, explanatory variables account for 14.3% (adjusted explained variation is 12.9%)

Summary Table:

Statistic	Axis 1	Axis 2	Axis 3	Axis 4
Eigenvalues	0.0970	0.0460	0.2005	0.1312
Explained variation (cumulative)	9.70	14.30	34.35	47.47
Pseudo-canonical correlation	0.7049	0.5125	0.0000	0.0000
Explained fitted variation (cumulative)	67.85	100.00		

Figure 14-4 Lower part of the *Summary* page of the *Constrained* analysis notebook.

◀ **Comparison** Summary Log Cases (1) RespVars

Following table allows you to evaluate importance of individual axes of constrained and unconstrained ordination applied to the same response data. Optional colouring of second row reflects comparison of unconstrained axes with broken stick model.

	1	2
Explained by constrained axis [%]	9.70	4.60
Explained by unconstrained axis [%]	23.92	19.66
Efficiency of constrained axis [%]	40.56	23.39

Figure 14-5 *Comparison* page comparing the axes of constrained and unconstrained analysis of the same data.

(which is by definition unconstrained because you have only two explanatory variables here) explains 20% of the total variability, which is more than both constrained axes explain together. This shows that there is a lot of variability in species composition that cannot be attributed to the two explanatory variables used in your analyses.

But based on your results so far, you can be confident that there is a highly significant relationship of species composition to the explanatory data. Now, you should decide which of the variables is important and how much does each explain, and how large is their shared effect. You will perform the **variation partitioning** to address these questions (see also Section 5.7). The Canoco Adviser provides several predefined analysis templates for the partitioning. You can either use the analyses testing the simple effects or analyses testing the conditional effects. Start with the latter option, corresponding to the

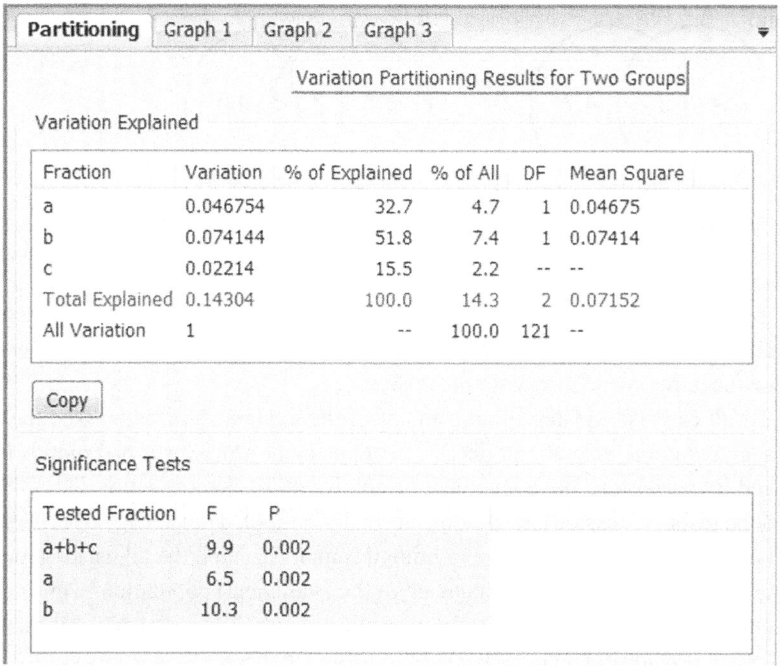

14.3 Changes in species richness and composition

Partitioning	Graph 1	Graph 2	Graph 3		

Variation Partitioning Results for Two Groups

Variation Explained

Fraction	Variation	% of Explained	% of All	DF	Mean Square
a	0.046754	32.7	4.7	1	0.04675
b	0.074144	51.8	7.4	1	0.07414
c	0.02214	15.5	2.2	--	--
Total Explained	0.14304	100.0	14.3	2	0.07152
All Variation	1	--	100.0	121	--

Significance Tests

Tested Fraction	F	P
a+b+c	9.9	0.002
a	6.5	0.002
b	10.3	0.002

Figure 14-6 Variation partitioning results without variance adjustment.

Var-part-2groups-Conditional-effects-tested analysis in the *Fertil.c5p* example project. In this analysis, you will not use the adjustment of explained variability, so un-check the default option (*Adjust explained variation*) in the Analysis Setup Wizard. Results of the partitioning can be seen in the upper part of the *Partitioning* page of that analysis' notebook, reproduced in Figure 14-6.

The most important results come in the *% of All* column of the first table. You already know how much the two variables explain together (14.3%). The simple effects of cover and dose are, respectively 9.6% ($b + c$ in the table) and 6.9% ($a + c$ in the table). The overlap is then 9.6% + 6.9% – 14.3% = 2.2%, and from this you can easily calculate the conditional effects (i.e. simple effect – overlap).[8] They represent 7.4% and 4.7% respectively (both highly significant, as seen in the second table of Figure 14-6).

The alternative analysis type using simple effects is represented by the *Var-part-2groups-Simple-effects-tested* analysis and provides identical estimates of the three fractions of explained variation, but also the tests of simple effects (they are again significant for the two predictors). Running both variation partitioning types enables better insight into the data, particularly when you inspect the ordination diagrams for individual analysis steps.[9] For example, it is interesting that in the simple analyses, *Galium aparine* is positively correlated with both the dose and the cover, whereas for the conditional

---

[8] Naturally, you do not need to do it here, as the conditional effects are directly represented by the *a* and *b* fractions quantified in the table.

[9] Each variation partitioning analysis in the example project contains three analysis steps, corresponding to required executions of the ordination algorithm. The first step uses both predictors as explanatory variables, while the second and third step use just one of the predictors as an explanatory variable (and for the analysis

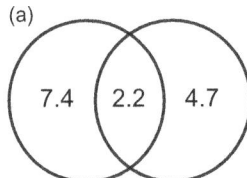 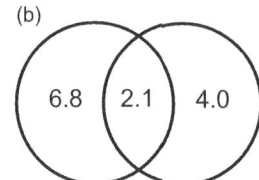

Figure 14-7 Partitioning of variance in weed species composition, explained by the *cover* (left circle in each pair) and *dose* (right circle in each pair) variables without (a) and with (b) variance adjustment.

effects, there is a strong correlation with the cover, but the diagram suggests very weak but negative correlation with the dose.

You can also opt for an adjustment of the explained variation (see the two *AdjVar*-analyses in the example project). The adjustment procedure corresponds in its method and its purpose to the adjustment of coefficient of determination (R^2_{adj}) as frequently done in the regression analysis: the normally estimated percentage of explained variation is biased (overestimated) due to limited sample size and the adjusted values provide an unbiased estimate of the percentage in the (statistical) population being sampled.[10] For our example data set, the pattern of explained variability is very similar, but the values are slightly lower (Figure 14-7), as one would expect.

From all the analyses, you can conclude that both simple and conditional effects of both explanatory variables are highly significant – i.e., each of the variables explains a significant proportion of species composition, both when taken separately as the only variable, and also in addition to the explanatory power of the other variable. There is the difference with the univariate analyses: the conditional (partial) effect of dose on the number of species was not significant in the regression for species richness (i.e., the dose does not add any explanatory power to the cover of the barley), but the dose improved the explanatory power of the model for species composition.

> It may sometimes happen (not in this data set) that the analysis with both variables used as explanatory variables is (highly) significant, whereas none of the analyses with one of the variables used as an explanatory variable and the other as a covariate is significant. This case is equivalent to a situation in multiple linear regression, where the ANOVA on the whole model is significant and none of the regression coefficients differ significantly from zero. This happens when the predictors are highly correlated. You are then able to say that the predictors together explain a significant portion of the total variability, but you are not able to say which of them is the important one.

The results of variation partitioning can be displayed in a diagram using Venn circles (Figure 14-7).

---

testing conditional effects, the other predictor is used as a covariate). The Graph Wizard then offers creation of a biplot diagram for each of these three analysis steps.

[10] Note, however, that this adjustment does not solve the issue with a different size of explained variation, a priori expected for groups with a different number of explanatory variables. This issue is, however, irrelevant for our data set.

When you compare the results obtained here with those of Pyšek and Lepš (1991), you will find important differences. This is caused by the omission of fertiliser type as an explanatory variable in your analyses. There were relatively large differences in cover due to fertiliser type within a dose. All these differences are now ascribed to the effect of barley cover. This also illustrates how biased the results can be when some of the important predictors in the data are ignored in a situation when predictors are not experimentally manipulated and they are interdependent.

14.4 Changes in species traits

Earlier analyses in this chapter demonstrated how the community composition is affected by the two correlated predictors and quantified their shared effect, as well as the unique contributions of each of them. Even a superficial inspection of the ordination diagram (Figure 14–1) suggests to a botanist knowing the species that the response of species depends on their properties. In particular, it seems that low plants are harmed most by fertilisation and the resulting increase of barley cover. This also supports a mechanistical explanation that the increase of available nutrients increases the intensity of competition for light and the tall plants are more successful in this competition. You will now test whether there are systematic changes in the species traits, related to fertilisation and the corresponding increase of barley cover. This section illustrates, with a relatively simple analysis the two analytical approaches used in analyses involving species traits, discussed in Chapter 9.

For this purpose, we have compiled (from various trait databases) the basic traits of the species included in the study (they are in the sheet *Traits* of the *FertilTraits.xlsx* spreadsheet file) – SLA [mm^2/mg], canopy height (*CanHei*) [m], seed weight (*sdweight*) [mg], and life form (LF). Four life forms were present – geophytes, therophytes, hemicryptophytes, and chamaephytes – and because some species can occur in two life forms, we decided to use fuzzy coding. Individual categories (life forms) are therefore coded as separate dummy (indicator) variables in the file (*LF.Ther*, *LF.Hemi*, *LF.Geop*, and *LF.Cham*). For the analyses of trait data, we have created an example Canoco 5 project *FertilTraits.c5p*, based on the two matrices that were used for the analysis of community composition and, in addition, the matrix with species traits.[11]

First, you will analyse the community weighted means (CWM) of traits, preparing the new table using the *Data | Add new table | Trait averages* menu command, accepting all the suggestions in the dialog box, and naming the table *CWM*. Start by computing an RDA using both predictors (this is the *Constrained* analysis in the example project). The steps are the same as for a corresponding analysis with species composition, but the focal table will be the CWM table with trait averages, and, because the traits are measured on different scales, you must centre and standardise by variables (as correctly suggested in the setup wizard). The relation between trait means and the two explanatory variables is highly significant (pseudo-F = 9.7, P = 0.002 with 499 permutations) and

---

[11] Note that when importing the traits, both NA and blank cells are interpreted by Canoco as missing values.

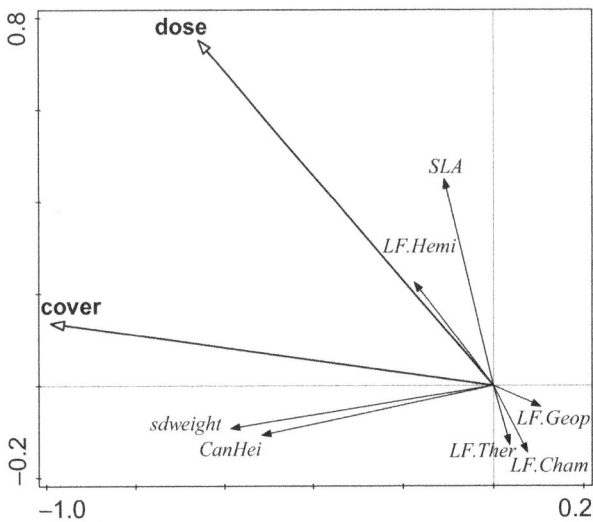

Figure 14-8 Ordination diagram with the first two axes of RDA, using barley *cover* and fertiliser *dose* as predictors and community-weighted means of trait values as response variables. The relationship is highly significant (P = 0.002 with 499 permutations).

the ordination diagram (Figure 14-8) suggests positive correlation of dose particularly with SLA, and of barley cover with canopy height and seed weight.

A more detailed overview of significant relations revealed in this analysis can be obtained with t-value biplots (see Section 11.5) that can be created using the *Graph | Biplots | t-value biplot* menu command – not shown here, but present as *Graph 2* and *Graph 3* in the example analysis. These biplots suggest that the canopy height and seed weight are positively and SLA negatively correlated with the cover of barley, whereas seed weight and canopy height are negatively and the proportion of hemicryptophytes and SLA positively correlated with the fertiliser dose.

Now, you can also carry out the variation partitioning similarly as for the species composition data. Again, the only difference is that the CWM table is used instead of the table with species composition data, and you must centre and standardise by CWM variables. The results (shown in the analysis *Var-part-2groups-Conditional-effects-tested*) demonstrate that the model with both variables is significant and also the two conditional effects (i.e. unique contributions of individual predictors) are related to trait averages with a high significance, with the total percentage explained variability being 12.6%, unique contributions of cover and dose being, respectively, 6.7% and 3.9%, and their shared effect represents another 2.0% of the total variance. The chosen analysis template uses partial ordination to perform variation partitioning, so it allows us to visualise the relation of species traits to unique effects of the two predictors (see *Graph 2* and *Graph 3* of the example analysis).

You can also explore the response of individual species and relate it to their traits. Here you will focus just on a simple graphical exploration of the hypothesis that only

14.4 Changes in species traits

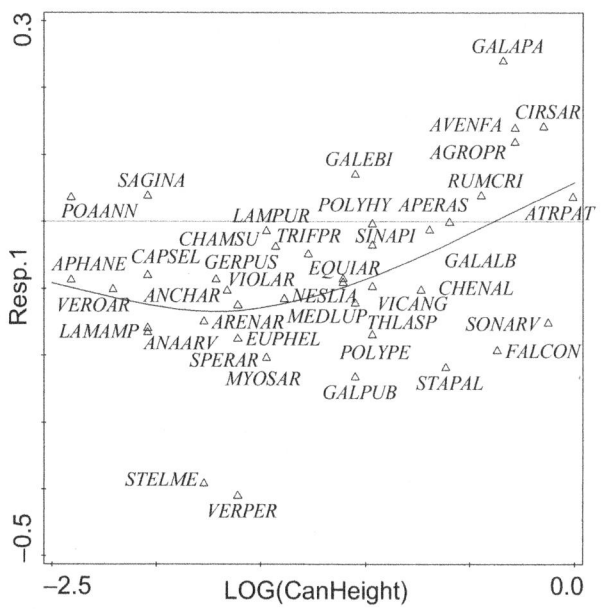

Figure 14-9 Relationship of species position on a partial constrained axis (representing response to increasing cover) to (log-transformed) species canopy height, fitted by a GAM (with df = 3, selected based on AIC_c, p = 0.0034).

tall weed species are able to cope with a dense cover of barley.[12] For this, you must first calculate the positions of species on the only constrained axis in a partial RDA, using cover as the explanatory variable and dose as a covariate (example analysis *Constrained-partial-cover*). These positions therefore represent the response of species to varying barley cover, after the possible confounding effect of dose (as a correlated predictor) was partialled out. Then, you can try to predict this response by the species canopy height. To do so, create a graph using the *Graph | Attribute plots | XY(Z) diagram* menu command and selects the canopy height (*CanHei*) from the *Traits* table as X and the species scores on the constrained axis (*Analysis 'Constrained-partial-cover' | Response variable scores | Axis 1*) as Y. In the example analysis, we fitted the relationship using a *GAM model*, with an automatic selection of model complexity (use the *Model Options* in the upper right corner of the dialog), and with log-transformed height.[13] The Resulting graph (Figure 14–9) shows that the short weeds are mostly suppressed and only the taller ones (with height over 0.3 m, corresponding to log value –0.5) are not negatively affected by increasing barley cover.

---

[12] Additional techniques of species-based approach to trait analyses are illustrated in Case study 5, Section 16.8.2.

[13] If you tick the *Show transformation and axis labels dialog* option in the dialog shown by *Edit | Settings | Graphing options*, you will get the possibility of selecting the desired transformation before the graph is plotted or re-plotted.

15 Case study 4: Evaluation of experiments in randomised complete blocks

15.1 Introduction

Randomised complete blocks design is probably the most popular experimental design in ecological studies, because in many cases it controls in a powerful way the environmental heterogeneity (see also Section 3.2). For a univariate response variable (e.g. number of species, total biomass), the results of experiments set in randomised complete blocks are evaluated using a two-way ANOVA without interactions. The interaction mean square is used as the error term – the denominator in the calculation of the F-statistic. In the following tutorial, you will use the program Canoco in a similar way to evaluate the community response (i.e. a multivariate response of the species composition of the vegetation).

This case study is based on an experiment studying the effect of dominant species, plant litter, and moss on the composition of a community of vascular plants, with special attention paid to seedling recruitment. In this way, some of the aspects of the importance of regeneration niche for species coexistence were tested. The experiment was established in four randomised complete blocks, the treatment had four levels, and the response (community composition) was measured once.

The experiment is described in full by Špačková et al. (1998), but here we provide its simplified description. The experiment was established in March 1994, shortly after snowmelt, in four randomised complete blocks. Each block contained four plots, each with a different treatment: (1) a control plot where the vegetation remained undisturbed; (2) a plot with the removal of litter; (3) a plot with the removal of dominant species *Nardus stricta*; and (4) a plot with the removal of litter and mosses.[1] Each plot was a 2 m × 2 m square. The original cover of *Nardus stricta* was about 25%. Its removal was very successful, with nearly no re-growth. The removal in the spring caused only a minor soil disturbance that was no longer apparent in the summer.

15.2 Data

In each central 1 m² plot, the cover of adult plants and bryophytes was visually estimated in August 1994. At that time, a new square (0.5 × 0.5 m) was marked out in the centre

---

[1] It seems that one treatment is missing – removal of mosses only; however, for practical reasons it was impossible to remove mosses without removing the litter.

of each plot and divided into 25 0.1 m × 0.1 m subplots. In each subplot adult plant cover and numbers of seedlings were recorded. In this case study, you will use the seedling totals in the 0.5 m × 0.5 m plots. The data are stored in the Excel spreadsheet file *seedl.xlsx*: response data, i.e. the numbers of seedlings of individual species, are in the *seedlspe* sheet and the design of experiment (the explanatory data) is in the *seedldesign* sheet. The latter sheet contains also the total number of seedlings (summed over all species) that you will use in the univariate analysis, but you should not import it to the Canoco project.

When importing the data into the Canoco 5 project (instead of using example project *Seedlings.c5p*), do not forget to state that all the explanatory variables in the second table (i.e. *treatment* and *block*) are factors. Otherwise Canoco will decide that *treatment* is a factor, because it does not contain numeric values, but it will interpret the *block* as a quantitative variable (this could be of course changed later in Canoco 5).

15.3 Analysis

You may ask whether is it necessary to use a multivariate method. Would it not be better to test the effect on each species separately by a univariate method, either by ANOVA or by a test on a generalized linear model? The answer is that it is much better to use a multivariate method. There is a danger in using many univariate tests. If you perform several tests at the nominal significance level $\alpha = 0.05$, the probability of Type I error is **0.05 in each univariate test**. This means that when testing, say, 40 species, you could expect two significant outcomes just as a result of Type I error ($0.05*40 = 2.0$). This can lead to 'statistical fishing', when one picks up only the results of significant tests and tries to interpret them. You could overcome this by using the Bonferroni correction[2] or perhaps some better method (see Section 5.8), but this leads to an extremely weak test. Another important deficiency of the separate-tests approach is that it completely ignores the fact that the tested variables (abundances or presences of individual species) are usually much correlated, so the statements made about them based on hypothesis tests are not independent. Another advantage of the multivariate test is that it potentially tests composition per se. If you are a community ecologist you are more interested in aggregate effects than population effects.

We consider it possible (but some other statisticians probably would not) to use the univariate methods for particular species when we find the community response significant; however, you should keep in mind that the probability of a Type I error is α in each separate test. You should be further aware that if you select the most responsive species according to the results of constrained ordination, those species very probably will give significant differences, and such a univariate test does not provide any further

---

[2] Bonferroni correction means, in this context, dividing the nominal significance level (e.g. 0.05) by the number of tests performed and performing the particular tests on the resulting level. This assures that the overall probability of the Type I error in at least one of the tests is equal or smaller than α (see Rice 1989 or Cabin & Mitchell 2000).

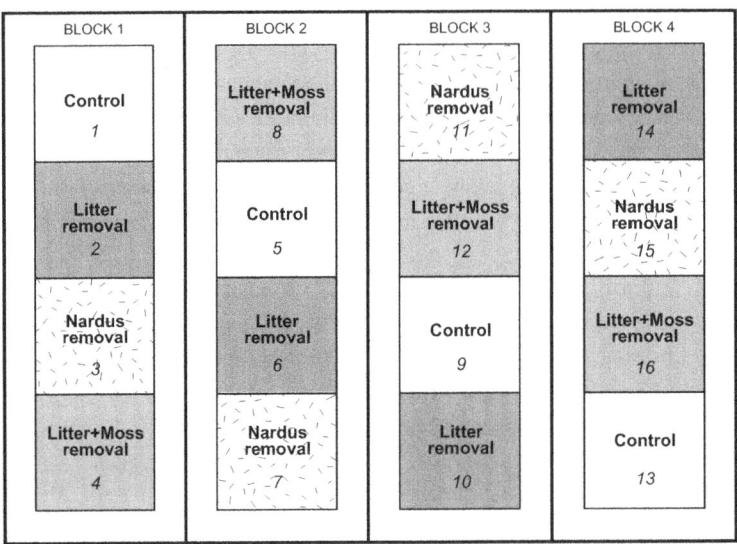

Figure 15–1 The design of the field experiment analysed in this case study.

information in addition to the ordination results. Even in the case when there are no differences between the treatments at all, some species will be **in your sample** more common in some of the treatments, just by chance. When you then plot the ordination diagram, select the 'most responsive' species (by eye) and test them by ANOVA, the test result will very probably be significant.

The design of the experiment is shown in Figure 15–1. Each quadrat is characterised by (1) the type of the treatment, and (2) the block, in which it is located.

Performing the univariate analysis in R is quite simple, as illustrated by the following code (make sure the *block* variable is set to factor type during or after import, as its values in Excel are numeric). In the analysis of randomised complete blocks design, the interaction between treatment and block is missing and it becomes the error term.

```
seedl.aov <- aov(seedlsum~treatment+block,data=seedl)
summary(seedl.aov)
            Df Sum Sq Mean Sq F value Pr(>F)
treatment    3  13540    4513   4.223 0.0403 *
block        3    647     216   0.202 0.8926
Residuals    9   9620    1069
```

The results show that the treatment effect is significant ($F_{3,9} = 4.223$, $p = 0.0403$), but there are hardly any differences among the blocks. Because *block* explains a very low amount of variability, the analysis is weaker than it would be if you disregarded the block effects (then $F_{3,12} = 5.275$, $p = 0.015$). This is a general pattern: using randomised complete blocks is more efficient than completely randomised design only when there are differences between individual blocks.[3]

---

[3] But to disregard the existence of blocks *ex post*, in your analyses, based on their non-significant effects, is from a statistical point of view (at least) questionable. On the other hand, this situation illustrates that the

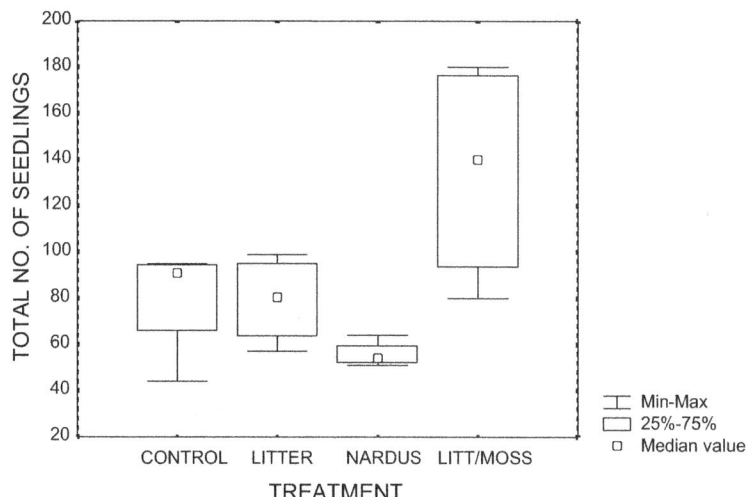

Figure 15-2 Box-and-whisker plot of the number of seedlings in individual treatments.

Now, you can use the multiple comparisons for testing pair-wise differences, and/or inspect a box-and-whisker plot (Figure 15–2).

```
library(multcomp)
summary(glht(seedl.aov, linfct=mcp(treatment="Tukey")))
        Simultaneous Tests for General Linear Hypotheses
Multiple Comparisons of Means: Tukey Contrasts

...

Linear Hypotheses:
                    Estimate Std. Error t value Pr(>|t|)
Li+Mo  - Cont  == 0    54.75      23.12   2.368   0.1533
Litter - Cont  == 0    -1.00      23.12  -0.043   1.0000
Nardus - Cont  == 0   -24.50      23.12  -1.060   0.7208
Litter - Li+Mo == 0   -55.75      23.12  -2.412   0.1441
Nardus - Li+Mo == 0   -79.25      23.12  -3.428   0.0316 *
Nardus - Litter == 0  -23.50      23.12  -1.017   0.7444
```

You can see that only the removal of both litter and moss has a significant effect on the total number of seedlings.[4]

In analogous multivariate analysis, *block* is used as a covariate and *treatment* as the explanatory variable.[5]

---

use of randomised complete blocks in homogeneous study areas is not necessary and might be even viewed as decreasing the power of statistical tests performed on the data.

[4] Using the Duncan test, the combined removal treatment is different from all the other treatments, while Tukey's test, which is usually recommended (and shown in the R output), shows that this treatment differs only from the *Nardus* removal.

[5] As a matter of fact, each of the variables is expanded into four 'dummy' variables, representing the four categories, when they are used in ordination methods, see Section 2.2.

The Canoco Adviser provides a template for a *Constrained-partial* analysis (in the *Advanced Constrained Analyses* folder in the last page of the New Analysis Wizard), where you can easily set the roles for both variables. This analysis is also present in the example project (*Seedlings.c5p*). As there is limited compositional variation in the vegetation of individual plots, you will use the redundancy analysis (RDA), whcih is based on the linear model. The length of the gradient on the first axis of a DCA is 1.98; on the second axis, 1.41.[6] Even more importantly, RDA (unlike CCA) enables you to carry out the analyses on both the standardised and non-standardised data. Now, you can test (at least) two hypotheses:

(1) The first null hypothesis can be stated as follows: *there is no effect of the manipulation on the seedlings*. To reject this hypothesis, it is enough if the total number of seedlings differs between the treatments, even if the proportion of individual seedling species remains constant. When the proportions of seedlings of different species change or when both the proportions and the total number of seedlings change, the null hypothesis will obviously also be rejected.[7] This hypothesis will be tested by RDA without standardisation by case norm.

(2) The second null hypothesis can be stated as follows: *the relative proportions of species among the seedlings do not differ between the treatments*. Rejecting this hypothesis means that the seedlings of different species differ in their response to treatments. The first hypothesis can be tested only when you do not standardise by cases (the default in the Canoco setup wizard) and when you use a standardisation by cases (usually by the case norm or case total), then you test this second null hypothesis. The test of the first hypothesis above is usually more powerful, but the rejection of the second hypothesis (by RDA or CCA[8]) is more ecologically interesting: the fact that seedlings of different species respond in different ways to particular treatments is a good argument for the importance of the regeneration niche for maintenance of species diversity. You will test this hypothesis by RDA with the standardisation by case norm.

The specification of RDA proceeds in a standard way. Just take care with the following issue. When performing the Monte Carlo permutation test, you can ask for permutation within blocks, indicated by the *block* covariate. Each permutation class will then correspond to a block in the experiment. The permutation within blocks is shown in Table 15–1. This is a **design-based** permutation test and this is the approach available in all versions of Canoco (under the null hypothesis, the treatments are freely exchangeable within a block). In the newer versions (since version 3.0), the residuals obtained after removing the effect of covariates can be permuted. In this case study, we recommend using unrestricted permutations (the increased strength is achieved without inflating the Type I

---

[6] See Section 2.3 for the explanation of the use of the length of gradient in turnover units.
[7] The univariate analysis has already shown that there are differences in the total number of seedlings, but this analysis can highlight the response of individual species.
[8] Note that the standardisation is part of the weighted averaging algorithm in canonical correspondence analysis. Consequently, a CCA will not find any difference between the plots differing in total number of seedlings, but with constant proportions of individual species.

Table 15–1 Permutations within blocks. An example of three possible permutations of four cases within each of four experimental blocks.

Original order	Block number	Permutation 1	Permutation 2	Permutation 3	...
1	1	2	4	1	
2		4	3	4	
3		3	2	2	
4		1	1	3	
5	2	7	5	7	
6		8	8	6	
7		5	7	8	
8		6	6	5	
9	3	11	9	9	
10		9	12	12	
11		10	10	11	
12		12	11	10	
13	4	14	16	14	
14		15	13	15	
15		16	15	13	
16		13	14	16	

error rate; see Anderson & Ter Braak 2003) – this is a **model-based** permutation test (see Section 5.4). In cases like this one, where the low number of experimental units leads to a low number of blocks, a test based on the within-block permutations is usually weak.

Of the two suggested analyses, only the RDA on data not standardised by case norm yields significant results (P = 0.028 for the test on the first axis, P = 0.056 for the test on all constrained axes, using 499 unrestricted permutations in each test). The fact that the test of first axis is much stronger[9] suggests that there is a strong univariate trend in the data. Accordingly, the second constrained axis is not significant (see Section 5.3 for a description of how to test the significance of higher axes). Also the ordination diagram (biplot with species and explanatory variables, see Figure 15–3) confirms that the most different treatment is the removal of moss and litter, with the plots with this treatment separated on the first constrained axis from the others.

In the *Summary* page, you should notice the striking difference between the eigenvalues corresponding to the first and to the second ordination axes (0.288 and 0.031, respectively). This again suggests that the explanatory power of the second axis is very weak. With such a difference between eigenvalues, there is also a great difference between graphs produced with scaling focused on species correlation and on the inter-case distances[10] (click on the *Edit scaling options* icon in the toolbar to change the graph scaling – the graph must be then re-created). Because the centroids of treatments

---

[9] This is also reflected by the computed pseudo-F statistic values for the two tests: 4.7 for the test on first axis, 2.0 for the test on all constrained axes.

[10] See the introductory part of Chapter 11 and the Canoco 5 manual, section 5.6.2, for additional information about the score scaling.

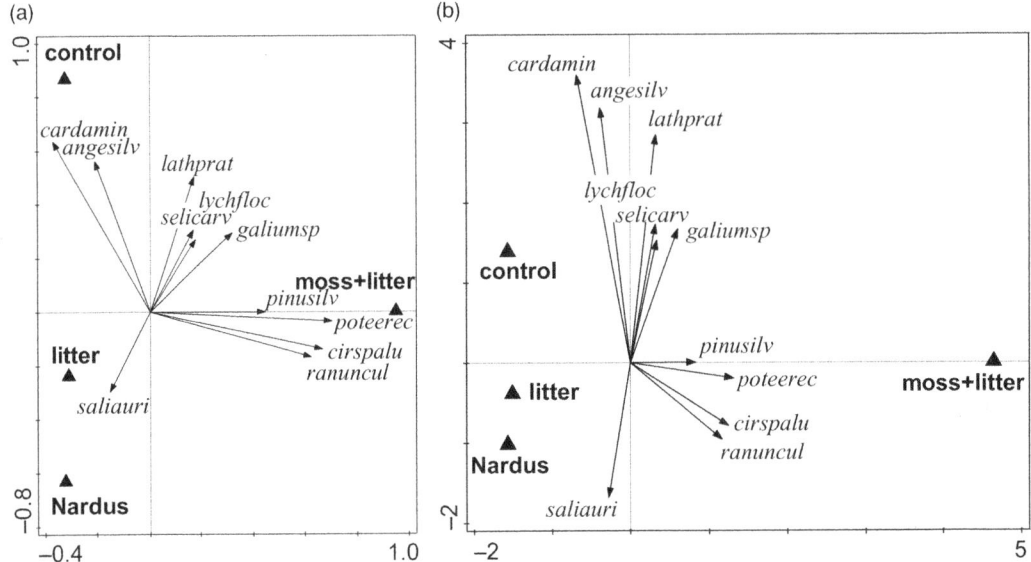

Figure 15-3 Species-explanatory variables biplot of RDA on non-standardised data. Seedling species are shown by arrows and the centroids of treatments by triangles. Both diagrams show the scores on the first two axes of identical analysis, but the left one was produced with score scaling focused on correlations among response variables and the right one with scaling focused on inter-case distances. The larger separation of the plots with *moss+litter* treatment is more obvious in the latter diagram.

are calculated as averages of individual case positions, the scaling focused on species correlation exaggerates the differences between treatments on the second axis, as you can see when comparing the two diagrams in Figure 15–3.

These diagrams are slightly different from what you will get from an analysis with default settings – in particular, the length of their species arrows is much larger. Because the length of an arrow has a meaning only in comparison with the length of other arrows, you can change the relative scaling of treatment centroids and species arrows to create a more readable graph. We recommend that you check the *Show scores rescaling dialog* option in the dialog shown by the *Edit | Settings | Graphing options* menu command, which enables changes in the relative scaling of the individual components of composite graphs.[11] Also, do not forget that you can change the number of species in the diagrams, as they can be selected on the basis of their fit.

If you need to make multiple comparisons,[12] you have to calculate separate analyses with pairs of treatments (omitting all the cases with the other treatments than the two

---

[11] This relative scaling of different components of an ordination diagram is not related to the scaling of the scores of the same type across different ordination axes that we have discussed in the preceding paragraph, although they interact in the resulting look of the diagram.

[12] I.e. testing which treatment pairs differ, as you have done in the univariate ANOVA.

compared; this change can be performed during the project setup), and then use some correction for multiple testing (e.g. false discovery rate, see Section 5.8).

The ordination diagram also shows that *Ranunculus* spp., *Potentilla erecta*, and *Cirsium palustre* are the species most responsible for the differentiation and all of them prefer the treatment with the removal of moss and litter. Although it seems that individual species prefer different treatments, the standardised RDA was not significant (P~0.5), so you are not able to confirm that there are differences between species in their seedlings' reaction to the treatments.

There is an important difference between ecological implications of significant treatment effect in the standardised-by-cases and the non-standardised analyses. The theory of regeneration niche (Grubb 1977) suggests that differential establishment of seedlings of various species in various microhabitats is a mechanism supporting species coexistence. The described experiment tested this suggestion. If there was differential establishment, then we could expect the proportions of species to be different in different treatments. Unfortunately, we were not able to demonstrate this. We were only able to show that mosses and litter suppress seedling recruitment, but this conclusion does not directly support Grubb's theory.

15.4 Calculating ANOVA using constrained ordination

At the beginning of your analysis, you have computed classical ANOVA for randomised complete blocks, using the total number of seedlings as the response variable. We will use this example to demonstrate how you can use the Canoco program to calculate classical univariate ANOVA with the Monte Carlo permutation test replacing the standard parametric test. One goal of this exercise is to demonstrate the equivalence of ANOVA (in this case an ANOVA for randomised complete blocks) with a corresponding RDA. But this approach has also practical applications. The non-parametric counterparts for more complicated ANOVA models are often not available. So the best solution for data that are not suitable for parametric ANOVA might be to calculate ANOVA, but instead of comparing calculated F values with the F distribution (which assumes normal distribution of the model residuals), you can use an appropriate Monte Carlo permutation test.

How to proceed? First, prepare a response variable table containing a single column – the number of seedlings in plots (the response variable in ANOVA). To do so in your existing project, select the *Data | Add new table(s) | Statistics of compositional table* menu command and in the dialog box uncheck all choices in the right-hand list except the *plot total*.[13] Then use the New Analysis Wizard to define a non-standardised RDA with the just created statistics table (*Data Statistics* in example project) selected as the focal table together with the *Design* table. You can then choose, in the last page of the

---

[13] Note that the word *plot* appears in the item name only if the term chosen for cases when creating the project was 'plot[s]', as in the example *Seedlings.c5p* project.

New Analysis Wizard, either the *Constrained-partial* or *Constrained* analysis template, depending on whether you want to take the blocks into account or not.

In the example project, the *ANOVA* analysis is based on the *Constrained-partial* template using *block* as covariate and *treatment* as explanatory variable. In the *Ordination Options* page, RDA is predefined and cannot be changed (because the table with response data has just one variable). On the following page, keep the *All constrained axes test* and *Unrestricted permutations* options, but increase the number of permutations to 4999. Then the test on all axes provides the correct test statistic (note that the F-ratio is exactly the same as in ANOVA, 4.2). If you have used the model-based permutations (i.e. freely exchangeable cases, without permutation blocks – as in the example *ANOVA* analysis), the significance estimate is P = 0.0376, and with the design based permutations (permutation within blocks), P = 0.0480, with both estimates based on 4999 permutations. These values are very similar to the P-value obtained from classical ANOVA, 0.0403.

Multivariate analyses corresponding to other ANOVA models (or other general linear models), including those with factor interactions, can be also carried out: you need to specify an appropriate combination of explanatory variables and covariates, with each model term tested using a separate analysis.[14]

---

[14] To fully represent the tests of main effects in the presence of interaction terms, not only the other main effects, but also the interaction terms must be used as covariates and the factors must be coded using the orthogonal (Helmert) contrasts, rather than using the standard dummy variables (called treatment contrasts in R).

16 Case study 5: Analysis of repeated observations of species composition from a factorial experiment

16.1 Introduction

Repeated observations of experimental units are frequently used in many areas of ecological research. A special case is the replicated BACI (before after control impact) design, in which the units (plots) are sampled first **before** the experimental treatment is imposed on some of them. In this way, you obtain 'baseline' data, i.e. the data where differences between the sampling units are caused solely by random variability.[1] After the treatment is imposed, the units are sampled once or several times to reveal the difference in the **development** (dynamics) of manipulated and control units.

To analyse a univariate response (e.g. number of species, or total biomass) in this design, you can usually apply the repeated measurements model of ANOVA. There are in fact two possibilities for analysing such data. You can use a split-plot ANOVA with time, i.e. the repeated measures factor, being the 'within plot' factor or you can analyse the data using MANOVA. Although the theoretical distinction between those two approaches is complicated, the first option (often called a 'univariate repeated measurements ANOVA') is usually adopted, because it provides a stronger test. But it also has stronger assumptions for its validity (see e.g. von Ende 1993; Lindsey 1993), which are not always fulfilled. The interaction between time and the treatment reflects the difference in the development of the units between treatments.

Canoco can analyse repeated observations of multivariate data (e.g. community composition) in a way equivalent to the univariate repeated measurements ANOVA. Whereas all the model terms (all main effects and interactions) are tested simultaneously in ANOVA, in Canoco you must run a separate analysis to test each of the model terms. We will illustrate the approach with an analysis of a factorial experiment applying fertilisation, mowing, and dominant removal to an oligotrophic wet meadow. The description of the experiment is simplified; a full description is in Lepš (1999).

16.2 Experimental design

In this experiment, we tested the response of an oligotrophic wet meadow plant community to various management regimes (mowing, fertilisation) and their combinations, and

---

[1] From the experimenter point of view, of course.

the effect of the dominant species, tested by its removal. We were interested in the temporal dynamics of species richness and species composition under various treatments, and also in which species traits are important for the species response.

The experiment was established using a factorial design with three replicates of each combination of treatments in 1994. The treatments were fertilisation, mowing, and removal of the dominant species (*Molinia caerulea*). This implies $2 \times 2 \times 2 = 8$ combinations in three replicates, yielding twenty-four 2 m × 2 m plots. The fertilisation treatment is an annual application of 65 g/m² of commercial NPK fertiliser. The mowing treatment is the annual scything of the quadrats in late June or early July. The cut biomass is removed after mowing. *Molinia caerulea* was manually removed (using a screwdriver) in April 1995 with a minimum of soil disturbance. New individuals of this species were removed annually.

Plots were sampled in the growing season (June or July) in four consecutive years, starting in 1994. The initial sampling was conducted before the first experimental manipulation in order to have baseline data for each plot. The cover of all vascular species and mosses was visually estimated in the central 1 m² of each 2 m × 2 m plot.

16.3 Data coding and use

Data are in the form of repeated measurements. For a univariate characteristic (the number of species) you will use the repeated-measurements ANOVA model (von Ende 1993). For species composition, you will use redundancy analysis (RDA): this method was chosen because the species composition in the plots is rather similar (see the end of Section 4.3). Because *Molinia* cover was manipulated, **this species was specified as a supplementary response variable in the analyses**. This is very important because otherwise we would show (with a high significance) that *Molinia* has a higher cover in the plots from which it was not removed. By using various combinations of explanatory variables and covariates in RDA, combined with appropriate permutation scheme in the Monte Carlo test, you will be able to construct tests analogous to the testing of significance of particular terms in ANOVA models (including repeated measurements).

Because the data form repeated observations that include the baseline (before treatment) measurements, the **interaction of treatment and time** is of the greatest interest and corresponds to the effect of experimental manipulation. When you test for the interaction, the plot identifier (coded as the *PLOT* factor in Canoco 5 project) is used as a covariate. In this way you subtract the average (obtained by averaging over years) of each plot,[2] and only the changes **within** each plot are analysed.

Values of time were 0, 1, 2, and 3 for the years 1994, 1995, 1996, and 1997, respectively. This corresponds to a model where the plots of various treatments do not differ in 1994 and the differences between treatments increase linearly with time.[3] We already know

---

[2] Separately for each species; the plot differences are also removed from the explanatory variables.
[3] This approach is analogous to using linear polynomial contrasts rather than the ordinary testing of effects in a repeated measurement ANOVA.

16.3 Data coding and use

Table 16-1 Values of interaction term (using bold style; before centring and standardisation) between Time and Treatment factors, when time values start with 0 (left part) or with the year value.

Time			0	1	2	3	1994	1995	1996	1997
Treatment	yes	**1**	**0**	**1**	**2**	**3**	**1994**	**1995**	**1996**	**1997**
	no	**0**	**0**	**0**	**0**	**0**	**0**	**0**	**0**	**0**

that the explanatory variables are centred and standardised before the analysis and consequently, it seemingly does not matter whether the values of time (as a quantitative variable) are 0, 1, 2, 3 or whether they are 1994, 1995, 1996, 1997. Whereas this is true if time is used directly (after centring and standardisation, you will get −1.162, −0.387, 0.387, and 1.162 in both cases), when you define interactions, they are calculated first (as the product of the interacting variables) and only then they are standardised. How different the interaction values can be is apparent from Table 16-1.

If time is coded as 0, 1, 2, 3, the plots have the same value of interaction at the start of the experiment (i.e. at the baseline), and then the differences linearly increase with time. On the contrary, if the baseline time is some high number (as in the right half of Table 16-1), then the values of interaction are very different in treated and non-treated plots, even at the baseline time. In summary, only coding time as 0, 1, 2, 3 correctly reflects the fact that the treatment and control plots are not different at the start.

The other possibility is to consider time as a categorical (nominal) variable and to code it as a factor (each year would be a separate category – factor level). In a 'classical' analysis using constrained ordination and diagrams, both approaches can be used (but only the analyses using time as a quantitative variable will be shown in this tutorial). Another method of visualising the results of repeated measurements analysis, the principal response curves (PRC), has been suggested relatively recently (Van den Brink & Ter Braak 1999). It is an extension of constrained ordinations and time is used as a categorical variable (see Section 10.1). In this chapter, we will demonstrate the univariate analysis, the classical constrained ordinations, and the PRC.

The original data are in Excel file *ohraz.xlsx*. The community composition data are in the sheet *Ohrazspe*, and the design of the experiment is in the *Ohrazenv* sheet. These two sheets are also imported into the example Canoco 5 project file, *Ohraz.c5p*, as the *OhrazSpecies* and *OhrazDesign* data tables. In the sheet *Ohraz-R*, there are numbers of species in individual relevés for the demonstration of univariate repeated-measurement ANOVA. In the *Ohrazspe* and *Ohrazenv* sheets, the cases are arranged in the following order: relevés from 1994 have numbers 1 to 24, relevés from 1995 are numbered 25 to 48, etc. The knowledge of this ordering will be important for the description of the permutation scheme. The names of cases are constructed in the following way: *y94p1* means a relevé recorded in 1994 at plot 1.

In the *Ohrazenv* sheet, the first three variables (*MOWING, FERTIL, REMOV*) describe which treatments were applied, using the following values: 1 for treated, 0 for non-treated. The next variable, *YEAR*, is the time from the start of experiment, i.e. time as a quantitative variable. The last column, *PLOT* determines the identity of individual plots.

After importing the worksheets into Canoco 5 project, remember to convert the three treatment variables and the plot identity into factors – click variable heading with the right mouse button and select *Convert... to factor* from the context menu. It will be useful for the ordination diagrams if you use short, self-explanatory names for individual factor levels (as *UNM* and *MOW* for unmown and mown plots). Short names will be also useful when you define the interaction with time, so that the short names of the interaction terms are understandable to the readers. This has been already done in the sample Canoco project *Ohraz.c5p*.

16.4 Univariate analyses

First you will use the R program to perform univariate analysis of the number of species in plots, using a split-plot (repeated-measurements) ANOVA with two error strata (among-plot and within-plot variation). Required data are in the worksheet *Ohraz-R*, the necessary command is shown next.[4]

```
summary(aov(NSP~mowing*fertil*remov*year+Error(plot),data=ohraz))
```

Table 16–2 summarises command output.

Altogether, 15 various tests were performed.[5] Of greatest interest here are the interactions of treatments with time, highlighted in Table 16–2, and they are all significant. Note that time (*year*) is treated here as a categorical variable (factor). The main treatment effects and their mutual interactions are also of interest, because sometimes they may provide a stronger test than the interaction with time.

16.5 Constrained ordinations

For most of the univariate tests shown in Table 16–2, it is possible to construct a corresponding Canoco analysis by a combination of explanatory variables and covariates. However the table, although devoted to the analysis of a single response variable, is rather long and not very lucid. One of the goals of multivariate analyses is to present the results in a comprehensible form simultaneously for many species. Consequently, we suggest simplifying the structure of the tests, and to test only selected interesting hypotheses. They correspond to the analyses presented in the Table 16–3 (example project *Ohraz.c5p* has all these analyses, using the *A1* to *A5* names). In addition, we use *year* as a quantitative variable so as to detect directional changes in time and simplify ordination diagrams.

---

[4] This command performs a split-plot ANOVA with the plots used as 'whole-plots' and the individual time points as split-plots.

[5] Interestingly, when you perform several t-tests without adjustment for multiple comparisons, many journal reviewers will complain, but when you perform 15 tests in a multiway ANOVA, the type I error rate is also given for each individual test, and typically nobody complains. (We realize that there **is** a difference between these two situations, but both provide an opportunity for 'statistical fishing'.)

16.5 Constrained ordinations

Table 16-2 Results of the univariate repeated measurements analysis of variance performed in R. Two-way interactions of treatments with time are emphasised using grey background.

Error: plot	DF	Sum Sq	Mean Sq	F value	Pr(>F)
mowing	1	65.0	65.0	1.59	0.22511
fertil	1	404.3	404.3	9.90	0.00624
remov	1	114.8	114.8	2.81	0.11296
mowing:fertil	1	0.3	0.3	0.01	0.93734
mowing:remov	1	213.0	213.0	5.22	0.03637
fertil:remov	1	75.3	75.3	1.84	0.19343
mowing:fertil:remov	1	6.5	6.5	0.16	0.69495
Residuals	*16*	*653.3*	*40.8*		

Error: Within	DF	Sum Sq	Mean Sq	F value	Pr(>F)
year	3	263.9	88.0	11.84	0.00001
mowing:year	3	226.6	75.5	10.17	0.00003
fertil:year	3	522.9	174.3	23.46	0.00000
remov:year	3	124.4	41.5	5.58	0.00229
mowing:fertil:year	3	44.0	14.7	1.98	0.13024
mowing:remov:year	3	34.4	11.5	1.55	0.21490
fertil:remov:year	3	7.7	2.6	0.35	0.79266
mowing:fertil:remov:year	3	10.6	3.5	0.48	0.70035
Residuals	*48*	*356.7*	*7.4*		

The analyses present in Table 16–3 test the following null hypotheses:

A1: There are no directional changes of the species composition in time that are common to all the treatments or specific for particular treatments. This hypothesis corresponds roughly to the test of all within-subject effects in a repeated measurements ANOVA.[6]

A2: The temporal trend in species composition is independent of the treatments, i.e. the individual treatments do not differ in their temporal dynamics.

A3, A4, A5: Fertilisation (or removal or mowing) has no effect on the temporal changes in the species composition. This corresponds to the tests of particular terms in a repeated measures ANOVA (the three highlighted rows in Table 16–2).

Note that when *PlotID* is used as a covariate (see Table 16–3), the main treatment effects (i.e. *M, F*, and *R*) cannot explain any additional variability and it is meaningless to use them, either as additional covariates or as additional explanatory variables.

In these analyses, all the interactions among the three factors representing the treatments are omitted for simplicity,[7] so that we, in fact, expect additive effects of mowing, fertilisation, and removal. During the analysis set-up in Canoco 5, you should not forget to tick the check-boxes *Select response variables ("species") to use* (and then choose

---

[6] And, more precisely, in Table 16–2 it corresponds to the highlighted terms tested together with the *year* term (but all with reduced degrees of freedom, because the *year* is treated quantitatively).

[7] But note that Canoco can test e.g. three-factor interactions by using all implied main effects and two-factor interaction terms as covariates.

Table 16-3 Results of the analyses using RDA applied to cover estimates of the species in 1 m² plots. Data are centred by species; no standardisation by case norm was done.

Analysis	Explanatory variables	Covariates	% expl. of total	% expl. of remainder	F ratio	P
A1	Yr, Yr*M, Yr*F, Yr*R	PlotID	12.9	24.0	5.4	0.002
A2	Yr*M, Yr*F, Yr*R	Yr, PlotID	3.2	10.8	2.8	0.002
A3	Yr*F	Yr, Yr*M, Yr*R, PlotID	2.6	6.1	4.4	0.002
A4	Yr*M	Yr, Yr*F, Yr*R, PlotID	1.5	3.5	2.5	0.002
A5	Yr*R	Yr, Yr*M, Yr*F, PlotID	0.8	2.0	1.4	0.084

% expl.of total – percentage of total community variability explained by all explanatory variables; this value can be found as *Canonical eigenvalues* (after multiplication by 100) in the dialog shown by the *Details* button on the *Summary* page;*% expl. of remainder* – percentage of variability that remained after partialling-out the effect of covariates; this value is shown (as the *account for* value) above the *Summary Table* in the *Summary* page. *F ratio*: the pseudo-F statistic for the test on all constrained axes (trace). *P*: corresponding significance value obtained by the Monte Carlo permutation test, using 499 random permutations. *Yr*: serial number of the year, *M*: mowing, *F*: fertilisation, *R*: *Molinia* removal, *PlotID*: identifier of each plot. The asterisk (*) between two variables indicates their interaction.

to ignore *Molinia* species), *Select explanatory variables*, *Select covariates*, and *Define interactions* of explanatory variables and/or covariates, where necessary.[8] Interactions between time and the individual treatment factors are then created in a separate wizard page, by selecting one in the upper left list, another in the upper right list, and clicking the *Create* button.

The null hypothesis for *A1* is a little more complicated and difficult to interpret ecologically. This analysis is useful for a comparison with the other analyses in terms of the explained variability. For the other analyses (*A2–A5*), under the null hypotheses, the dynamics of individual plots are independent of the treatments imposed. This means that if the null hypothesis is true, then the plots are interchangeable; but the records from the same permanent plot should be kept together. Technically speaking, the records done in different years in the same plot are subplots (*within-plots*) of the same main plot (*whole-plot*) and the main plots are permuted. To do this, you must use the options, described in the following two paragraphs, during the analysis setup in Canoco 5.

Testing and Stepwise selection page: Many people use *Both above tests performed*, but the procedure when one performs both tests and then uses the one which gives a better result **is not correct**. It may be expected that the test of the first ordination

---

[8] Note that these check-boxes are present at two pages of the Analysis Setup Wizard that are visible only after you switch off the QuickWizard mode (see Section 2.3).

16.5 Constrained ordinations

axis is stronger when there is a single dominant gradient; the test using all constrained axes is stronger when there are several independent gradients in the data. With a single explanatory variable, 'the first axis' and 'all the axes' are the same. The permutations should be set up as a *Hierarchical design*.

Then in the next setup wizard page named *Split-plot Arangement Permutation*, select the *Number of split-plots in each whole-plot* to be *4* (i.e. you have four records from each plot, taken in different years) and specify that the split-plots are selected by the rule *TAKE 1 SKIP NEXT 23*. This corresponds to the order of records in the data table: in this case, the records from the same plot are separated by 23 records from the other plots.[9] The whole-plots are freely exchangeable and 'No permutation' is recommended at the split-plot level.[10]

After running the analysis, it is advisable to check the permutation scheme in the *Log* page of the analysis notebook.[11] In your case, the output is:

```
*** Case arrangement in the permutation test ***
  Whole plot 1:
      1     25    49    73
  Whole plot 2:
      2     26    50    74
  Whole plot 3:
      3     27    51    75
etc...
```

This shows that the whole-plots are composed of the records from the same field plot, which is correct.

The relationship of particular species to experimental manipulations can be visualised by ordination diagrams. Probably the best possibility is to display the results of the analysis *A2* by a biplot with explanatory variables and plant species (Figure 16–1).

Because you have used year as a continuous variable, the interactions of time with the treatments are also continuous variables and are shown by the arrows. Because the time was used as a covariate, the trends should be seen as relative to the average trend in the community. For example, the concordant direction of a species arrow with the *FER*Yr* arrow means either that the species (e.g. *festprat,* i.e. *Festuca pratensis*) cover increases in the fertilised plots or that it decreases in the non-fertilised plots (or both).

Due to the balanced design and each factor having just two levels, the arrows for complementary treatment levels (e.g. FER*Yr and UNF*Yr) are just opposite and of the

---

[9] Take care: if you use a permutation within blocks – which is **not** your case here, then the number of plots to skip is the number of plots to skip **within a block**.

[10] According to the Canoco 5 manual (p. 356): 'Optionally the time points could be permuted as well, but this ... has a less secure basis than the permutation (randomization) of sites'. Permutation of time points uses the within-plot error. Permutation of (whole-)plots only makes less stringent assumptions and is more comparable with the MANOVA approach.

[11] The *Log* page is shown only for notebooks in the non-brief mode. To set this mode, choose the *Edit | Settings | Canoco5 Options* menu command, uncheck the *Show brief version of notebooks* ... box in the *General* page, and then *Hide* and again *Show* the notebook.

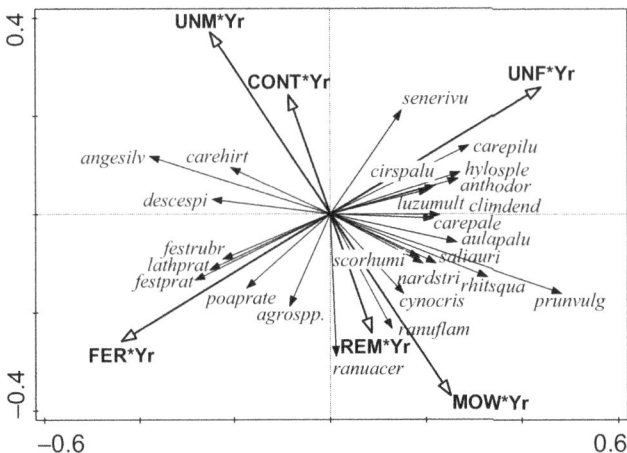

Figure 16-1 Results of the *A2* analysis (the first two RDA axes are plotted). The significance for the arrows of individual explanatory variables can be obtained from analyses *A3–A5* in Table 16–3.

same length. Saying this in the figure caption, you can further simplify the diagram by keeping just one of them (usually showing the positive one), yet we are not sure whether such a change really improves the clarity of the figure. The simplest way to do so would be to leave the three treatment variables as quantitative ones, with values 0 and 1 (as originally imported), and you will get just one arrow. Note here that when you define e.g. fertilisation as a factor with two levels (FER and UNF), two *dummy* variables are formed for Canoco calculations and named by the factor levels, each with values 0 and 1 (see Section 2.2).

The use of time as a quantitative variable is clearly a simplification, because in this way you look just for a linear trend of increasing differentiation. We nevertheless believe that it is a reasonable approach in the first four years of the experiment, when ongoing differentiation can be expected. Now we have available close to 20 years of data from this experiment and for such an extended data set, we cannot recommend the use of time as a quantitative variable, because we cannot expect the trend of increasing differentiation to continue after such a long time.

When you compare the explanatory power of individual treatments, the effect of fertilisation is most pronounced, the effect of mowing is weaker, and the effect of removal is the weakest (and only marginally significant with $P = 0.084$). It is interesting to note, however, that the corresponding test for the number of species provided a significant result (Table 15–1, $P = 0.0023$). This is strange, because it is clear that the number of species cannot change without a change in species composition. The change in species richness was likely caused by relatively rare species, that are also most sensitive to competition and prone to competitive exclusion, but they do not affect much the results of RDA without standardisation by species.

The fact that the variability explained by all the three factors is only slightly higher than the effect of fertilisation itself (see Table 16–3, analyses *A2* and *A3*) shows that fertilisation was the most important determinant of the vegetation composition development. Further, the amount of explained variability is considerably higher if you include time (i.e. the common temporal trend). This shows that there is some trend that is rather independent of the treatments applied and you will check it in Section 16.7.

16.6 Principal response curves

In previous analyses, time was considered a quantitative variable, and you looked for a linear trend. If you want to check the validity of this assumption and look at the temporal dynamics independently of this assumption, use the Principal Response Curves (PRC, see Section 10.1). This method provides an alternative presentation of the data analysed in this case study. The resulting response curves show you the extent and directions of development of grassland vegetation under different experimental treatments, compared with the control treatment. Additionally, we can interpret the directions of such compositional changes by the response of individual plant species using a simple 1-D plot, which can be well integrated with the PRC diagram.

The vertical scores of PRC curves (such as shown in Figure 16-2) are based on the scores of explanatory variables from a redundancy analysis (RDA), where the factor for sampling time is used as a covariate and the interaction between the treatment and sampling time is the explanatory variable (see Section 10.1).[12]

For the construction of PRC in Canoco 5, you need two explanatory variables coded as factors: one will represent the treatments and the other the time. For our example, you must first construct a single variable describing the treatments, i.e. each of the eight treatment combinations will be one level of the same factor. Technically, you will construct an interaction of the three imported treatment variables – Mowing, Fertilisation, Removal.[13] To have a full range of visualisation tools available for this analysis, you cannot create this interaction only in the analysis context (using setup wizard), but you must create it in the data table before you define the analysis.[14] You must also convert the *YEAR* variable into a factor. Or, preferably, copy it to a new *YEARFac* variable and then convert it, to have year available for further analyses both as a quantitative variable and as a factor, as we did in the example project.

The Canoco Adviser offers a template for PRC: in the New Analysis Wizard go to the *Advanced Constrained Analyses* folder and select one of the *principal-response-curves* templates (the resulting analysis is named *Principal-response-curves-first-set* in

---

[12] So this makes the analysis most similar to *A2* in Figure 16-3, except that the plot identifier is not used as a covariate and the main effect of treatment therefore stays in the model.
[13] Note that interaction variables in the Canoco sense include both the main effects of interacting factors and all the implied interaction terms.
[14] This interaction of three factors must be created in two steps. In the *OhrazDesign* table, first choose the *MOWING* and *FERTIL* factors (the first two columns), right-click one of them, and select the *Create factor interaction* command from the context menu. Then select *REMOV* factor and the newly created interaction variable and repeat the same command for them. The created variable is a single factor with eight levels, representing the eight possible treatment combinations. If you do not see the whole names of the variables originating from interactions, ask for full labels.

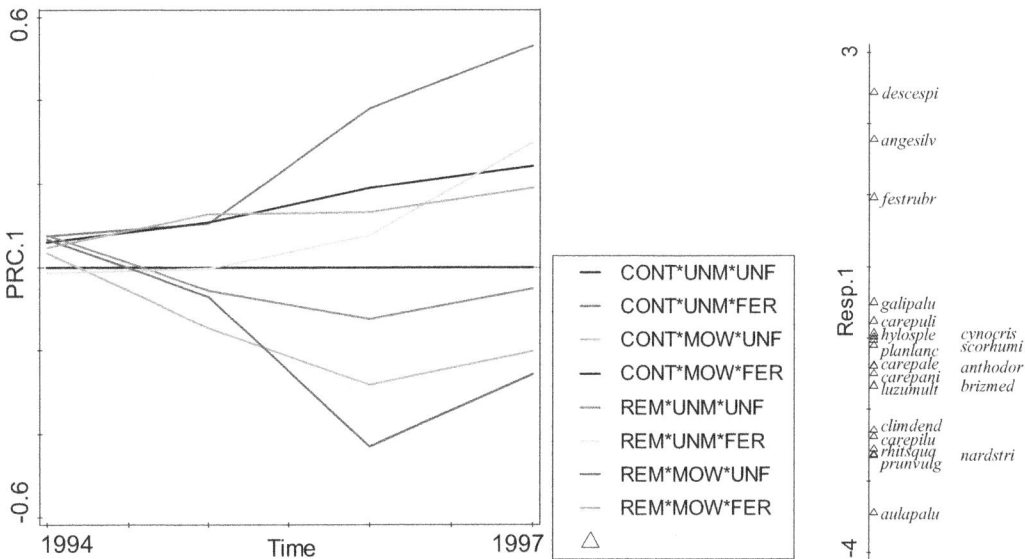

Figure 16-2 PRC diagram, the first attempt.

the example project). The first axis PRC will be demonstrated in this tutorial (the other template – *principal-response-curves-N-sets* – can be used to compute and test additional PRC axes).

In the setup wizard for the PRC template, you just select the treatment variable (the three-way interaction – you will not see its name correctly unless you ask for full labels) and temporal factor (*YEARFac* in the example project). Check that the time levels are in correct order, and you can adjust level names to reflect the real years, when the observations have been made. The reference level of the treatment factor is used to define the reference (horizontal) line in the PRC diagram.[15] All the other options will be suggested automatically by the Canoco Adviser, but as you will need to go through all the relevant pages, you can check which options have been chosen. We recommend you to keep suggested options unchanged, except you ask for the log-transformation of the cover values (this might improve quantitative interpretations of PRC diagram, see Section 10.1).

After you close the Analysis Setup Wizard at its *Finish* page, the PRC analysis is executed and the Graph Wizard offers creation of the PRC diagram. On the second page of the Graph Wizard, you should (at least for our example data set) decrease the number of plotted response variable scores (use e.g. 20 best-fitting species). After you close the Graph Wizard, you can check the PRC diagram with the species scores on its right side,

---

[15] To check or change the identity of the reference level for a factor, right-click the factor variable heading in the Canoco 5 data table and select the *Modify levels of...* command from the pop-up menu. The current reference level is shown with cyan background and if you choose a different level, the *Set Base Level* button becomes enabled.

stored in the analysis notebook and illustrated in Figure 16–2 (and stored in the support data for this chapter as a separate Canoco 5 graph file – *PRC1.c5g*).[16]

This is a fine graph when you do not need black-and-white presentation as we have for this book. The easiest way to adopt settings distinguishing the principal response curves for different treatments without colour is as follows. (1) select the *Edit | Settings | Visual Attributes* menu command; (2) in the *Visual Attributes Manager* window select the *Black & white* choice – but not in the upper *Defined sets* list, but lower in the control labelled as *Active attribute set*; (3) re-create the diagram from the toolbar or using the *Graph | Recreate graph* menu command.[17]

After re-creating the PRC diagram, you will note that the vertical plot with species scores at the right side became almost unreadable, because the scores are shown for all species present in the data set. To change this, you must specify the selection of best-fitting species (your earlier choice applied only to the Graph Wizard work). After you choose the *Analysis | Plot creation options* command, use in the *Species Selection* (the first word varies depending on the term you have chosen for response variables) page the control labelled *On horizontal axis: / at least*. This is a rather paradoxical choice, given the scores are plotted in the vertical direction, but this choice refers implicitly to the first ordination axis, which is the one plotted in the PRC diagram. You cannot set the number of species directly here, but the text field in the lower part of the page informs you about the number of passing species, as you modify the percentage value of fit for individual response variables. To display 20 species as before, the value must be set between 7.1 and 7.4% and after closing the dialog box with *OK*, the graph must be again re-created. The resulting graph is shown in Figure 16–3 and also stored as *PRC2.c5g* file in the example data.

The PRC diagram shows that there are two directions of departure from the vegetation composition on the reference plots (which are not mown, no fertiliser is added to them, and *Molinia caerulea* is not removed from them). The mown plots (with negative PRC scores) have higher abundance of *Nardus stricta*, of several small sedges (*Carex pallescens*, *Carex panicea*), and also of many forbs. Also the mosses (like *Aulacomnium palustre* or *Rhytidiadelphus squarrosus*) have a much higher cover. The extent of the mowing effect is somewhat smaller than the oppositely oriented effect of fertilisation, shown by the lines directed to the upper (positive) side of the vertical axis. Fertilised plots became dominated by competitive grass species (*Deschampsia cespitosa*, *Festuca rubra*, but others as well) with only a few forbs (e.g. the tall forb *Angelica silvestris*) increasing their cover. Finally, there is only a limited effect of dominant grass removal, and it is similar to the effect of mowing.

---

[16] The actual look depends to some extent on your current Canoco 5 settings and we have also adjusted the positions of the species labels at the right side of the diagram. This can be done either by dragging labels with the left mouse button or by selecting them and using the arrow keys on your keyboard.

[17] You can further fine-tune the line type and thickness using the Attribute Editor (shown with a toolbar button or using a command in the *Graph* menu). Note however that this modification is not reflected in the displayed legend, so you must change each line type also there. To be able to do so, you must first select the *Unlock* command from the *Graph | Legend* submenu and then select the legend rectangle, right-click it again and select *Lock selected*.

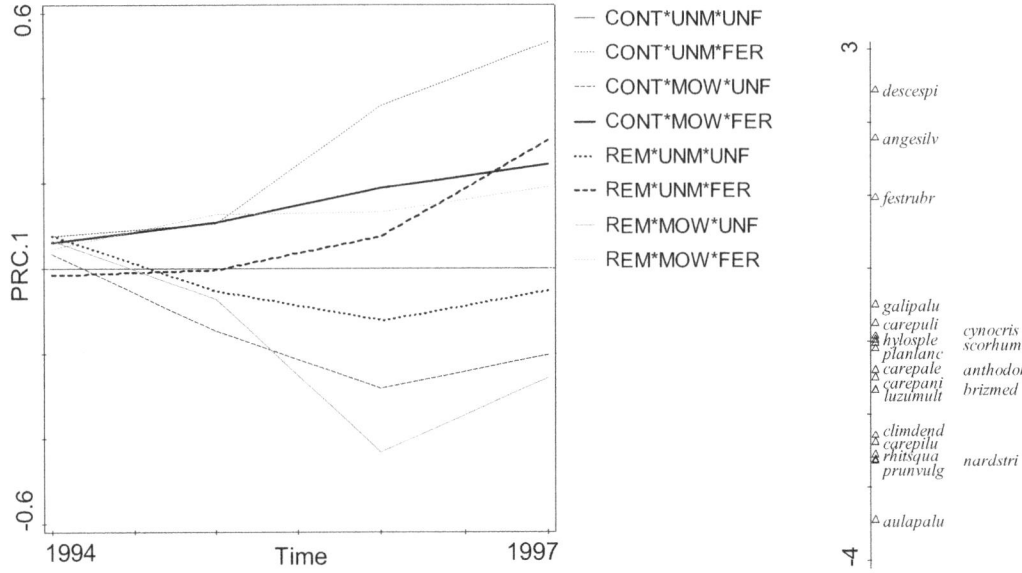

Figure 16-3 Final diagram with principal response curves.

The extent of increase in cover of grasses like *Festuca rubra* (*festrubr*) in fertilised plots can be quantified using the rules described in section 6.3.4.2 of the Canoco 5 manual. The score of that grass species is roughly +1.0, as seen in the vertical subplot of Figure 16-3. If you look up the PRC score of the fertilised-only plots (*CONT*UNM*FER* in the diagram) in the year 1997, you can see that it is approximately +0.5. The estimated change is, therefore, $\exp(1.0 \cdot 0.5) = 1.65$, so the grass *Festuca rubra* is predicted to have, on average, 65% higher cover in the fertilised-only plots compared with the control plots.

The variability of PRC scores in the year 1994, visible at the left side of the diagram in Figure 16-3, cannot be caused by the experimental treatments (which were started afterwards), so it provides a 'yardstick' to measure the background variability among the plots. This suggests (because the second PRC axis is not significant here) that there is probably no real difference among the three treatments where fertilisation was combined with mowing and/or with the removal of *Molinia*. Likewise, the mown plots did not differ from those where the removal treatment was added to mowing.

The unmown plots may not be the best choice for control treatment. If you want to use the **mown** but unfertilised plots without *Molinia* removal as the reference level, the only adjustment you must make in the Canoco project is to change the reference (base) level for the *REMOV*MOWING*FERTIL* factor (in the way described earlier in this section, in footnote 15). You must then click the *Modify* button below the list of analyses (choosing probably the *Replace in existing analysis* option), go through the Analysis Setup Wizard pages (without any change) and then re-run the analysis (using the *Re-analyze* button). You will then create the PRC diagram as described above. The

16.6 Principal response curves

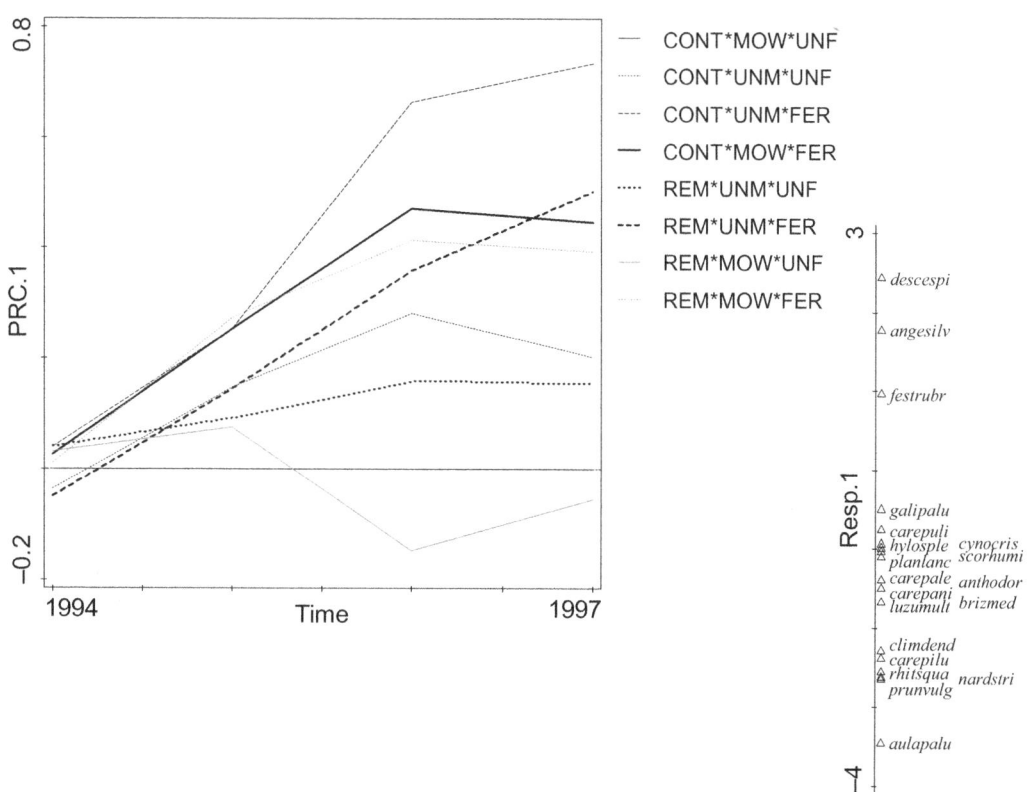

Figure 16-4 PRC diagram for the first RDA axis. The mowing-only regime was taken as the reference (control) treatment.

resulting graph is illustrated in Figure 16-4 and also stored as a standalone graph file *PRC3.c5g*.

If you compare the new PRC diagram with the original one (in Figure 16-3), the only difference is that the identity of the reference line, which is 'rectified' (flattened) along the horizontal axis, has changed. The scores of species on the first RDA axis (the right side of the plot) have not changed at all and neither have the statistical summaries of the analysis, and this gives an assurance that the interpretation of the PRC curves has not changed, either. But as the horizontal zero line of any graph represents a natural reference for readers, we believe that the graph in Figure 16-3 (with all fertilized treatments above the reference line and other ones below it) is easier to read and understand than the graph in Figure 16-4.

Finally, you can compare the information you obtained by summarising the standard constrained ordination (as presented in Section 16.5) using an ordination diagram (e.g. the biplot diagram in Figure 16-1) with that provided by the PRC diagram (e.g. Figure 16-3). We should first note that the analyses and corresponding diagrams are not the only possible ones and you should probably run more of them when evaluating your research data. For example, the constrained analyses with time coded as a categorical variable could also be graphed, and a partial PRC analysis can be constructed for each of the factors separately, using the other ones as covariates. Also, you could compare the species scores on the first principal response axis with species traits (and you would find that species with positive scores, i.e. those most positively affected by fertilisation, are the tall species – see also Section 16.8).

However, in a research paper, only the most instructive diagrams can be included. The PRC diagram is superior in its display of temporal trends. In our example, the PRC diagram clearly shows that the development of plots diverges according to the treatment during all the years, particularly for the fertilised plots (and in this way confirms that the use of time as a quantitative variable is a good approximation). Now, we have data for a much longer time (Lepš, submitted) and it is clear that the fast (nearly linear) divergence finished after five years (and so the use of time as a quantitative variable would not be a good approximation for a longer time series), but we observe a continuous shift in species composition in all treatment combinations and continuous loss of species in fertilised plots.

On the other hand, the classical diagram is better at showing the affinities of individual species to the treatments (which one is more responsive to mowing, and which to fertilisation), and also the mutual relationships of the effects (e.g. the similarity of effects of mowing and dominant removal). So the information presented in the two alternative diagrams is partially overlapping, and partially complementary.

Based on the combined information from all the analyses, you can conclude that many species (particularly the small ones) are suppressed by either the fertilisation or by the absence of mowing or by a combination of both factors. Only a limited number of (mostly tall) species is favoured by the fertilisation. This suggestion is further explored in Section 16.8.

16.7 Temporal changes across treatments

The analyses presented in the preceding two sections focused on the differences in temporal development of community composition among the experimental treatments, removing any eventual temporal change shared by all plots (*YEAR* or – in PRC analyses – *YEARFac* variable was used as a covariate, except in *A1*). This is a very natural approach, as this study is primarily concerned with the experimental manipulation and the changes shared by all the plots are of less concern, representing perhaps some wider-scale changes in the system or even undesirable artefacts of the experiment.

But it might still be interesting to see the whole picture of changes in the permanent plot composition and address questions like 'Is there any overall development of the vegetation?' or 'How much were the plots with different treatments different at the start of the experiment?'. In this section, you will start with a test of overall temporal change and then focus on visualising the changes in an enhanced ordination diagram, in which the pathways of temporal development can be compared across the treatments.

16.7 Temporal changes across treatments

Before testing for an overall change in plots, we must first define our task more precisely. The majority of plots were modified by the imposed treatments and you have already demonstrated (in analyses performed in the preceding two sections) that the treatments really affected the community and that the magnitude of these effects increased through time. You can therefore reasonably expect (before performing the test) that a straightforward test of time effect, by permuting the four records within each plot, will yield a significant result, as it subsumes the impact of treatments. So, you need to focus here on the changes that cannot be interpreted as the effect of treatments, already ascertained earlier.

The most obvious approach is to focus on temporal changes within control plots, but with this you are back to the problem of what is the appropriate control here (i.e. whether mown or unmown plots). For simplicity, we suggest that you use the unmown–unfertilised plots without *Molinia* removal as the reference (the area was not mown for some time, before the experiment started) and work with the corresponding three permanent plots, each sampled over four years.

Before creating the analysis, it will be useful to define a group of cases representing these three plots, using the *Project | Groups | of relevés*[18] menu command. In the dialog box click the *By Rule* button and then modify the settings as illustrated in Figure 16–5. This uses the *REMOV*MOWING*FERTIL* variable, which you have created in the preceding section before performing PRC, and selects the cases with the particular combination of the treatments (the control plots).

After group definition, close both dialogs, switch off the QuickWizard mode (if set), and create a new analysis with the New Analysis Wizard, using the *Constrained-partial* template in the *Advanced Constrained Analyses* folder. Make sure that in the setup wizard you check the *Select cases ("relevés") to use* option at the second page[19] and in the *Selection of cases* page click the *From group* button in the lower right corner and after selecting the group you have created earlier, click *OK*. This selects the cases that are **not** members of the group, so that you can move them to the left-hand list of ignored relevés (using the << button).

In the following *Definition of Groups* page use *YEAR* as the only explanatory variable and *PLOT* as a covariate.[20] Keep the suggested choices (identical with those used in *A1* to *A5* analyses) in the *Ordination Options* page, as well as those in the *Testing and Stepwise selection* page (with the *Hierarchical design* choice set). But changes are required at the next page, called *Split-plot Arrangement Permutation*. The correct choices are illustrated in Figure 16–6.

---

[18] Note that the 'of relevés' part of the command name varies depending on the term you have chosen for your cases.
[19] No need to omit *Molinia caerulea* this time, as you will retain only the control plots.
[20] Note that given the design nature and consequently the permutation options chosen as described below, you must use the numerically coded year descriptor here, to test for a directional change. Using the factor version (*YEARFac* in the example project), each permutation would just 're-label' the years in each permanent plot and the tested within-plot variation among years would therefore remain identical in the permuted data, yielding identical pseudo-F statistic values and the *P* equal to 1.0.

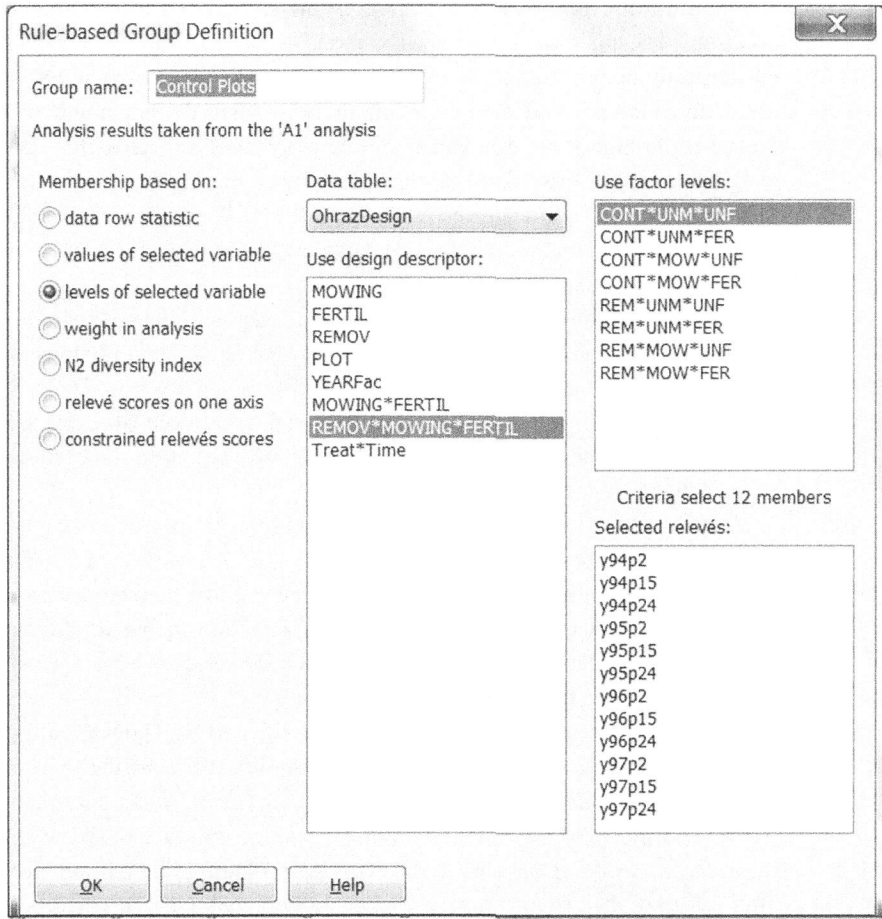

Figure 16-5 Dialog with a rule-based selection of cases representing particular combination of treatments.

First, you should take into account the fact that the *TAKE – SKIP* option works only with the 12 retained relevés, so 'TAKE 1 SKIP NEXT 23' is no longer correct, as we have only three plots left for each year. So this must be changed to 'TAKE 1 SKIP NEXT 2'. Second, you test the main effect of time within permanent plots, so the whole-plots are not permuted here and the split-plots (four repeated measurements on each permanent plot) are permuted at random. Finally, the permuted order of years must be applied in parallel for all three permanent plots, because the identity of four years has the same meaning across all of them (see Section 5.5 for additional details). This is specified by the choice of *Dependent across whole-plot* option in the *Split-plot Permutations* area.

After you close the setup wizard and execute the analysis (named *Temporal-change-on-controls* in the example project), you will see in its *Summary* page that the temporal change within the control plots is non-significant ($P = 0.172$). The pseudo-F statistic

16.7 Temporal changes across treatments

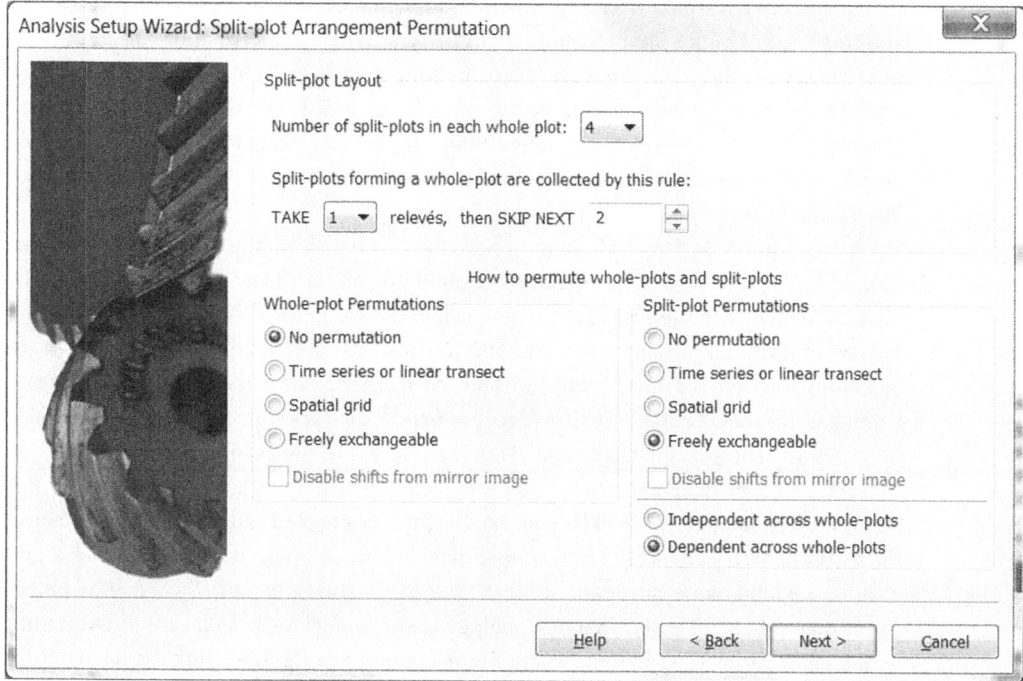

Figure 16-6 Correct choices for split-plot design-based permutation test when testing for time effect with only three permanent plots.

value (representing effect size) is quite high (3.2), suggesting that our inability to reject the null hypothesis is due to the low power of the test, which is performed with a small number of plots, rather than due to non-existent change. In fact, the minimum significance you can achieve with this data subset is about 0.083, given that you are permuting four cases (split-plots) in the same way for all three permanent plots (whole-plots).[21]

Despite this non-significant outcome, it is perhaps instructive to create a biplot with the one explanatory variable and the plant species best fitted by the *YEAR*.[22] The resulting graph (not reproduced here) suggests that the most pronounced changes were the decrease of *Nardus stricta* cover and increase of species like *Anthoxanthum odoratum*, *Festuca ovina* or *Festuca pratensis*.

You will now visualise the temporal changes in a way that will enable you to interpret both any shared trends and possible differences among plots with different treatments. For this, you will create a constrained ordination (RDA) using the interaction of experimental treatment with time as the explanatory variable. Because you will use neither time nor

---

[21] This means that the number of distinct permutation sequences is the factorial of 4 (i.e. 24), but you are constraining the ordination axis by a (linear) effect of time, so half of the permutations of the four split-plots yields the same constrained axis (with the same pseudo-F value) as they mirrored counterparts, only with an opposite sign. Hence, if all split-plot arrangements different from the original one (1, 2, 3, 4; or its mirror) produce smaller pseudo-F, you still get a P = 1/12 (if all arrangements are generated with exactly the same frequency, which cannot be guaranteed for a Monte Carlo permutation test).

[22] Only the fit on the horizontal (first) axis should be considered here, see Section 12.2 for details.

plot identity as covariates here, this 'interaction' will include also the main effects of time and treatment. Unlike the preceding analysis, you will use the categorical description of year here, so that each combination of treatment and year will be plotted as a single centroid in the diagram. By connecting the four centroids corresponding to the same treatment combination by line segments, you will plot a temporal trajectory of vegetation change, representing averages of the three plots available for each such combination (see Figure 16–8 for the final result).

First you must start with defining a new factor variable in the data table containing design descriptors (*OhrazDesign* in the example project). Bring forward the table (by clicking its name in the *Data tables* list), select the variables *YEARFac* (created earlier to represent time in categorical form) and *REMOV*MOWING*FERTIL* (created in the preceding section for PRC analysis) and then choose the *Create factor interaction* command from the context menu. We recommend that you rename the resulting variable to something shorter, such as 'Treat*Time', as we did in the example project.

Next, you can create a series collection effectively describing which centroids for this newly created factor (with 8*4 levels) shall be inter-connected and in which order. Select the *Project | Series | of design descriptors* menu command (the 'design descriptors' part of the command name can again differ in your project) and then the *By Selection* button in the dialog box. This shows a new dialog – *Specify Series Collection*. In it, you must define eight series (one by one, using the *Add* button in the lower left corner) and for each select the four levels of the factor *Treat*Time* (this name precedes the level names in the rightmost list) that start with a different year, but have the same combination of treatments.[23] In our example project, we have named the series using single-letter acronyms for each specific treatment that was applied, so e.g. the 'RF' name refers to the plots without mowing, but with the removal of *Molinia* and addition of fertiliser. Figure 16–7 illustrates the look of the dialog box in the middle of the job, with four series defined.[24] After you define all eight series, close the *Specify Series Collection* dialog and in the *Series Collection Manager* check the *Use selected collection in graphs* box before closing the dialog with the *OK* button.

Finally, define the analysis (named *Temporal-changes-by-treatments* in the example project), which is a simple constrained ordination (RDA) with the newly created *Treat*Time* factor as its only explanatory variable. The permutation test for this analysis would attempt to handle at once the main effects of time and treatments, as well as their interaction. But with different positions of time and treatment in the model hierarchy, the permutations cannot be properly set up and so you should not perform the test or at least ignore its results. But remember to remove the *Molinia* species from the analysis. After executing the analysis, ignore Graph Wizard suggestions (using *Cancel* button) and create a graph using the *Graph | Scatterplots | Design descriptors* menu command. To

---

[23] The dialog boxes in Canoco 5 can be enlarged by dragging their corners or edges and this will be particularly useful here.
[24] As it happens, the four treatment–year combinations in each series are already in the correct order, but note that you might need to re-order them in other data sets and if so, you can use the small up and down buttons at the left side of the central list.

16.8 Changes in composition of functional traits

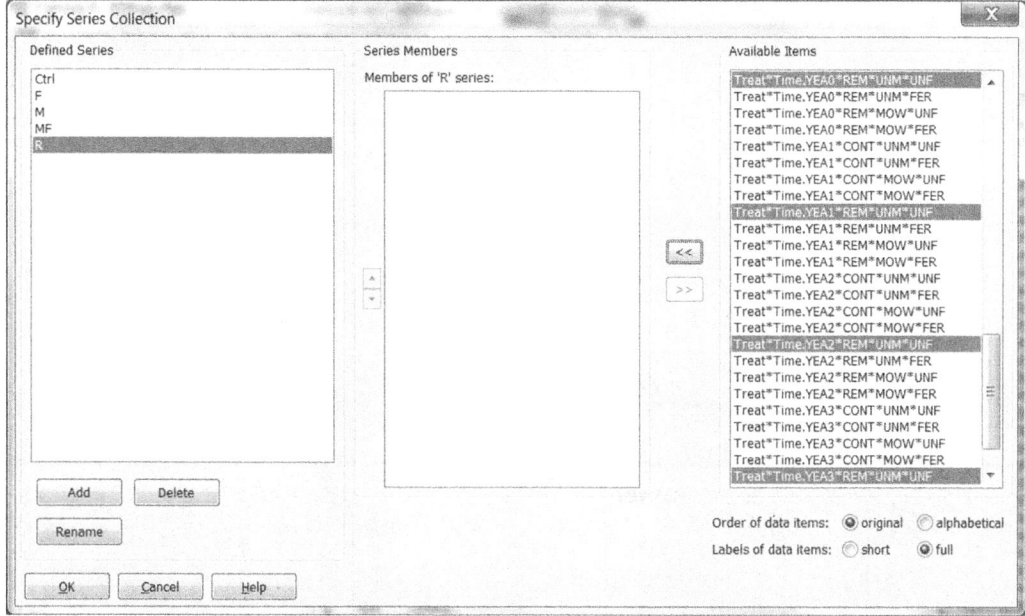

Figure 16-7 Specify Series Collection dialog at the stage of four series defined and the four cases belonging to the fifth series (R) selected for their inclusion.

produce the graph illustrated in Figure 16–8, we have further adjusted it by (a) choosing a black and white set of visual attributes (*Edit | Settings | Visual attributes*); (b) removing the triangles representing the position of centroids (and duplicating the positions implied by the series lines); and (c) removing centroid labels except the last one for each series, with its description simplified to reflect the combination of treatments and final year.

The diagram nicely summarises the general trend of community development (seen to a limited extent also for the control plots) from the left to the right side of the diagram.[25] Fertilised plots moved further than those without fertilisation and the mown and non-fertilised plots followed a somewhat different trajectory.

16.8 Changes in composition of functional traits

The response of species composition to management is often interesting, but results are rather specific to the investigated site or, at best, to the investigated vegetation type. The response in terms of functional traits might, on the contrary, provide more general results and also, because the functional traits have known ecological function (as, e.g. Westoby's 1998 LHS system), they might be more useful for the interpretation

---

[25] The example project also offers a diagram with plant species, selecting those with a high fit on the first axis, which illustrates the species most correlated with this dominant gradient of shared change in time.

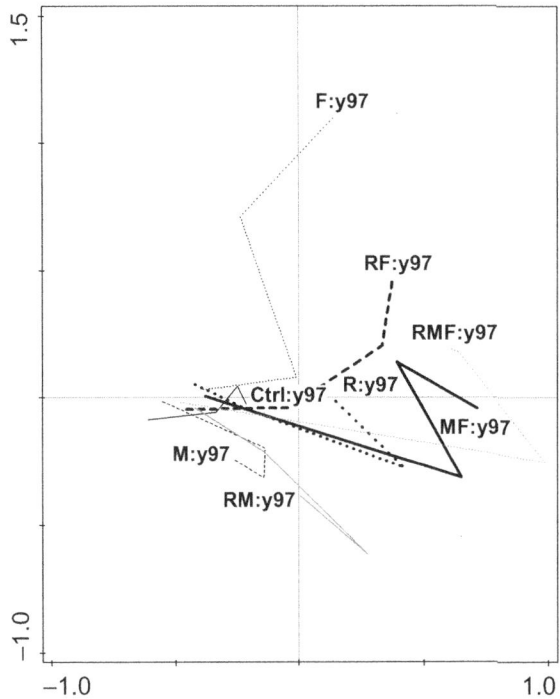

Figure 16–8 Change of average composition in time for individual treatment combinations, based on the first two axes of RDA. The position of the last year for each treatment combination is labelled.

of ecological mechanisms (Chapter 9). Here we show some of the possible analyses using the species traits.

The data are stored in the *OhrazWithTraits.xlsx* file, but you can also start from the example project (*OhrazTraits.c5p*), which already includes the analyses discussed below. We have added a species x trait matrix and to avoid some uncertainties with the traits, we have slightly simplified the species composition data. For this tutorial, we have also decided to use the species traits just as fixed values (i.e. the values of each species are fixed and do not change according to treatments – see Section 9.1). Analyses reflecting both the between- and within-species variability are presented by Lepš et al. (2011). For the purposes of this case study, we have collated measured values (averaged over treatments) with published data and in some cases, amended the data with 'best professional guesses'.[26] In this way, we have obtained a data table typical for trait data, i.e. with values for some traits available for all the species, but also quite a few traits with many missing values.

The following traits were used. Factors: *functional group* defined as grasses, sedges (including rushes), and forbs (includes every species which is not a grass or sedge), *life*

---

[26] Such a mixture of measured and published traits should be used with great caution and should be preceded by some checks of consistency of individual sources.

form (according to Raunkier, i.e. hemicryptophytes, geophytes, chamaephytes, therophytes), *presence of rosettes* (rosettes, semirosettes, erosulate), and *life cycle* (monocarpic, polycarpic). Quantitative variables: *seed weight* (in μg), *Onset of flowering* (as a Julian day), *foliar C:N*, *foliar P*, *specific leaf area* (SLA, mm^2/mg), *leaf dry matter content* (LDMC, dry weight/fresh weight * 1000), and *plant potential flowering height* (in cm). If a species can potentially belong to more than one category of a factor, we have selected the most likely category at our site. Alternatively, you can code each factor as multiple dummy variables representing individual categories and use the fuzzy coding (see Section 2.2).

16.8.1 Response of community weighted means of traits to treatments

In the first analysis, you will answer the question of how the community trait composition (characterised by community weighted means of individual traits, called CWMs below) changes according to the treatments applied. In the first step, you will calculate the CWMs of all the traits (for factors, each level has a separate CWM variable). Use the *Data | Add new table(s) | Trait averages* command. This first shows the dialog box illustrated with the required settings in Figure 16–9.

In our example project, we have decided to use the cover values as weights without any transformation (so we unchecked the *Use the implicit transformation Log(x+1)* option). Consequently, the CWMs are affected mainly by dominant species. Further, we have omitted the manipulated species, *Molinia caerulea* (using the *Subset* button in the central part of the dialog). Both these decisions affect the analysis, but in both cases, alternative choices can be justified. If you decide to log-transform the cover values, then the CWMs will be more affected by the subordinate species and by comparing the analyses based on transformed/non-transformed cover values, you can see in which case the CWMs are more affected by the treatments and hence whether the dominant or subordinate species are more sensitive.[27]

By omitting *Molinia,* we consider only the remaining part of the community in non-removal treatments (similarly as in the species data analyses in Section 16.5). This should be done, because the removal was performed to demonstrate the competitive effect of *Molinia*; having *Molinia* included in the analyses, you would probably find a large effect of the removal treatment, manifested by decreasing proportion of grasses in removal plots, immediately in the first year after the removal. But if you are interested in further development of the community after many years, you can also analyse CWMs based on all the species including *Molinia.* This will be particularly useful if you ask whether the plots with *Molinia* removal developed similar functional structure as the original plots.

---

[27] You can even choose a binarising transformation for the *Species* data table (turning its content into species presences and absences) before computing CWMs and due to such a choice the trait averages computed here would be no longer weighted ones, but rather plain averages of the trait values for all present species. Further, the transformation of trait values is also often appropriate.

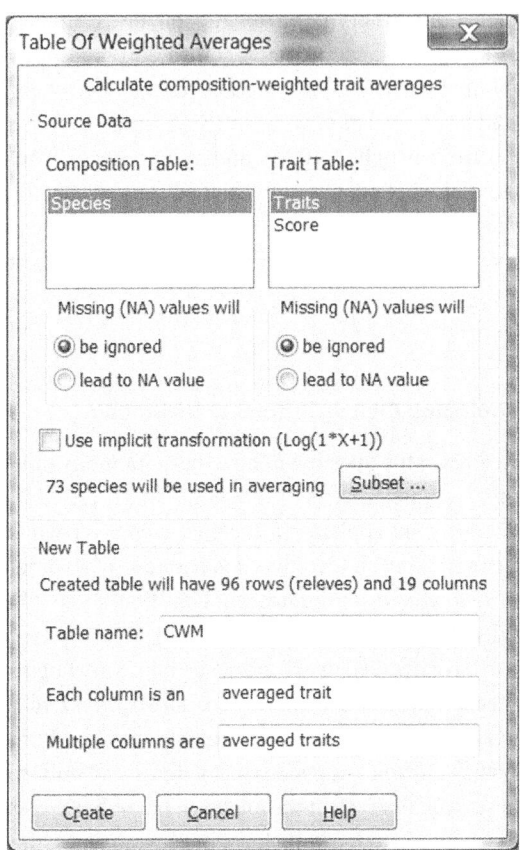

Figure 16-9 Dialog box for creating the data table with trait CWMs.

The CWM data table can be analysed in a similar way as the original community composition data (instead of individual species, you have individual species traits). Because the CWM data table is not compositional (the traits are measured in different units), you are restricted to linear methods, and you should always use the option *Center and standardize* (by averaged traits). With these restrictions, the CWM table can be now analysed using the same combinations of explanatory variables and covariates as the *Species* table; we will demonstrate the RDA for repeated measures (corresponding to analysis *A2* on species composition data and shown as the *RDA-traits-on-treatments* analysis in the *OhrazTraits.c5p* project), but alternatively PRC can be computed for CWM table as well.[28]

For the repeated measures RDA, you must use again the interactions of all three treatments (*MOWING, FERTIL, REMOV*) with time (*Yr*) as explanatory variables, and plot identity (*PLOT*) and year (*Yr*) as covariates, and choose the permutation type

---

[28] You cannot, however, interpret the resulting PRC diagram using the supplementary vertical plot in the quantitative way which was demonstrated for the PRC diagram based on compositional data in the preceding section.

16.8 Changes in composition of functional traits

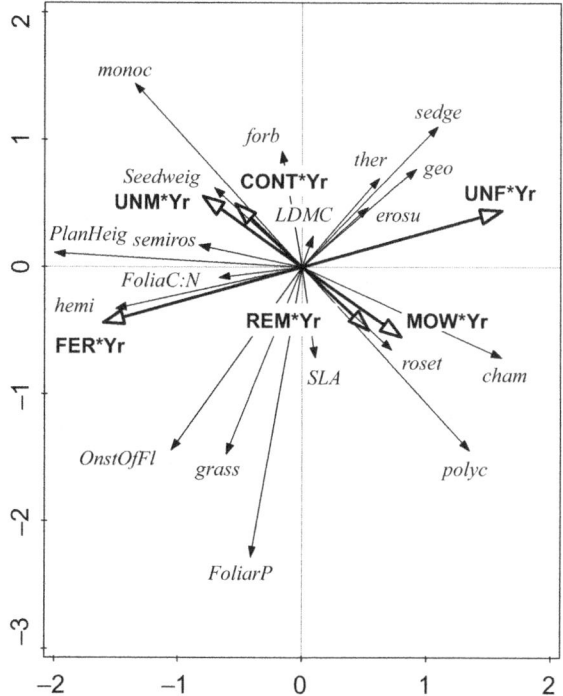

Figure 16–10 RDA for repeated measurement on the trait CWM data (the first two axes are plotted).

appropriate for the hierarchical design (exactly as in the *A2* analysis). The permutation test is highly significant (P = 0.004) and the results show that the CWMs of traits diverged in time according to the treatments applied (Figure 16–10, reproducing *Graph 1* in the analysis notebook). You should be aware, however, that the significant outcome means that the community weighted means changed – this change might happen simply by a prevalence of a single species (the log-transformation, if used, would partially downplay such an effect), and you should be aware that the traits that are responsive in this way need not always be the functional ones (e.g. these might be traits that are correlated with another trait, which is the actual cause of the species response).

The diagram in Figure 16–10 shows that the plant height CWM increases with fertilisation and decreases with mowing,[29] that both fertilisation and mowing support the increase of average foliar P concentration, etc. Now you can perform analyses analogous to *A3–A5* to see, which of the three factors has significant effects on trait CWMs (FER*Yr: P = 0.002, MOW*Yr: P = 0.106, REM*Yr: P = 0.356, illustrated, respectively, in the analyses *RDA-traits-on-treatments-fert*, *RDA-traits-on-treatments–mow*, and *RDA-traits-on-treatments–rem*). So, in fact, only the fertilisation affected the trait

---

[29] Remember that we have used fixed values for all traits, e.g. the potential height which a species can reach, obtained from published sources.

CWMs in a statistically significant way – the results are fairly less significant than those obtained for species composition.

This is probably because we have included some traits that cannot be expected to respond to our treatments, particularly in a short-term experiment. For example, you can hardly expect that the seed weight will have an effect on species performance in a community of mostly clonally spreading plants (which is our case), at least in the relatively short time span of four years. But you at least have a take-home message that the inclusion of all the traits that can be downloaded from databases will likely decrease the power of your test when you include all of them irrespectively of your research question.

Next, it might be useful to check, which CWM values (i.e. which traits) respond significantly to the treatments. For this, you should return to the first analysis (*RDA-traits-on-treatments*). You will create a t-value biplot: select the *Graphs | Biplots | t-value biplot* menu command and choose the FERTIL.FER*Yr interaction in the dialog box. In a t-value biplot, the length of the arrows is inversely related to the strength of the relationships: the longer the arrow, the weaker is the relationship; only the variables that fit within one of the two van Dobben circles have a significant relationship with the selected explanatory variable (see Section 11.5). Here you will get a graph with many long arrows, showing that most of the traits behave independently of the treatments. Consequently, the picture is dominated by these non-responsive traits, and the circles are rather small.

To see which variables really fit into the circles, you should rescale the picture (*Graph | Range of axes*) and you will see that the fertilisation is positively related to potential plant height, foliar P, proportion of hemicryptophytes, grasses, and late flowering species, whereas in the unfertilised plots, sedges, geophytes, and chamaephytes are more abundant (Figure 16–11 and *Graph 2* in the example analysis notebook). If we were to construct a similar diagram for mowing, we will also get some significant relationships, but their interpretation would not be reliable, because the test for the partial effect of mowing*year interaction (see above) was not significant.

You can further explore the variation of trait values in the context of constrained ordination space of CWM data, using the functional diversity (FD) index estimated for each recorded case. You will calculate FD estimates and place them into a new data table in the *OhrazTraits.c5p* project. To do so, select the *Data | Add new table(s) | Functional diversity* menu command. But when the *Calculate Functional Diversity* dialog opens, you will see it includes a message '7 traits selected', and this does not match the number of traits (11) in the *Traits* data table. This is because the Rao FD can be calculated, in a straightforward way, only from numerical traits, but the *Traits* table contains four factors. To combine numerical and factor traits, you can transform the trait data into a matrix of distances, using e.g. the Gower distance that is able to combine numerical and factor variables (see Section 6.2.2).

So close now the dialog box with *Cancel* button, create a new analysis, keeping only the *Traits* table checked in the first page of the New Analysis Wizard and then selecting the *Principal-coordinates* analysis template from the *Specialized Analyses* folder in the next page. In the first page of the Analysis Setup Wizard (we assume you are working in the QuickWizard mode), select the Gower distance. You will note that the *PCO Options* page allows you to export calculated distances into a file, so, in principle, you are able

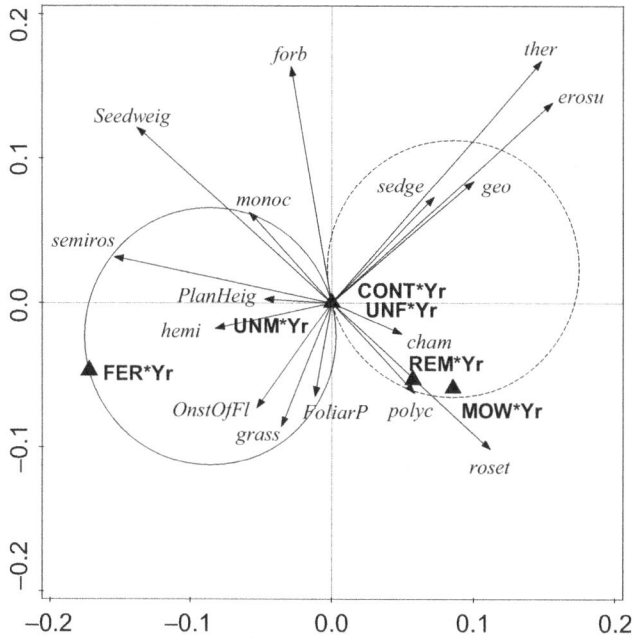

Figure 16-11 T-value biplot (first two axes) with van Dobben circles suggesting a significant relationship of particular traits with fertilisation.

to use them directly with Rao's formula of functional diversity (as given in Section 9.3). Canoco 5, however, does not allow you to include an external matrix of distances in the functional diversity calculations, so you must proceed using a workaround described in the next paragraph. But first execute the analysis.

Next you need to transfer the calculated scores of plant species (representing differences in the trait values among individual species) from the PCoA results into a new data table. Create a new data table first, using the *Data | Add new table(s) | Empty tables* menu command, choosing that *Each table row represents a* 'species', using 'PCoA score'/ 'PCoA scores' for *Each table column represents a / multiple columns are called* options, and e.g. 'PCoA Scores' as the table name. Then switch to the new (PCoA) analysis notebook (re-opening it in non-brief mode, if not already shown in that way, see Section 2.3), go to its *Cases (1)* page, select all 21 PCO columns, and copy them (using the *Copy expanded* command in the context menu) to the Clipboard. Switch back to the *PCoA Scores* data table, right-click the heading of its first empty (orange-background) column and choose the *Append PCoA Scores from Clipboard* command from the context menu. Remove any variables preceding the first inserted (*PCO.1*) column. You can see that some species do not have PCoA scores defined at all – these correspond to species automatically excluded from the analysis due to missing trait values (see the following Section 16.8.2 for a discussion of how to adjust the handling of missing values in Canoco 5).

Now you can re-select the *Data | Add new table(s) | Functional diversity* command and in the *Calculate Functional Diversity* dialog box, select *PCoA Scores* as the data

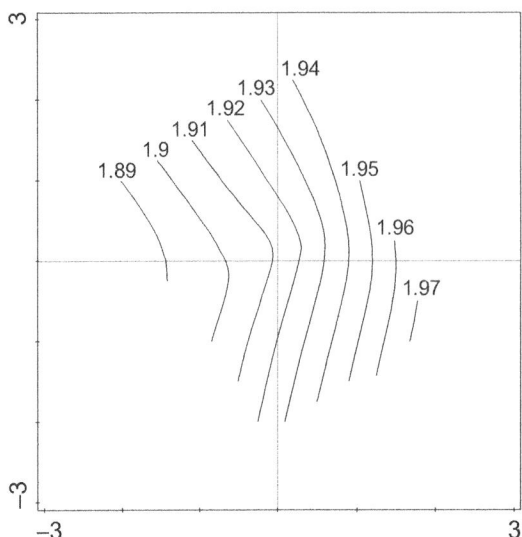

Figure 16–12 Change of the functional diversity of a plant community through the ordination space (using its first two axes) defined by the interactions of experimental treatments and time. Visualisation is based on a loess smoother model and uses the same ordination space as Figure 16–10.

table providing traits to use.[30] Click on the *Select columns* button and exclude the *molicaer* species (for the same reasons the *Molinia caerulea* was excluded in earlier analyses). Finally, you can change the default *center and standardize (Z scores)* choice in the *Standardization of functional traits* menu to *do not standardize*, because the PCoA scores already have proper scaling, reflecting the decreasing importance of consecutive axes by the smaller range of their score values. After clicking *OK*, a new data table with a single column representing functional diversity based on the Rao index is created (named *Functional Diversity* in the example project).

Now you can plot the diversity values into the ordination space of RDA on trait CWM data by selecting first this analysis (*RDA-traits-on-treatments* in the example project) and then choosing the *Graph | Attribute plots | In ordination space* menu command. In the *Create Attribute Plot* dialog, select *FD Indices | FD(Rao)* in the left-hand list, then choose the *Loess model* in the upper right corner and review (or change) the smoother options using the *Model Options* button before clicking the *OK* button. The resulting diagram is shown in Figure 16–12 (as well as in *Graph 5* of the example analysis notebook). Comparing the figure content (with functional diversity apparently increasing from the left to the right side of the ordination diagram) with the standard biplot for the same

---

[30] The values in the new *PCoA Scores* data table can be seen as representing the same information as the original *Traits* data table, but now on a fully quantitative scale. There are, however, two limits to the correctness of such a statement: (a) the variation is portrayed exclusively through the metric of Gower distance, and (b) the Gower distances cannot be completely embedded into a Euclidean space – if you check the information shown in the *Principal-coordinates* analysis' *Summary* page for the *PCO* step, you can see that a part of the variation in distance values corresponds to negative eigenvalues (ca. 0.6 out of 2.7+0.6 for the total variation). It might be, however, reassuring, that after using the seven numeric traits

ordination space (see Figure 16–10), you can see that the functional diversity decreases through time on the fertilised plots.

With FD values calculated, you can also export them and analyse in the same way you have already done with the number of species (i.e. by a repeated measurements ANOVA, see Section 16.4). This will tell you directly to which experimental factors the functional diversity responds.

16.8.2 Species-based approaches

Plant species responded significantly to fertilisation and mowing (see Table 16–3), so to demonstrate species-based approaches, you will first estimate species response to those two treatments. These two sets of response values will be calculated separately using partial constrained analyses, in the form of response variable (species) scores on a single constrained axis. This constrained axis is defined either by fertilisation*year interaction (analysis *Constrained-partial-fert*yr* in the example project) or by mowing*year interaction (analysis *Constrained-partial-mow*yr*) and these two analyses correspond to *A3* and *A4* in Table 16–3. To obtain species response values, open the analysis notebook in a non-brief view (see Section 2.3) and select the *RespVars* page. The first column (*Resp.1*) contains the scores characterising the extent and direction of the temporal change of each species with the treatment (fertilisation or mowing). Before copying the scores, however, create a new table for them (using the *Data | Add new table(s) | Empty tables* menu command).

When defining the new table, make sure that you choose in the initial dialog box that *Each table row represents a . . . species*, rather than *releve* and that you choose the appropriate terms for the table columns (e.g. *score/scores*). You can then copy the *Resp.1* scores from the two analyses into the new table by right-clicking the column, selecting the *Copy expanded* command, switching to the new table, right-clicking the first column (or the first non-initialised one) and selecting *Paste score* (or *Append score from Clipboard*).[31] We recommend that you rename the individual scores – e.g. to *RFert* and *RMow*, as used in the example project's *Score* data table. These scores will now become the response variables. Because the effect of removal (analysis *A5*) was not significant, a similar analysis for removal is not needed. Consequently, we can now use also the *Molinia caerulea* species, because we are able to estimate its response to both mowing and fertilisation.[32]

First, you can predict each type of response separately – you will use the response to fertilisation, which was most pronounced, but the same can be done for the response to

---

only, FD values calculated directly from these traits and FD values based on PCoA scores calculated from them using Gower distance have Pearson correlation value r = 0.974.

[31] Beware, however, of possible axis reversal. For example in our example project, *FERTIL*YEAR* variable in the *Constrained-partial-fert*year* analysis has a negative correlation with the constrained axis, as can be seen from ordination diagrams or from the negative sign of its *RegrE.1* score in the *ExplVars* page of the analysis notebook. In such a case, the species scores must be first copied to Excel, all multiplied by minus one and only then inserted into the *Scores* data table.

[32] But there is a slight danger that the response will be underestimated, because the species does not respond in the removal plots, as its cover is close to zero there, irrespective of mowing and fertilisation treatment.

mowing. However, before carrying out the analysis, it will be useful to exclude species with a low frequency – you can reasonably expect that the estimate of their response to a treatment will be imprecise. For this data set, for example, we recommend that you omit species with less than 9 occurrences in the table. It can be easily done by defining a group of *frequent species* (defined as species found in the whole data set at least in 9 relevés).

Start with the *Project | Groups | of species* menu command and select the *By Rule* button at the right side of the dialog. In a newly opened dialog, you should choose in the *Membership based on* option the *data column statistic* value and then in the *Use statistic* area choose *the count of occurrences*. In the right-hand area you specify that a species is included in the group *From 9*. Once the group is defined, you can use it in any analysis, where you ask for selection of species in the Analysis Setup Wizard. At the corresponding wizard page, you will see in the bottom right corner the *From group* option. After choosing it, you will ask to *Select non-members*, species with a low frequency will be highlighted, and you can move them to the left-hand column (i.e. the ignored species).

The analysis *Regress-fert-response-on-traits* in the example project represents a multiple linear regression model fitted using RDA. It is defined using the *Constrained* analysis template, with the variable *RFert* from the *Score* table used as a response variable and the variables from the *Traits* table as explanatory variables. The analysis was defined with the QuickWizard mode switched off, so that the *Select cases ("species")* and *Select response variables ("scores")* options could be checked at the second setup wizard page; the plant species were then subset using the *Frequent species* group (as described above) and *RFert* variable selected in the following page.

This analysis also uses the *Summarize effects of expl. variables* option in the *Testing and Stepwise selection* page. Before executing this analysis, you need to reflect the fact that the trait data contain many missing values. So uncheck the *Execute this analysis after Finish* box in the last setup wizard page, before clicking the *Finish* button. To specify how to cope with missing (NA) values, select the *Traits* data table first (clicking on its name in the upper left area of the Canoco 5 workspace) and then choose the *Data | Set handling of missing values* menu command, adjusting the settings in the dialog box as illustrated in Figure 16–13.

We suggest omitting species with at least five trait values missing and replacing the other missing trait values with the trait average. You can then execute the analysis.

In the original paper (Lepš 1999), where only a limited number of traits was available, we have looked first at bivariate scatterplots and fitted linear regression of species response to fertilisation on plant potential height. You can easily display such a scatterplot directly in Canoco, using *Graphs | Attribute plots | XY(Z) diagram*, and selecting *Plant height* for the X axis (from the *Traits* data table), and the *RFert* variable (from the *Scores* table) for the Y axis. You can also specify which type of model will be fitted. Here you can fit e.g. a generalized additive model (GAM) with two degrees of freedom complexity (see Figure 16–14). In the same dialog, check the *Use inclusion rules* box so that only the species active in the analysis are plotted (excluding species with low frequency).

16.8 Changes in composition of functional traits

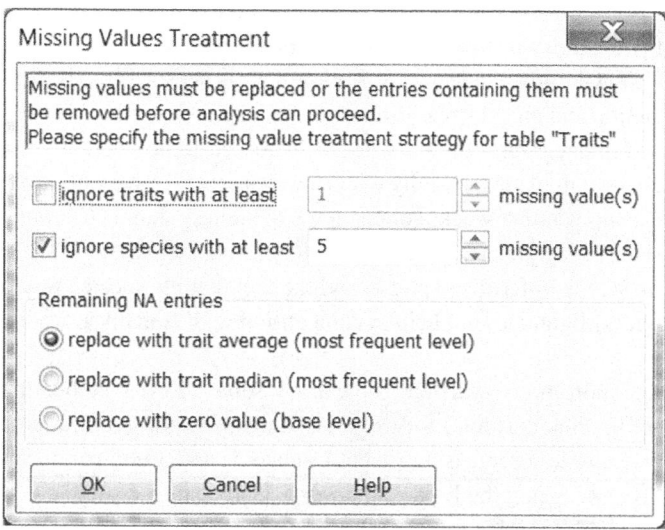

Figure 16–13 Dialog box specifying the missing (NA) values handling for the *Traits* data table.

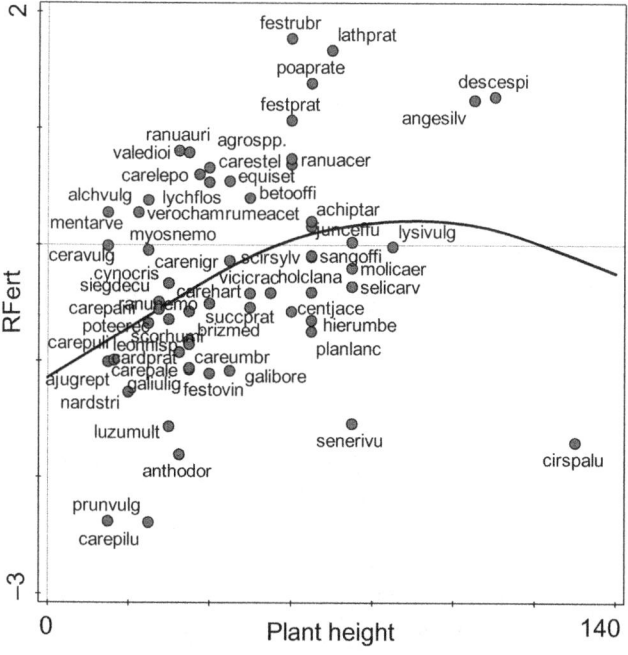

Figure 16–14 Relationship between plant species height and species response to fertilisation fitted with a generalized additive model with df = 2 complexity.

Using the labels plotted for individual species, you can directly see which species are outliers and you can try to figure out why.[33] So, you can see that on average, only tall plants are able to take advantage of fertilisation. This is because fertilisation removes the nutrient limitation and thus decreases the underground competition. This leads, together with increased aboveground biomass, to an increased importance of competition for light, and here the tall plants naturally win. However, not all tall plants are favoured, and so the relationship is rather weak. For example, the tallest plant is *Cirsium palustre*, but this species is monocarpic and so it is limited by microsites for seedling establishment (and these are scarce in fertilised plots), and the fact that the species would be superior in competition for light does not help in competition with clonally spreading polycarpic species.

Having now more traits available, we can try some of the methods that will select the best set of predictors (traits) for us. We will demonstrate two of them, the forward selection procedure in a regression (using Canoco 5) and a non-parametric regression tree method (CART, using the R package *rpart*). Classical regression can be defined, in fact, as an RDA with a single response variable and without standardisation and this is the analysis you have already defined above. Now you can review the provided overview of simple and conditional effects requested in the *Regress-fert-response-on-traits* analysis and shown at the bottom of the *Summary* page of its notebook (see Figure 16–15).

The first selected variable (with highest explanatory power) is the *Plant height*, which is only marginally significant when the false discovery rate (FDR) correction is applied ($P = 0.0442$). It is followed by the monocarpic/polycarpic dichotomy, but its conditional effect is significant only when no correction on multiple testing is applied ($P = 0.035$), otherwise it is non-significant ($P = 0.301$). Interestingly, the simple effect of this variable was highly non-significant, and the variable becomes significant only after inclusion of plant height, probably just to correct the prediction of a positive response of monocarpic *Cirsium palustre*.[34]

If you want to display the effect of individual traits on the response, it is better to use a regression biplot, because it displays the partial effect of individual explanatory variables, showing correctly here that the monocarpic species have a less positive response – *Cirsium*, in fact, a negative one. The classical biplot displays for the two *CARP* levels their centroids and they represent the independent effects, which would in this case suggest a positive effect of monocarpy on species response to fertilisation. You can find additional information about regression biplot in the Canoco 5 manual, p. 220.

The regression tree models (as one type of the CART family) perform binary partitioning of the set of cases and for each partitioning, one (best) predictor is selected and the threshold value of a quantitative predictor or subset of factor levels, best separating two groups of cases with distinct values of the response, is found. So, for each node in the tree, you get a condition (rule). If the condition is fulfilled, you will follow

---

[33] In the final graph for publication, you will probably delete most if not all the labels, as the figure is rather messy, but for your inspection the labels are invaluable.

[34] There are very few monocarpic species in the whole data set, so a single species has in this case a large weight.

16.8 Changes in composition of functional traits

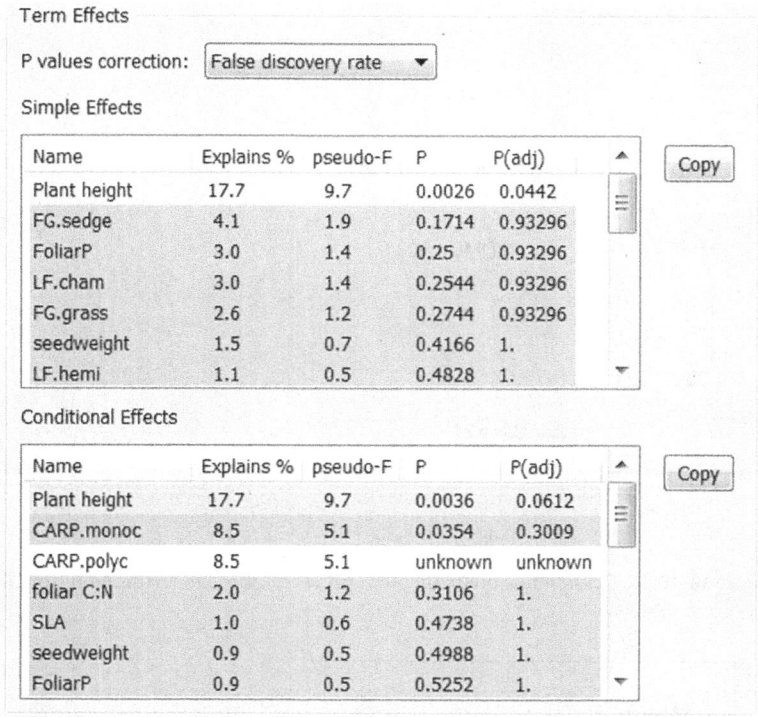

Figure 16-15 Simple and conditional effects of plant traits explaining the species response to fertilisation.

the left branch, if not, you will follow the right branch (see Section 8.7 for additional details).

The regression tree shown in Figure 16-16 suggests that the most important trait for species response to fertilisation is the plant height, concordantly with the results of all the methods above. The species taller than 36 cm should be favoured by fertilisation (at least their majority, with high LDMC values or with lower foliar C:N ratios) and the lower species should be disadvantaged. From the group of short species, the grasses and sedges are negatively affected more than other species. Nevertheless, the lower divisions are, in this case, not sufficiently supported by the tree size cross-validation procedure and, honestly speaking, are also difficult to interpret ecologically.

You will predict now both sets of scores (i.e. the response to fertilisation and the response to mowing) simultaneously in a single RDA (analysis *Constrained-both-responses-by-traits*). You will follow the same procedure as for the analysis with the multiple regression for the *RFert* variable described above, just with the difference that both variables in the *Score* data table will be included as response variables. The only significant predictor is again the *Plant height* variable (if you apply the false discovery rate adjustment, then P=0.09), other variables are non-significant. But the resulting ordination diagram (Figure 16-17) is interesting, showing that the responses to fertilisation and mowing are negatively correlated, i.e. in opposing direction on both the

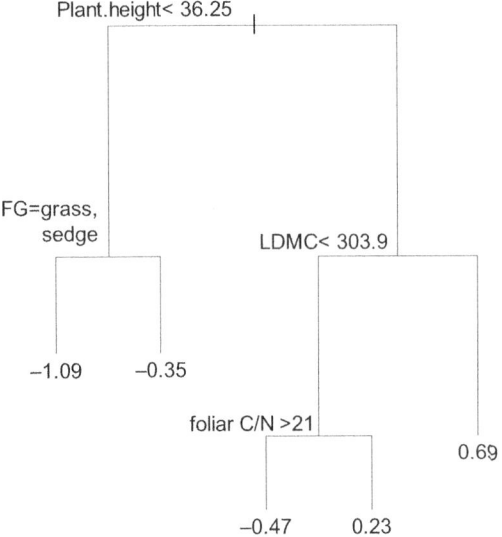

Figure 16–16 Regression tree predicting the species response to fertilisation from trait values.

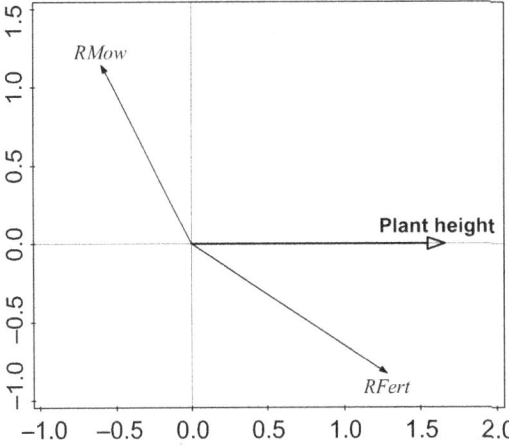

Figure 16–17 Plant species responses to fertilisation and mowing, explained by plant height in an RDA with forward selection of predictors (traits). The first two RDA axes are plotted.

constrained and also the first unconstrained axis. This means that not only are the tall species supported by fertilisation and harmed by mowing, but the negative relationship of both response characteristics is stronger than can be explained by their discordant relationship to plant height.

You can also carry out a reverse analysis (see *Constrained-traits–by-scores* in the sample project), i.e. try to explain the trait composition by species responses to fertilisation and mowing (the analysis suggested by the Canoco 5 manual, p. 299). This analysis shows the correlation among individual traits (each category of a factor is taken

16.8 Changes in composition of functional traits

as an independent quantitative variable). But the relationship is highly non-significant (P~0.7), and the t-value biplots reveal just one significant response, namely the relation of plant height to fertilisation.

All the analyses show concordantly that of the three factors, fertilisation had the largest effect on all evaluated community characteristics: species number, species composition, and trait composition. Plant height (i.e. potential plant height) is the most important trait for the species response, being positively related to the response to fertilisation in all analyses. The relationships of other species traits are rather weaker. A more significant relationship with traits was found in the plot-based methods and this is not surprising. Significance can be caused there by the pronounced response of one (or several) of the most abundant species to a treatment.

There are also some lessons learned for designing similar studies and selecting the measured traits. In the original paper (Lepš 1999), we have used just the plant height (and demonstrated that mycorrhizal status has no effect at all): this was not only because the trait databases were not available at this time (and so we retrieved the information on height from a local flora), but also because the decline of low statue species was already evident in the field and we had a plausible ecological explanation for it. Now, we have measured several species traits, and additional basic characteristics are easily available for most species in various databases. As a matter of fact, if you were to check first the significance of the full model with all presently available traits as predictors, the result would be non-significant. Similarly, when you use any of the corrections for multiple testing (e.g. the false discovery rate approach, adopted in the above analyses) during stepwise selection of effective traits, then with so many non-responsive traits present, even the plant height is only marginally significant (with P~0.08). Such a situation can lead to three alternative analytical strategies.

1. You can start with just the traits that are likely to respond (or that you see to respond in the field). In this case, you could get significant results, but many of them might be trivial or just confirming what you already know from the field. Moreover, the statistical test is strictly invalid if you have selected the variables from the effects you already saw in the data (*prophecy ex eventu*, Jongman et al. 1987, p. 24). Statistical tests need to be selected before you saw the data. This selection approach would be fine, however, if the traits are selected from prior (literature) knowledge only.
2. You can use many traits, including those whose response is not so self-evident or expected, and a correction for multiple testing (and/or guarding the stepwise selection with a preliminary test of the whole model). But this will result in a very low power of the test and possibly nothing will be found significant.
3. You will use all the traits as above, but you will not apply any multiple testing correction and run the selection even in the case that the full model is not significant. In this case, you have a good chance to get new interesting results, but also a high chance that some (if not all) of the significant results are simply results of 'statistical fishing' – i.e. the result of the inflation of Type I error probability.

The situation is nicely described in a paper criticising the Bonferroni type of Type I error corrections (Nakagawa 2004). Yet there seems to be no simple solution, and your decision should be based on the properties and aims of the individual studies. Undoubtedly a two-step solution, as recommended by Hallgren et al. (1999) – see Section 3.7, i.e. data diving with cross-validation (or other types of cross-validation) might help, but it would also decrease the power of the tests – particularly in the species-based analyses, the number of species cannot be increased to increase the power of the test.

Finally, it should be noted that the relatively weak relationship of the traits with experimental factors in our study is partially caused by the relatively short observation period reflected in the analysed data. In a community of perennial plants, the competitive exclusion is relatively slow, and so are the changes in trait composition. Now that we have data available for a longer period, it seems that the trait responses are more pronounced. This again highlights the importance of choosing an appropriate temporal scale in ecological experiments.

17 Case study 6: Hierarchical analysis of crayfish community variation

Biotic communities vary at multiple spatial (and temporal) scales, often hierarchically related. To understand this variation, including its relation to biotic and abiotic factors, we must start with testing the variation's existence and quantifying its extent at particular scales. In this chapter, you will study the hierarchical components of spatial variation of the crayfish community in the drainage of Spring River, north central Arkansas and south central Missouri, USA. The data were collected by Dr Camille Flinders (Flinders & Magoulick, 2002). The statistical approach used in this study was already described in Section 10.4.

17.1 Data and design

The primary data table consists of 567 records of the crayfish community composition. There are 10 variables, which actually represent only five crayfish species, with each species divided into two size categories, depending on carapace length (above or below 15 mm). The data matrix is quite sparse. In the 5670 data cells, there are only 834 non-zero values; 85% of the data cells are empty. This would suggest high beta diversity and, consequently, use of a unimodal ordination, such as CCA. There is a problem with that choice, however. There are 133 records without any crayfish specimen present and such empty cases are uninformative with respect to species composition. They cannot be compared with the other records using the chi-square distance, which is implied by unimodal ordination methods. If you use a unimodal ordination, the empty records will be automatically omitted from the analysis and some of the permutation tests shown in this case study will not work, due to the introduced unbalancedness of the data. You must, therefore, choose the linear ordination and rely on its robustness.[1]

The sampling used for collecting data has a perfectly regular (balanced) design. The data were collected from seven different watersheds (*Watershed* or *WS*). In each, three different streams (*Stream* or *ST*) were selected, and within each stream, three reaches (*Reach* or *RE*) were sampled. Each reach is (within this data) represented by

---

[1] In case you wish to use analyses reporting only the compositional changes (i.e. in the proportions of individual taxa), as would be the case with unimodal methods, you can increase the robustness of linear methods by standardising by cases (either standardising by the case norm or using Hellinger standardisation – see Section 2.3). The analyses reported in this case study do not adopt this approach, however.

Table 17-1 Analyses needed to partition the total variance in the crayfish community data.

Variance component	Explanatory variables	Covariates	Permuting in blocks	Whole-plots represent
watersheds	WS	none	no	ST
streams	ST	WS	WS	RE
reaches	RE	ST	ST	RU
runs	RU	RE	RE	none
residual	none (PCA)	RU	n.a.	n.a.
Total	none (PCA)	none	n.a.	n.a.

three different runs (*Run* or *RU*) and, finally, each run is represented by three different samples (cases). This leads to the total of 567 cases = 7 WS · 3 ST · 3 RE · 3 RU · 3 replicates. These abbreviations (identical with the brief versions of the variable labels in the *Design* data table) will be used throughout this chapter.

Both data tables are stored in the *Scale.xlsx* Excel file, but they were also imported into a Canoco 5 project stored in the *Scale.c5p* file and you can start the work with it.

When performing the variance decomposition for crayfish community data, you must partition the total variance into five different components, using the method summarised in Table 17-1.

In the permutations tests assessing the effects of spatial scales (watersheds, streams, and reaches) upon the crayfish community, the individual cases cannot be permuted at random. Rather, the groups representing the individual instances of the spatial level immediately below the tested level should be held together. This can be achieved using the split-plot design permutation options.

17.2 Differences among sampling locations

You will start your analyses with a 'standard' PCA, in which you will not constrain the ordination axes and you will not use any covariates. Instead, the factors representing the watersheds and streams are passively projected in the ordination space to visualise the degree of separations at the two highest spatial levels. You could also use the factors describing the lower spatial levels, but the resulting diagram would be too overcrowded. The corresponding analysis in the example project has the name *Unconstrained-suppl-vars*.

Click the *New* button below the *Analyses* list (in the lower left corner of the Canoco 5 workspace), keep the default choices in the first two pages of the New Analysis Wizard and select the *Unconstrained-suppl-vars* analysis template at the last page. After you click the *Finish* button and the first page of the Analysis Setup Wizard appears,[2] keep only the *WS* (*Watershed*) and *ST* (*Stream*) factors in the right-hand list and move

---
[2] Presumably in the QuickWizard mode, otherwise you must progress through the pages not referred to here.

17.2 Differences among sampling locations

Summary of Results

Method: PCA with supplementary variables

Total variation is 5296.649, supplementary variables account for 31.1% (adjusted explained variation is 28.6%)

Summary Table:

Statistic	Axis 1	Axis 2	Axis 3	Axis 4
Eigenvalues	0.3818	0.1924	0.1280	0.0963
Explained variation (cumulative)	38.18	57.42	70.22	79.85
Pseudo-canonical correlation (suppl.)	0.4723	0.7451	0.6398	0.5987

Figure 17-1 Summary of PCA results.

the others to the *Ignore supplementary variables* list. At the following page (named *Ordination Options*), change the recommended *DCA* choice to PCA (see the previous section for an explanation), progress to the next page, and click the *Finish* button there.

In the Graph Wizard page click the *Cancel* button, as you will need a finer control over the graph contents. But look first at the analysis summary, shown in the *Summary* page of the analysis notebook (see Figure 17–1).

The sum of all eigenvalues is, of course, equal to 1.0, because Canoco standardises in this way the response data table in all linear ordination methods. This makes it easier for you to read the explained variance fractions in linear analyses (without covariates). You can simply multiply the eigenvalues with 100 to obtain the percentage of crayfish community variance explained by individual ordination axes. The same information is then summarised in the *Explained variation* row in a cumulative way, so you can see that the first two ordination axes (principal components) summarise about 57% of the total variance and this value also represents the quality of conclusions drawn from the default ordination diagrams, in which the first two axes are used.

You start with a graph where only the watershed positions are shown. For that, you must suppress the display of streams. To do so, select the *Project | Suppressed in graphs | spatial units* menu command and in the dialog box move the *ST* factor to the left-hand list before pressing *OK*. Now select the *Graph | Biplots | Crayfishes + spatial units* menu command and in the resulting plot adjust the positions of labels as needed (see Figure 17–2 for a possible look of the resulting graph).

The content of the diagram is dominated by the difference of the *Bay* watershed from the other ones, so most of the location centroids are concentrated in the middle part of the diagram. You will therefore create also another diagram (Figure 17–3), which shows only the central part of the previous ordination diagram, and shows centroids not only for the watersheds (black squares), but also for each of the three streams within each watershed (grey squares), except the *Bay, Bay2,* and *Wf3* centroids, cut off from the diagram.

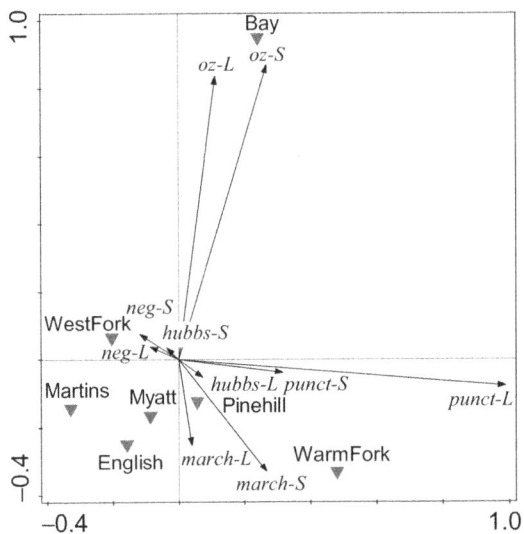

Figure 17-2 Biplot of crayfish taxa with centroids for individual watersheds; first two PCA axes are shown, explaining 57% of the total variance.

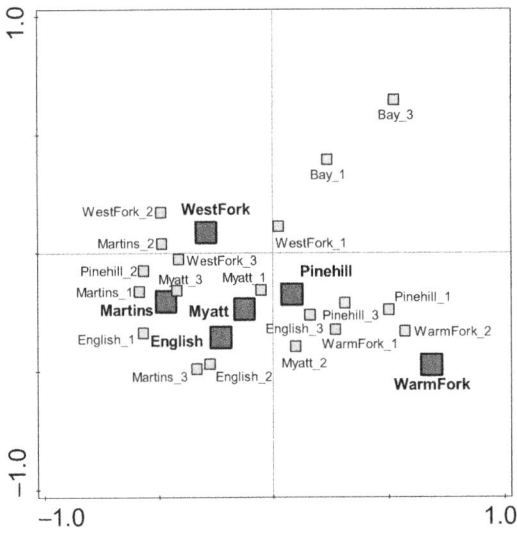

Figure 17-3 Diagram with the first two PCA axes, with projected centroids of watersheds and streams.

To create such a diagram, you must first classify the supplementary variables into two classes (one with the 7 watersheds, the other with the 21 streams; the remaining factor variable levels may remain non-classified) using the *Project | Classifications | of spatial units* menu command. In the first dialog select the *By Selection* button and then define in a new dialog (*Specify Classification*) the two classes using the *Add* button and add members to them by moving factor levels from the right-hand list to the central one. When back in the *Classification Manager* dialog, make sure you check the *Use selected classifications in graphs* option. You might also like to adjust the symbol types and colour for the two defined classes by selecting the *Edit | Settings | Visual attributes* menu command and in the manager dialog clicking the *Edit* button and then changing symbol type, outline, and fill attributes for the *Supplementary vars. | Symbols* with *Class 1* and then *Class 2* selected in the lower left dialog corner. Finally, you must select *Project | Suppressed in graphs | spatial units* and undo earlier suppression of the *Stream* factor.

You can then create e.g. a scatter of centroids using the *Graph | Scatterplots | Spatial units* menu command and then focus into its central part e.g. by selecting the *Graph | Range of axes* menu command, setting the *Use the range specified below* choice and selecting an appropriate range for both axes. The choice used in Figure 17–3 cuts off some of the stream centroids, namely *Bay_2* and *WarmFork_3*, but makes the differences among streams more obvious than when the full range was retained.

You can see that while the positions of some stream triples are clustered in the proximity of their parental watershed, the difference among the streams is much larger for others (like *Bay*, but also *Pinehill*).

17.3 Hierarchical decomposition of community variation

Now you should create a separate Canoco analysis for each of the spatial levels you want to evaluate (see Table 17–1). In fact, you can omit the calculations for one of the levels, because the fractions sum up to 1.000. Probably the best choice for omission is the residual variation (within-runs variance), as there is no meaningful permutation test and defining the indicator variables (to be used as covariates in the partial PCA) for individual runs is quite tedious.[3] We will not provide detailed instructions for all the analyses you can set up here, but if you feel confused you can always look at the corresponding analysis defined in the example project file (see Appendix B).

Start with the analysis described in the first row of Table 17–1 (in the example project, this is the *Watersheds* analysis). This analysis has no covariates and the factor variable *Watershed* is used as the explanatory variable. To define this analysis, select the *Constrained* analysis template in the last page of the New Analysis Wizard, and in the first page of the Analysis Setup Wizard (named *Selection of explanatory variables*) keep only the *Watershed* variable in the right-hand list. As in the previous analysis, change to a linear ordination on the next page (i.e. to RDA).

At the following page (*Testing and Stepwise selection*) choose the *Hierarchical design* in the *Permutation Test Parameters* and on the following wizard page (named *Split-plot Arrangement Permutation*) select value 27 for the *Number of split-plots in each whole plot*. This value follows from the fact you want to permute individual streams, keeping all the reaches of a particular stream together. Therefore there are 27 split-plots

---

[3] You will need to do exactly this anyway in the analysis of the effect at runs level, where the runs will represent explanatory variables.

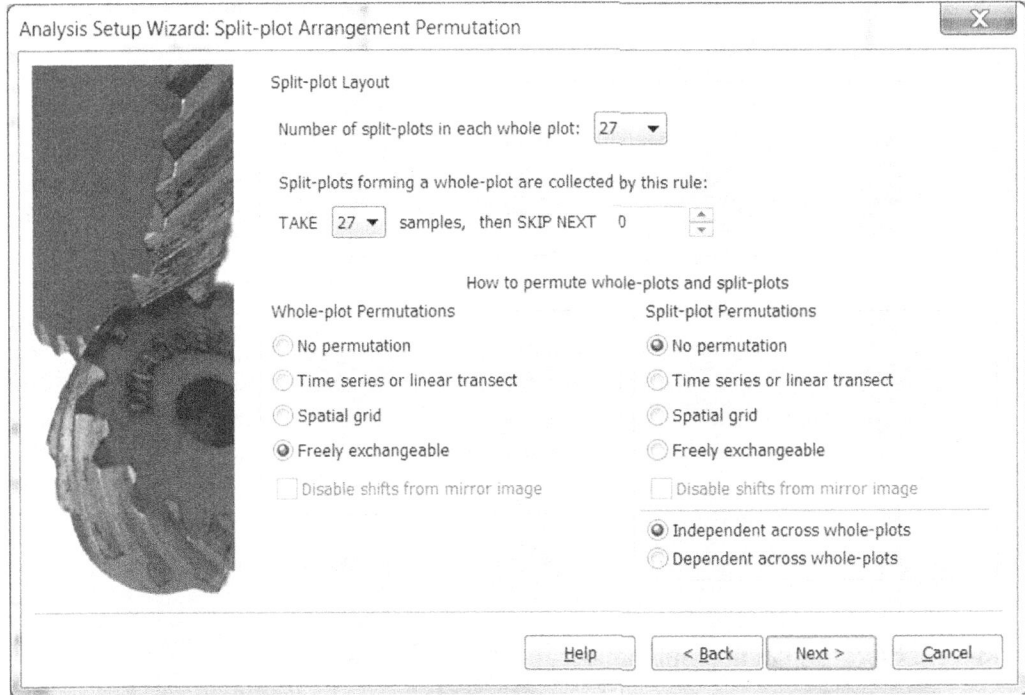

Figure 17-4 Split-plot permutation settings for the *Watersheds* analysis.

(3 replicates * 3 RU * 3 RE). The cases representing one whole plot (one watershed) are contiguous in the data table, so you can keep the default setting for *TAKE* and *SKIP NEXT* parameters (i.e. 27 and 0, respectively). In the lower part of the page, you should choose to freely exchange the whole-plots and not to permute the split-plots, as illustrated in Figure 17-4, representing the resulting state of that page.

After you execute this analysis, click *Cancel* in the Graph Wizard and now you can determine the amount of variability explained by the streams from the *Summary* page, looking at the information provided above the *Summary* table (*explanatory variables account for 18.9%*), But the information present there will not be usable for the other analyses, where covariates are present,[4] so instead click on the *Details* button and look at the value labelled *Canonical eigenvalues* (0.189 after rounding).

The watersheds explain 18.9% of the total variability in crayfish community data. Canoco also reports in the lower part of the *Summary* page about the permutation test results. These results show that none of the 499 permuted (randomised) data sets produced as large an amount of variability explained by watersheds as the true data configuration, because you have achieved the smallest possible Type I error estimate (i.e. $(0 + 1)/(499 + 1) = 0.002$).

---

[4] This is because the percentage of explained variation is in partial analyses calculated from the total variation after removing the part explained by the covariates, and this makes it impossible to compare this number across multiple analyses with different covariates.

17.3 Hierarchical decomposition of community variation

The next analysis (the second row in Table 17–1) quantifies the variability explained at the streams level (this is the *Streams* analysis in the example project). You will use the ST (or *Stream*) factor as the explanatory variable and the WS (or *Watershed*) factor (representing the next higher spatial level) as the covariate. In the New Analysis Wizard, therefore, you must select the *Constrained-partial* analysis template from the *Advanced Constrained Analyses* folder. Because you do not want to involve the differences between watersheds in the permutations simulating the null model, you must permute the whole-plots (the next lower spatial level, i.e. the reaches) randomly, but only **within** each of the watersheds. Therefore, you will not only have the split-plot design restrictions on the permutations, but you will also permute in blocks defined by the covariates.

So the important difference in the new analysis is (in addition to specifying a different explanatory variable and newly the covariate) that you check, in the *Testing and Stepwise selection* wizard page the *Blocks defined by covariates* option (and you keep the *Hierarchical design* option chosen, of course). After clicking the box, new content is added to the wizard page where you should specify which of the covariates define the permutation blocks by their values. In your case, with only a single covariate available, the task is simple – move the *WS* (*Watershed*) factor to the right-hand list using the ≫ button. In the following page, change the selection (remembered by Canoco from the previous analysis) of the *Number of split-plots in each whole plot* from 27 to 9 (i.e. 3 replicates · 3 runs, in every reach). The other settings should be already correctly set (*Freely exchangeable* whole-plots and *No permutation* for the split-plots).

You can see – by consulting the *Canonical eigenvalues* value in the dialog displayed with the *Details* button in the *Summary* page – that the streams explain ca 12.3% of the total variability and (back to the *Summary* page) that this is a significantly non-random part of the total variation (pseudo-F = 6.9, P = 0.008).

The next analysis (the *Reaches* analysis in the example project) is similar to the one you just performed, except you will move one level down in the hierarchical spatial structure: the placement of cases in reaches, coded by the *RE* (or *Reach*) variable, is used as the explanatory variable, and the *ST* (*Stream*) factor is used as a covariate. You should permute within the blocks defined by this covariate and use the split-plot structure, where the whole-plots are the individual runs (each with three records – the split-plots).[5] The variability explained at the reaches level is estimated as 15.8% of the total variability and the Type I error of the permutation test is estimated as P = 0.002 (with pseudo-F statistic value 3.6).

The last variance component you will estimate is the variation between the runs (see the *Runs* analysis in the example project). Beside changing the identity of the explanatory variable (now the *Run* (*RU*) variable) and of the covariate (now the *Reach* (*RE*) variable), another important change is that in the *Testing and Stepwise selection* page, you change the *Hierarchical design* choice to *Unrestricted permutations*. This is because there is no internal structure within individual runs – just three replicate cases. The *Reach* covariate must still define the permutation blocks, however. The variation

---

[5] Due to its increased complexity, the permutation tests for this and the following analysis might run somewhat longer.

Table 17-2 Results of variance decomposition for crayfish community over multiple spatial levels.

Component	Explained variability [%]	DF	'Mean square' value	Significance
WS	18.9	6	3.150	0.002
ST	12.3	14	0.879	0.008
RE	15.8	42	0.376	0.002
RU	19.5	126	0.155	0.002
residual	33.5	378	0.089	n. a.
Total	100.0	566	0.177	n. a.

explained at the runs level is 19.5% of the total variance. The significance level estimate is P = 0.002, pseudo-F = 1.7.

Finally, you can deduce the amount of variability explained by the differences between the cases within runs by a simple calculation: 1.0 − 0.189 − 0.123 − 0.158 − 0.195 = 0.335.

The results of all the analyses are summarised in Table 17–2.

The table includes not only the absolute fractions of the explained variation, but also the values adjusted by the appropriate number of degrees of freedom. These DF values are calculated (similar to a nested design ANOVA) by multiplying the number of replicates in each of the levels above the considered one by the number of replicates at the particular level, decreased by one. By the number of replicates we mean the number **within** each replicate of the next higher level. For example, there are seven watersheds, each with three streams, and each stream with three reaches. Therefore, number of DFs for the reach level is 7 · 3 · (3–1) = 42.

It is, of course, questionable which of the two measures (the variation adjusted or not adjusted by degrees of freedom) provides more appropriate information. On one hand, the reason for adjusting seems to follow naturally from the way the sum of squares is defined.[6] On the other hand, it seems quite natural to expect an increasing variation at the lower hierarchical levels: there must be always more reaches than there are streams or watersheds, so a higher opportunity for crayfish assemblages to differ.

In any case, the variation at the watershed level seems to be relatively high compared with the other spatial scales, and all the spatial scales seem to say something important about distribution of crayfishes, given the fact their 'mean-square' terms are substantially higher than the error mean-square (0.089).

---

[6] It does not matter here that we use the data which were standardised to total sum of squares equal to 1. All the components are calculated with the same standardisation coefficient – the total sum of squares of the data does not vary across the individual analyses.

18 Case study 7: Analysis of taxonomic data with discriminant analysis and distance-based ordination

In this chapter you will work with plant taxonomy data. This type of data does not strictly belong to the field of ecological research, but we hope it provides an easy-to-understand illustration of how to apply the demonstrated methods and how to interpret their results. You will learn how to analyse non-compositional (general) data tables with PCA, how to find which variables separate a priori specified classes (using discriminant analysis), and how to perform unconstrained and constrained ordination with an a priori chosen dissimilarity (distance) measure using, respectively, principal coordinates analysis and distance-based RDA.

18.1 Data

The data come from a not-yet-published taxonomic study (Štech et al., unpublished) of several closely related taxa of the genus *Melampyrum* (hemiparasitic plants in the family Orobanchaceae), where the differences on a wider geographical scale were evaluated using both morphological measurements and molecular data, obtained using the AFLP method (Vos et al. 1995). In this case study, you will be using a subset of 71 individuals originating from 19 populations to address questions relating the variability in morphological characters or in genome to an a priori classification of individuals into four taxa coded as *DEG*, *HOE*, *NEM*, and *SUB* (this classification was based on other molecular data, namely chloroplast DNA). The questions concerning the definition of well-separated groups from scratch, based on the collected data, are not asked here, but you will be informally evaluating the consistency of a priori defined groups with the variation in the data.

The data used in this chapter can be either imported from the Excel file *Melampyrum.xlsx* or you can start from the *Melampyrum.c5p* example project, which already contains the analyses discussed in this chapter. Twenty morphological characters were measured on specimen flowers (characters starting with *B* letter were measured on inflorescence bracts, those starting with *K* on calyx, and those starting with *C* on corolla). They are stored in the *Morphology* table of the example project, together with *Taxon* and *Locality* factor variables, distinguishing individual taxa as well as sampled populations. This data table is marked as a general type, not only due to the presence of the two factors, but also due to the fact that the values of individual measured characters cannot be meaningfully summed even though some of them use the same measurement units.

The matrix of 126 AFLP bands was imported as a separate data table (*AFLP*) and the table is marked as compositional, representing just the presence/absence of individual DNA fragments in a sample from a particular specimen.

18.2 Summarising morphological data with PCA

You will start with summarising the variation in morphological characters and visualising the difference among individual plants. When interpreting the results, you will focus on the distinction between the plants assigned to different taxa.

Before creating the analysis (called *PCA* in the example project) using New Analysis Wizard, make sure that the QuickWizard mode is switched off (see Section 2.3), because you will need to eliminate first two variables (*Taxon* and *Locality*) from the *Morphology* table, before the data are analysed by Canoco. In the New Analysis Wizard (started with the *New* button), keep only the *Morphology* table checked at its first page and select the *Unconstrained* analysis template in the next page. In the second page of the Analysis Setup Wizard (shown after you close the New Analysis Wizard), check the *Select response variables ("characters")* box and in the later *Selection of response variables* page move the *Taxon* and *Locality* variables to the left-hand list, so that they are ignored in the analysis. You can see in the *Ordination Options* page that PCA is the only ordination method available, as the unimodal (D)CA is not appropriate for this kind of data.[1] Also the centring and standardisation are preselected for the measured characters, to ensure that all variables are considered equal, with the same impact on PCA results.

You will also note that the setup wizard does not offer transformation of the morphological variables, although it would perhaps seem as appropriate. This is because here you can set the transformation only for a whole data table and such shared transformation is supported only for compositional data tables. You will fix this problem later, before executing the analysis. To do so, you must uncheck the *Execute this analysis after Finish* box at the following page, before clicking the *Finish* button. The analysis is now defined, but not yet executed, so its notebook is hidden.

Now you can specify required transformations for individual characters by selecting the *Morphology* table and then executing the *Data | Default transformation and standardization* menu command. Canoco informs you first that the log transformation was suggested for one of the characters and after pressing *OK*, the *Variable Transformations* dialog box appears. Quick inspection of displayed variables reveals (see Figure 18–1) that the log transformation in the form $log(10*x+0)$ was suggested for the *BD* character.

While the Canoco Adviser uses heuristic rules (concerning the minimum value and the ratio between the largest and smallest positive values of each variable) to recommend log-transformation for just a single character, we suggest to log-transform all of them, using the $log(1*x+0)$ transformation. All the measured characters are quantitative with only positive values and when you ask questions about their difference among specimens or taxa, you expect the answer to be in a multiplicative form, such as 'the bract is

---

[1] They do not represent compositional data – see Section 1.2.

18.2 Summarising morphological data with PCA

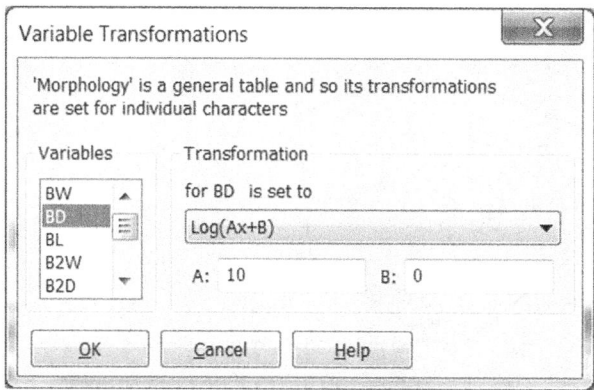

Figure 18-1 Log-transformation of *BD* variable as suggested by the Canoco Adviser.

1.5-times longer in A than in B' (see Section 1.3.1 for additional discussion of this approach). To set the log-transformation for all 20 characters, you must select the variables one by one in the left-hand box, choose the *Log(Ax+B)* transformation type, and then set the *A* and *B* values, respectively, to 1 and 0.[2]

Now you can proceed with the PCA. Select the analysis in the list and then click the *Perform* button. It is quickly executed and you will see the Graph Wizard offering the creation of a diagram with characters and specimens. If you decide to create it, you will find that the cases (specimens) cannot be excluded. Luckily, they can be placed in a separate subplot of the graph and this option is selected by default. After finishing with the Graph Wizard, you are asked whether to display the analysis notebook, so select *Yes* to inspect the results. Starting with the *Summary* page, you can see that the first PCA axis summarises about a third of the total variation, while the first two axes together summarise 46% of the total variation.

The composite graph (not shown here) created by the wizard shows in its left part that most of the measured characters increase their values in the partly shared direction (from right to left), representing the often-seen pattern of positive correlations among the morphological parameters characterising various aspects of the size of individuals (or here flowers). The right-hand subplot shows the position of individual plants on the first two principal components, but to make more sense of it, this diagram needs to be improved by classifying the specimens based on the taxon, to which they belong.

To do so, select the *Project | Classifications | of specimens* menu command, click the *From Data* button in the *Classification Manager* box, and as it happens, just click the *OK* button in the following *Classify from Data* box (because the *Taxon* variable is the first one in the first data table and for factors, the *distinct values* definition strategy is preselected). After you close this dialog, another one is shown that allows you to inspect the membership of individual specimens in classes, but you can close it now with the

---

[2] Note that the dialog box retains the previously chosen *A* and *B* settings, when you move to a new variable, except for *BD* where the 10 and 0 values were defined earlier.

Analysis of taxonomic data

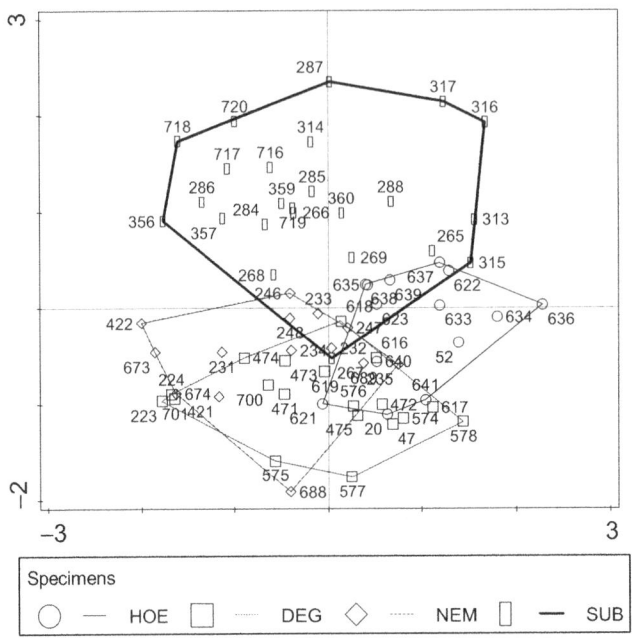

Figure 18-2 The first two axes of PCA ordination space computed from morphological data. The first (horizontal) axis explains 33.2% of the total variation, the second (vertical) axis adds another 12.8%.

OK button. Back in the *Classification Manager*, make sure you check the *Use selected classification in graphs* option before closing the box with *OK*.

You can re-create the existing graph, but we suggest that you create a new one, containing only the scatter of specimens and adding a few extras. So first make sure that with the specimen symbols classified, a legend will be added to the graph. Select the *Edit | Settings | Graphing options* menu command and in the *General* page make sure that the *Create legend in the graphs* option is checked. Following options control the placement and internal layout of the legend and we recommend for this graph to place the legend near the *bottom* edge, with item layout *Horizontal with headings*. After you close this dialog with *OK*, choose the *Analysis | Plot creation options* command and check the *specimens* box in the *Plot envelopes for* area at the *General* page. Finally, use the *Graph | Scatterplots | Specimens* menu command to create the graph, shown in Figure 18-2.

You will note that your graph differs from the one displayed in Figure 18-2, which is not a colour graph merely transformed to grey levels, as evidenced e.g. by the patterned envelope lines. Canoco 5 maintains two default sets of graphing attributes (called visual attributes), namely one for on-screen viewing and for inclusion into presentations and another one for creating graphs for printed publications. You can also create a third set of visual attributes and adjust it to your needs (see *Canoco 5 manual*, p. 395). To quickly change to the *Black & white* set of visual attributes, select the *Edit | Settings | Visual attributes* menu command and select *Black & white* in the *Active attribute set* field (not in the list present near the upper left corner!). After closing the dialog with *OK*, you must recreate the graph to apply the new attributes.

The ordination diagram in Figure 18–2 shows very limited separation of the *HOE*, *DEG*, and *NEM* taxa, with only the *SUB* taxon standing apart. The separation of the four taxa using morphological characters is nevertheless possible, but an unconstrained ordination (here PCA) is not the right method for such a task. From its results, you have now learned that the two major gradients separating the specimens in their morphospace are not very good for separating the a priori taxa. Finding the best way to distinguish the four taxa using (a subset of) measured characters is a different task and you will focus on it in the following section.

18.3 Linear discriminant analysis of morphological data

The linear discriminant analysis (LDA) that we introduced in Section 10.3 can be used to find a combination of morphological characters, most effectively separating a priori defined classes – here the four taxa. In the context of ecological studies, it can be used e.g. to find the differences in environmental conditions among community types distinguished, say, based on their species composition.

But there are other questions, where the application of LDA seems to be a good fit, yet the method usually does not provide satisfactory results. A good example might be an attempt to use discriminant analysis for the selection of taxa (species) indicating a particular type of community or habitat. The discriminant analysis then attempts to separate the community types using a linear combination of species abundances. However, because the species occurrences/abundances are often strongly correlated and each species alone provides little information, it is usually impossible to single out a few such diagnostic species, even if the species composition differs strongly among the distinguished types.

For your present case study, LDA is quite appropriate and you can create the corresponding analysis (named *LDA* in the example project) using the following steps. Click the *New* button to open the New Analysis Wizard, keep again only the *Morphology* table checked in its first page and select the *Discriminant-analysis* template in the *Specialized Analyses* folder. Assuming you are defining the analysis with the QuickWizard mode switched on again,[3] the first page you will see in the setup wizard is the *Discriminant analysis (CVA)* page with just two choices in its lower part.

In the left-hand list, make sure that the *Taxon* variable is chosen as defining the classification of specimens, and then choose the *filtered by stepwise selection* option at the right side, so that you can get rid of the morphological characters not effective for predicting to which taxon a specimen belongs. When using stepwise selection in discriminant analysis (or in any other context), you should be aware of the fact that the selection procedure greatly inflates the Type I error estimates for the selected subset of predictors (see Section 5.6), but here you will protect your analysis by using adjusted P-values.

At the following page (*Definition of Groups*), move all the variables except *Locality* from the *Pool* to *Members* list and you can also verify that the *Taxon* variable is already

---

[3] If not and you have already started the Analysis Setup Wizard, simply skip the additional pages shown in it, using the *Next* button.

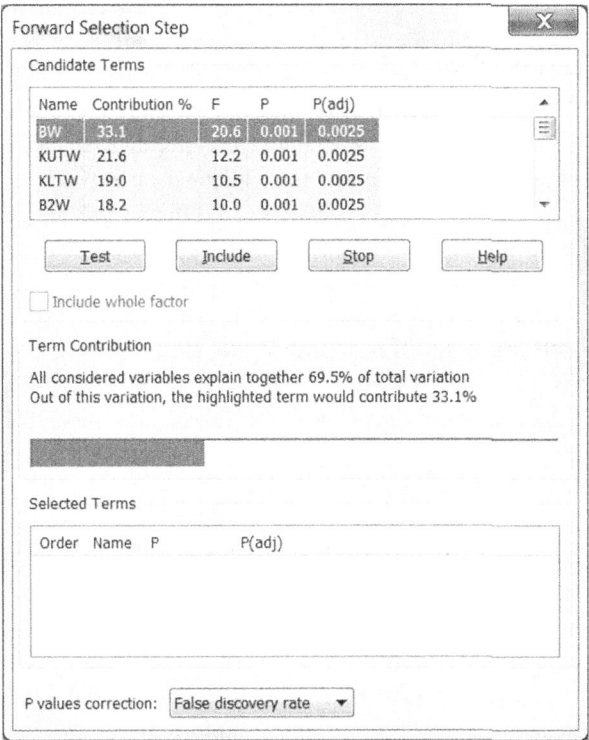

Figure 18–3 Stepwise selection of morphological characters effectively differentiating four *Melampyrum* taxa; the selection dialog is shown here in its initial stage, before any character was selected.

assigned to the *Classification* group. In the next setup wizard page, keep the *Unrestricted permutations* chosen, but possibly increase the *Number of permutations* to 999. After you click the *Finish* button in the last wizard page, analysis execution starts and you are soon confronted with a dialog box named *Forward Selection Step*, illustrated in Figure 18–3.

In the initial stage of stepwise selection, all morphological characters are present in the upper list of *Candidate Terms* (variables), together with percentage values representing the relative contribution of the term to the total variation that could be explained by selecting all the candidate variables. This percentage value is followed by the results of automatically performed permutation tests of each predictor contribution: pseudo-F statistic, significance estimate (P), and the P-value adjusted with respect to performing multiple tests with the same data set.[4]

---

[4] If you see the same *P* values, but different *P(adj)* values in your display, compared with the snapshot, please note that Canoco 5 does precompute the significance tests only for the variables visible in the list. While computing the remaining tests does not change the visible *P* values, it may change the *P(adj)* values, because during the adjustment procedure, all the *P* values are taken into account. To get the most accurate estimates of *P(adj)*, you must scroll through the full list of candidate terms, waiting for the permutation tests to be accomplished. If you do not see the *P(adj)* column at all, select the *False discovery rate* option from the *P value correction* control at the dialog bottom.

18.3 Linear discriminant analysis of morphological data

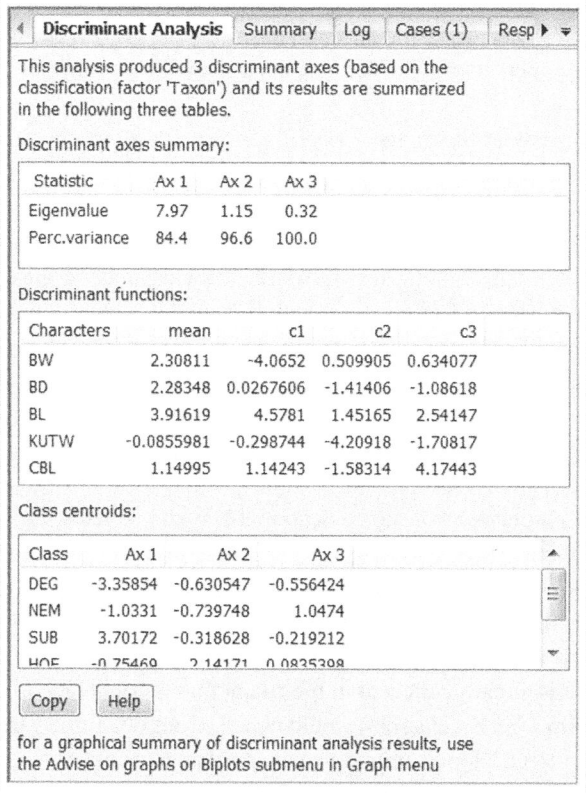

Figure 18–4 Discriminant Analysis page of the analysis notebook.

Next you should proceed with a repeated selection of topmost (best) candidate predictor by clicking the *Include* button, until the best candidate does not yield a significant value (i.e. stop when $P(adj) > 0.05$), clicking the *Stop* button then. This procedure will select, in the end, five morphological characters, shown in the LDA diagram in Figure 18–5. To create the diagram, keep the *Discriminant analysis diagram* checked at the first page of the Graph Wizard and progress through its other two pages.

Main statistical summaries are presented in a specialised *Discriminant Analysis* page of the analysis notebook, illustrated in Figure 18–4.

In the first table, the eigenvalues and percentage of variance explained by individual discriminant axes are shown with values calculated using formulas preferred in the context of discriminant analysis (differing from those provided on the *Summary* page, see the Canoco 5 manual, section 3.10, for additional details). There are three discriminant axes, because you are distinguishing four classes. The following table in the page presents discriminant functions for each of the three axes. These functions can be used directly to calculate the discriminant score e.g. for a new specimen, so that you can predict to which taxon it belongs. You first subtract from the measured values of *BW*, *BD*, *BL*, *KUTW*, and *CBL* their corresponding *means*, then you multiply each difference by the coefficient (e.g. in the *c1* column for the first discriminant axis) and finally you sum all five multiples together. You can then compare the resulting value with the class centroid coordinates, stored in the third table (ideally using all three discriminant axes). The centroid closest to the computed specimen coordinates corresponds to the taxon to which the specimen most likely belongs.

You might stop briefly at the *Summary* page of the analysis notebook. As the most important LDA results are summarised on the specialised *Discriminant Analysis* page, this page provides only a few additional details. Most useful among them is the report on stepwise selection, shown in the lower part of the page. You will also note that the *Method* is labelled as CCA (canonical correspondence analysis), because the discriminant analysis is calculated in Canoco 5 through this method, using the factor variable describing group membership as the response data. Another strange thing you will see is that the *Summary* page informs you that there are two steps in the analysis, but when you choose to see the summary of the second step (by clicking the scrollbar in the upper left area), no information is actually shown. This happens because the second step is present only to provide information used internally by the Canoco Adviser to adjust the length of the discriminant predictor arrows (see Section 10.3) and so its results are removed after analysis execution.

Now have a look at the graph produced by the Graph Wizard. Assuming you have started your work with the project by accomplishing the tasks in the preceding section, your specimens are already set up to be shown classified into four taxa classes, so this classification is already reflected in the subplot at the right side of the discriminant analysis diagram.[5] Yet the diagram would benefit from two further improvements: you should enable the envelopes around specimen classes again (this is an analysis-specific setting, see the previous section for a description how to set it) and the arrows for discriminating characters seem too short for you to fully understand the plot.

To upscale arrow lengths, you should first select the *Edit | Settings | Graphing options* and check the third option in the *General* page (*Show scores rescaling dialog...*) of the displayed dialog box. Then, after you close the dialog and re-create the diagram, a new dialog is shown with the name *Rescaling of ordination scores*. We suggest that you change the value 1 in the *Scaling factor* field for *Character scores* (in the lower part) to 5. The right-hand subplot can be further improved by removing the specimen labels (select one of the labels, right-click it, choose the *Select suchlike* command, and then press the *Delete* key). The resulting graph is shown in Figure 18–5.

With character arrows rescaled, you can see now more clearly that the largest differentiation is provided by the *BW* (first bract width) and, to a lesser extent, by *BL* (first bract length) variables, increasing (BW) or decreasing (BL) from the *SUB* taxon, over *HOE+NEM* taxa, to *DEG*. The second discriminant axis essentially separates the *HOE* taxon from the others by its lower values of *KUTW* (width of upper calyx tooth) and *CBL* (length of corolla basis), with further contribution of the *BD* character. There was no overall test of significance computed in this analysis, because it includes only the morphological characters that were shown as significantly contributing to the discriminating power during the stepwise selection.

---

[5] If not, follow the instructions in the preceding section to create and activate specimen classification.

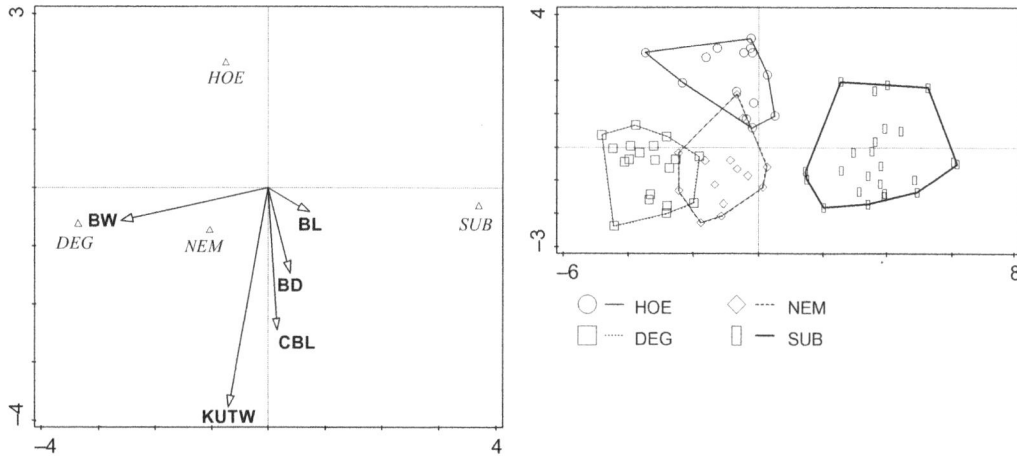

Figure 18–5 Ordination diagram of the first two discriminant analysis axes.

18.4 Principal coordinates analysis of AFLP data

You will now move to the AFLP data representing the presence and absence of fragments of particular length in a DNA sample, coming from a particular specimen and broken down with restriction enzymes. You will use these raw band-level data to calculate dissimilarities (distances) among specimens and use the resulting distance matrix (a) to summarise the similarity of individual specimens and (b) to test the difference among the four taxa (in the following section). This approach fits into the group of band-based statistical methods commonly used to analyse AFLP and similar data (Bonin et al. 2007). Task (b) corresponds partly to the AMOVA method widely used in molecular ecology, although we do not distinguish the population level in this analysis. See Section 18.6 for an overview of how to perform the analyses on population level and also check Chapter 17 for an example of how to perform hierarchical decomposition of variance with constrained ordination.

The two tasks outlined above correspond pretty closely to those addressed in preceding Sections 18.2 and 18.3 and you might be wondering why you cannot analyse the AFLP data with PCA and LDA methods. There are three main arguments for not doing so: you work here with non-numeric variables that have the nature of the simplest possible factors (with 0 and 1 levels), these band descriptors are too numerous for effective stepwise selection,[6] and their identity cannot be functionally interpreted. With the distance-matrix-based methods, you will summarise the patterns present in the band data using the principal coordinates analysis (PCoA) axes (for task (a)), and the specimen scores on these axes might be further subjected to a constrained ordination to test research hypotheses (such as the one in task (b), targeted in Section 18.5).

---

[6] And you cannot simply keep them all in the role of explanatory variables, due to their massive collinearity.

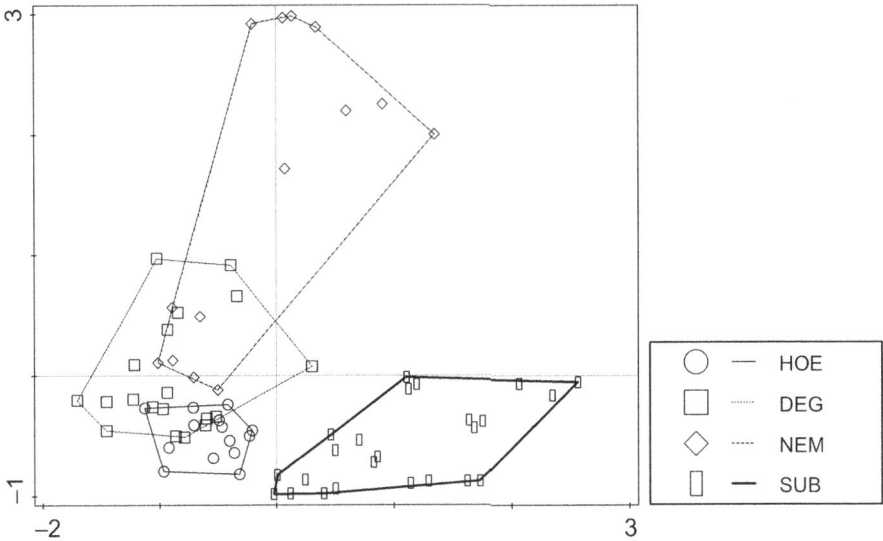

Figure 18–6 Ordination diagram displaying the first two axes of PCoA calculated from AFLP band data. The enclosing polygons separate plant specimens belonging to the recognised four taxa.

To create an analysis summarising the AFLP data (the *PCoA* analysis in the example project), start again with New Analysis Wizard, keep only the *AFLP* table selected in its first page, and then choose the *Principal-coordinates* analysis template (from the *Specialized Analyses* folder) in the second wizard page. Assuming you work with QuickWizard mode enabled, the first page you will see in the subsequently displayed Analysis Setup Wizard is named *PCO Options* and here you must select the options specific to PCoA.

First, in the upper part of the page, you need to keep the choice stating that the distance matrix must be calculated from data in the AFLP table (the other possibility enables you to import the distance or similarity matrix computed elsewhere). This choice is followed by the selection of distance measure to use. Following the suggestions of Bonin et al. (2007), we recommend that you choose the *SQRT(Simple matching coef.compl.)* measure, but the Jaccard or Sørensen coefficients would be also a good choice. When applying PCoA to, say, morphological data, the Gower distance is a good choice, as it allows easy combination of numeric and factor variables. Finally, keep the other options with their default values, including the use of the original variables from the AFLP table in the role of supplementary variables.

After you execute the analysis, a diagram summarising specimen similarity is offered by the Graph Wizard. It can be further adjusted by enabling the envelopes enclosing specimens belonging to a particular taxon (see Section 18.2 for more details) and by removing the specimen labels. The resulting graph (created with Black & white set of visual attributes) is shown in Figure 18–6.

It is interesting to note that similarly to the results of the PCA run on morphological data, even here only the *SUB* taxon is fully separated from the remaining three taxa that

18.4 Principal coordinates analysis of AFLP data

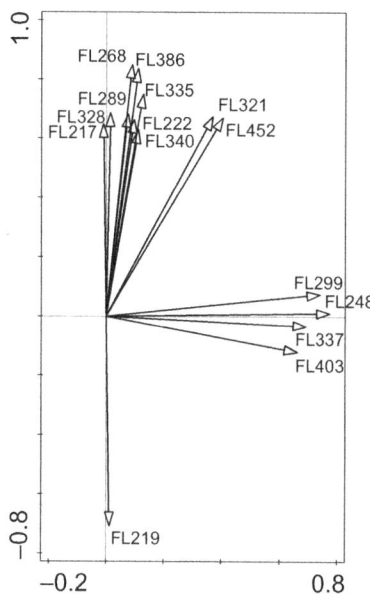

Figure 18-7 Scatter of 15 AFLP bands most correlated with the first two ordination axes of PCoA.

overlap each other to a certain extent. We have suggested earlier that the AFLP bands do not have any interpretable identity, but this is not necessarily the case with other types of data coming from molecular methods (such as the band 0/1 data from isoenzyme analyses), so we will illustrate here how to select among the numerous bands those most differentiated along the ordination axes. Please note that while the variables in the AFLP table were the basis of the distance matrix used as the response data by PCoA, they are projected into the resulting space as supplementary variables. Therefore, to select the 'best-fitting' AFLP bands, you must select them differently than normal response variables. Select the *Analysis | Plot creation options* menu command and choose the *Predictor Selection* page in the dialog box. You can select here the bands with the highest absolute value of their correlation with one of the displayed ordination axes (e.g. $|r| >= 0.60$). To do so, specify values -0.60 and 0.60 in the two fields in the upper row of spin controls, and this selects 15 bands most correlated with the axes. Then you can create a diagram using the *Graph | Scatterplots | Bands* menu command – see Figure 18-7.

The graph shows the bands most indicative for the *SUB* taxon (FL299, FL248, etc.) and those separating a large part of *NEM* specimens from the other taxa by their presence (FL268, FL386, etc.) or absence (FL219).

The suggested way of presenting the original variables (that were used to calculate the distance matrix) in the ordination space is controversial. When you plot response variables e.g. in PCA/DCA ordination diagrams, you know that the implied distance measure (Euclidean distance/chi-square distance) matches the chosen type of

> response model – i.e. linear/unimodal response, leading to arrows/symbols choice for response variables. You lose this firm ground when a different distance measure is chosen, like the simple matching coefficient. There are no known models of how the response variables change along the ordination axes for other distance types and the use of e.g. the linear change model, as we did in Figure 18–7, is an ad hoc choice.[7]

18.5 Testing taxon differences in AFLP data using db-RDA

As in the case of morphological data, unconstrained ordination (here PCoA described in the preceding section) is not appropriate for addressing questions about differences among recognised taxa. You will use the distance-based RDA (db-RDA, described in Section 6.6) instead. Because you will use the *Taxon* factor as the only explanatory variable, db-RDA represents here a multivariate ANOVA on PCoA scores, identical with those calculated in the preceding section.

To create the analysis (called *db-RDA* in the example project), click the *New* button to open the New Analysis Wizard, keep both data tables selected in its first page, choose the AFLP table as focal in the second page, and then select the *Distance-based-RDA* analysis template (without forward selection) in the *Specialized Analyses* folder in the final wizard page. In the setup wizard, keep in the *PCO Options* page the same choices as recommended in the preceding section (namely the use of the square-rooted simple matching coefficient) and leave only the *Taxon* variable in the right-hand list of the *Selection of explanatory variables* page. We recommend that you create only the second of the two diagrams offered by the Graph Wizard, as their content is very similar. The resulting graph is illustrated in Figure 18–8.

It is interesting to note that the diagram content suggests that the *HOE* and *DEG* taxa cannot be separated using AFLP data. But before you make such a conclusion, recall that to separate four classes of specimens, three constrained axes are needed, so the two overlapping taxa might be further separated by the third axis. And indeed, when you use the Graph Wizard (invoking it again e.g. by the *Graph | Advise on graphs* menu command) to plot the second offered diagram, but this time with the first and third axis,[8] it shows clearly that the third axis separates the *HOE* taxon from the other taxa (Figure 18–9). You can also try the technique described in the preceding section to display the AFLP bands most contributing to taxa separation.

Our choice of db-RDA (and of PCoA in the preceding section) was driven partly by the desire to use a distance measure (simple matching coefficient) widely adopted in the

---

[7] Not completely so in this particular case: square-root transformed simple matching coefficient complement can be viewed as a Euclidean distance computed on binary data, so the choice of arrows has some justification here.

[8] Use the *Add* button in the graph specification page to add this combination of axes and then delete the *Axis 1 & Axis 2* choice.

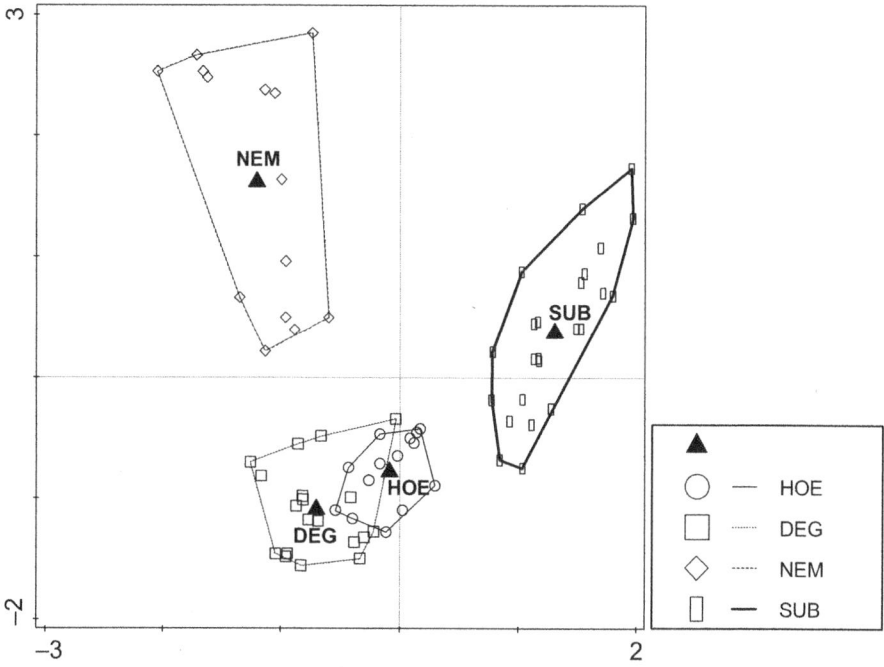

Figure 18–8 Ordination diagram with the first two axes of distance-based RDA, summarising the differences among the four taxa using AFLP data.

research field of molecular taxonomy and ecology. Alternatively, you might have used the canonical correspondence analysis (CCA) here, and the correspondence analysis (CA) for the task addressed in the previous section. The correspondence analysis is known as a method appropriate for summarising contingency tables (Greenacre 2007) and the 0/1 band data can be viewed as a very simple form of such tables.

A similar situation occurs when a community ecologist chooses to prefer a distance measure (such as Bray–Curtis distance) different from those implied by standard unimodal or linear ordination. Consequently, (s)he is required to use principal coordinates analysis (PCoA) or non-metric multidimensional scaling (NMDS) or distance-based RDA or PERMANOVA (for testing hypotheses). In our perspective, the uncertain gains in the adopted distance metric qualities do not sufficiently compensate the loss of the detail, because the distance-based methods do not provide good support to compare and evaluate responses of individual species constituting the community. Similar conclusions (albeit based on partially different grounds) were reached by Legendre and Fortin (2010, p. 842) who wrote that 'Scientists should use multiple regression (for a single response variable) or canonical redundancy analysis (RDA) when investigating response-environment relationships or spatial structures, unless the hypothesis to be tested is strictly formulated in terms of distances (or involves the variance of the distances).'[9]

---

[9] These authors show their preference for the linear constrained (canonical) method in the context of assuming that the data more suitable for unimodal ordination can be transformed (standardised) in a way to make

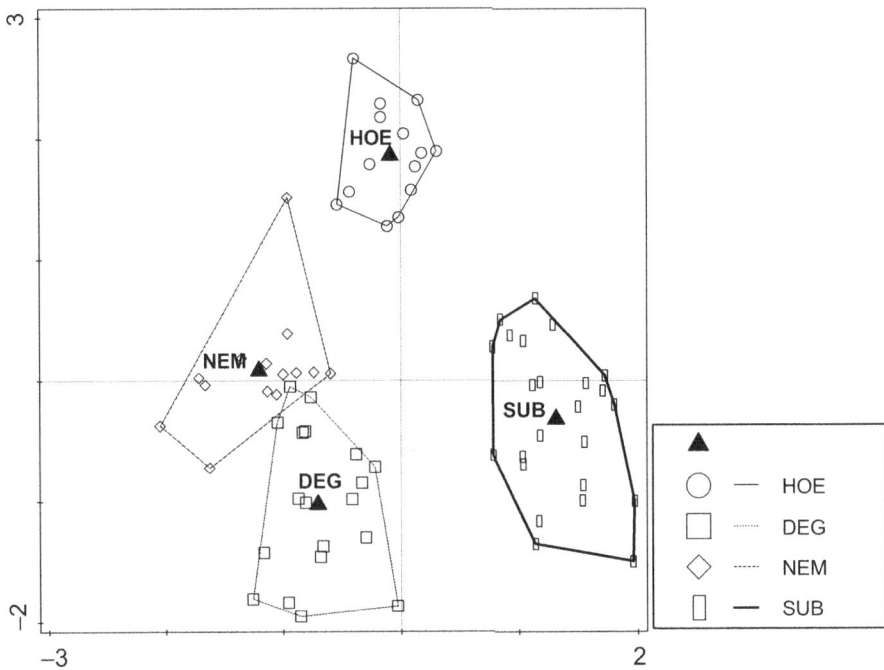

Figure 18-9 Ordination diagram with the first (horizontal) and third (vertical) axis of distance-based RDA, summarising the differences among the four taxa using AFLP data.

18.6 Taking populations into account

In the analyses demonstrated in the preceding sections, you have worked with plant specimens as independent units of the evidence concerning differences among taxa. But in reality, the specimens taken from the same local population are not independent and to properly test for differences among the species (for example), you must compare the between-species variation among population averages (i.e. among the populations) with the within-species variation of the same averages (i.e. variation among the populations of the same species).

Ideally, you would do so using the data set you have available for the specimens, simply taking the *Locality* factor into the account. But this is not possible with this data set (and the majority of other data sets), due to its unbalanced design. The number of specimens differs among populations, so you cannot set up a split-plot design (as you did in Chapter 17) with populations being whole-plots and the individual specimens

---

them acceptable in linear methods. Although we do not share the belief this is the best choice for highly heterogeneous data, we note that e.g. the Hellinger standardisation is also available among the Canoco 5 setup wizard choices.

18.6 Taking populations into account

the split-plots. The solution[10] is to aggregate the original data tables, representing each population by the averages of its specimens' values (see also Section 5.5).[11]

To work with averaged data, however, you need a new project where the data table rows represent the populations, not specimens. We will not guide you through the analyses here, but we will explain how to create this new project and note some differences of the analyses performed on the averaged data. The example project *Aggregated_Melampyrum.c5p* provides the equivalents of the four analyses discussed in Sections 18.2 to 18.5.

To create the new project, select the *Project | Create derived project | Aggregate cases* menu command and in the dialog box choose the *Locality* factor in the *Morphology* data table, before clicking the *Save* button. You must then select a name for the new project. After its creation, you can close the parent project and open the new one. This can be done easily using the *File | Recent files* submenu, where the new project name was put at the top of list. Canoco 5 creates the project data tables using the same names and the same terms for table columns. The rows are named, in our example, *aggregated specimen(s)*, so you should use the *Data | Change table name or terms* menu command to replace the row term with more appropriate *population(s)*. You can then repeat the analyses done in the preceding sections using the averaged data, just note that the term 'population' replaces the 'specimen' in the instructions.

Taking out the within-population variation affects the data in the newly created project and also some of the results. The most profound change occurs in the *AFLP* data table, where the binary (0/1) data are replaced by the fractions of specimens (in each population) with the presence of particular band. So in *PCoA* and *db-RDA* analyses, you should replace the (square root of) simple matching coefficient complement with the Euclidean distance, which is its natural counterpart, as we have already explained in footnote 7 of this chapter. In discriminant analysis, there are fewer characters selected for distinguishing the four taxa and you will see in *PCoA* analysis that the meaning of the first and second axis was swapped. The differences among the species, based on 19 population averages rather than on 71 specimens still remain obvious and significant.

---

[10] We do not recommend the alternative solution, namely the 'least common denominator' approach, unifying the number of specimens per population, as this would mean here retaining only 2*19 specimens out of the total 71.

[11] This feels like an immense loss of information, but it is not, when the among-population level is tested. As a parallel in univariate statistics, performing a split-plot ANOVA with the tested factor changing at the level of whole-plots produces a F-statistic and its degrees of freedom (and hence the resulting p value) identical with those that you obtain by first averaging the values of the response variable within whole-plots and then performing a simple one-way ANOVA on the averages.

19 Case study 8: Separating effects of space and environment on oribatid community with PCNM

In this case study, you will learn how to test for existence of spatial structuring of the variation in biotic community composition and how to correctly test for the effects of environment upon community composition when the spatial effects are present in your data, manifested as a spatial correlation among the sampling locations. We also show how to quantify the spatially structured and spatially non-structured fractions of the community variation explained by the environment and the spatial variation in the community composition not related to the environment, and how to visualise these fractions.

You will use a data set first published by Borcard et al. (1992) and describing oribatid mite community in a peat bog. The data set is based on 70 substrate cores, accompanied by measured environmental variables, characterising local conditions: peat density, water contents, substrate type, relief type, and shrub cover. These data are present in the first two data tables of the *Oribatida.c5p* project: the *Oribatids* data table with the counts of 35 oribatid taxa and the *Environment* table with environmental descriptors. The third data table (named *Coordinates*) stores the X and Y coordinates describing the spatial position of each core within the sampled area, which was 2.5 metres wide and 10 metres long. The additional, Xc and Yc variables represent the centred (i.e. with a zero average) versions of X and Y, and Xc^2 and Yc^2 are the centred second powers of Xc and Yc, respectively. Their use is explained in Section 19.2. Alternatively, you can import the data from the three sheets of the *Oribatids.xlsx* spreadsheet file. Please note that the data table in the first sheet (*Species*) must be transposed during the import, as the cores are represented by columns.

19.1 Ignoring the space

Your first analysis will not work with the spatial positions of the cores at all. It represents the common way of working with the field data, when the information about spatial location of individual cases is either not available or it is deemed unimportant. The *Oribatids-and-environment* analysis of the example project presents both unconstrained (PCA) and constrained (RDA) ordinations of the oribatid community and so it allows you to compare the overall variation in the community composition with the effects of measured environmental descriptors. To re-create this analysis using the New Analysis Wizard, click the *New* button, keep only the *Oribatids* and *Environment* tables selected

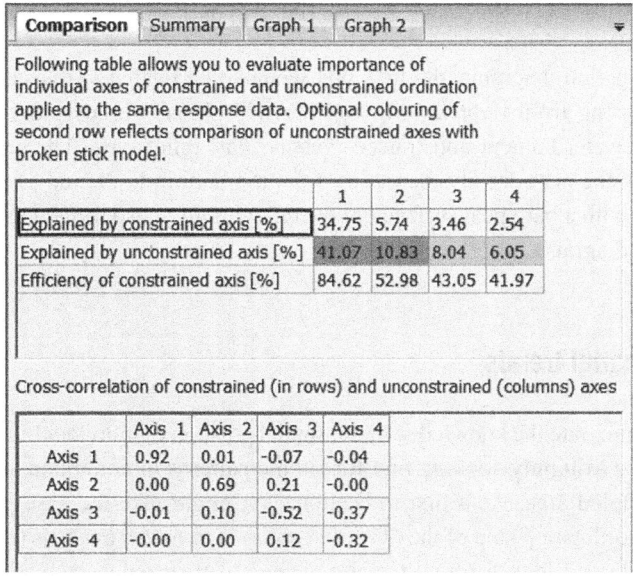

Figure 19-1 Snapshot of the Comparison page of the analysis notebook for the *Oribatids-and-environment* analysis.

in its first page, keep the *Oribatids* table as the focal one, and then select the *Compare-constrained-unconstrained* analysis template from the *Standard Analyses* folder in the third page. In the Analysis Setup Wizard, you must select all five environmental variables as Members. We have chosen *No* in the dialog box shown before the analysis execution which asked about the increase of the number of calculated axes from 4 to 11. Both suggested graphs were created, but only with the 10 best-fitting oribatid taxa shown.

When looking at the *Comparison* page in the analysis notebook (see Figure 19–1), you might note that the first two entries in the second row of the top table are green. They correspond to the first two axes of unconstrained ordination (PCA) and the green colour indicates that these axes explain more variation than expected based on the broken-stick model (see Section 12–1) and that they might be therefore worthy of interpretation. For constrained ordination (summarised in the preceding row of the table), you can see that its first axis explains a very large fraction of the total variation (34.8%) and it is quite efficient when compared with the unconstrained ordination reference.[1]

In addition, the first axis seems to report a similar kind of gradient in both constrained and unconstrained analyses, as can be seen from the very high correlation (+0.92) among these two axes, reported in the second table. Yet the correlation between the second axes is also quite high (+0.69), which can be seen also indirectly by comparing the directions of oribatid taxa arrows in the two graphs from constrained and unconstrained ordination

---

[1] The efficiency of constrained axes (shown in the third table row) is computed as the percentage of variation explained by a constrained axis divided by the percentage explained by the corresponding (same-order) unconstrained axis, multiplied by 100.

(*Graph 1* and *Graph 2*, respectively, in the example analysis notebook, not reproduced here).

In both ordination diagrams, the first axis seems to be related to the substrate water content (increasing from right to left) together with the shrub cover decreasing with the increasing water content and 'raised' versus 'flat' relief type. The second axis is correlated with the peat density increasing from the bottom to the top of the diagram, with the cores with a bare peat surface or covered with litter also separated at the top of the ordination diagram.

19.2 Detecting spatial trends

You will now integrate the knowledge of the spatial position of individual cores into your analyses. You start lightly, looking first for simple patterns of community composition across the sampled area. As a first indication, you might take the case scores of the unconstrained ordination step of the *Oribatids-and-environment* analysis (demonstrated in Section 19.1) and look for a systematic change of their values across the sampled area. This might be done, for example, using the case scores of the first PCA axis and summarising their relation to the X and Y coordinates with a loess smoother model. The first PCA axis represents the most important gradient of change in oribatid community composition, irrespectively of any of the measured environmental variables.

You can also handle the constrained case scores (CaseE) of the constrained ordination (RDA) step in the same way. Because these scores represent the best gradients in community composition that can be defined using known values of environmental variables, you will obtain in this way an indication of the spatial variation in the environmentally determined community variation and you can compare it with the spatial patterns revealed using the scores from the unconstrained ordination.[2]

To create a contour diagram combining smoothed changes of both PCA.1 and RDA.1 scores (see *Graph 3* in the analysis notebook of the *Oribatids-and-environment* analysis), you must use the *Graph | Attribute plots | XY(Z) diagram* menu command and in the dialog box choose the *Spatial Coordinates / x* and *Spatial Coordinates / y* items, respectively, in the upper left and upper right list. Then in the lower left list choose the *Analysis 'Oribatids-and-environment' / Step ConstrainedAnalysis / Constrained case scores / Axis 1* for plotting the RDA.1 case scores, and in the lower right list choose the *Analysis 'Oribatids-and-environment' / Step UnconstrainedAnalysis / Case scores / Axis 1* for plotting the PCA.1 case scores. In the upper right area of the dialog, choose the *Loess model* instead of the default *Symbol plot* choice, check the *Iso-scale X and Y* box and possibly uncheck the *Add points to model* box. The resulting graph shows clearly that the scores of individual cores on the first axis of both PCA and RDA change in a systematic way (they increase) from the top to the bottom.

---

[2] This approach has some limitations, however, because (a) we already know the resulting plots will be quite similar, given the high correlation between the PCA.1 and RDA.1 axes (see earlier discussion) and (b) for an easy comparison, we have to pick just one of the axes summarising the community variation in both unconstrained and constrained ordination.

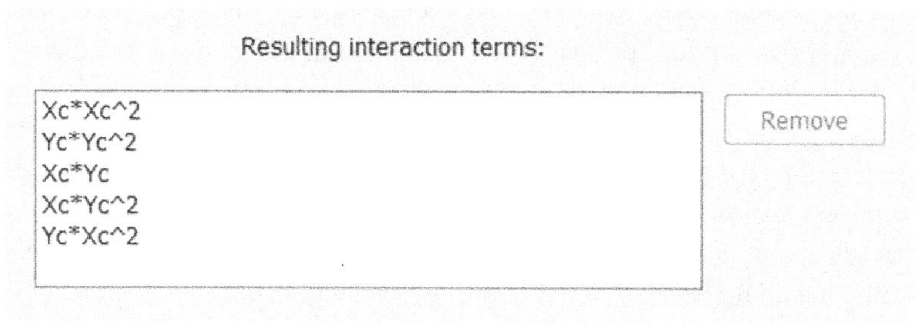

19.2 Detecting spatial trends

Resulting interaction terms:

```
Xc*Xc^2
Yc*Yc^2
Xc*Yc
Xc*Yc^2
Yc*Xc^2
```

Remove

Figure 19–2 Polynomial terms defined as interactions in the Analysis Setup Wizard.

You will now use a so-called **trend surface polynomial** to describe directly the effect of spatial position on the community variation and test its significance. To do so, you will use the X and Y coordinates, as well as the other terms of their third-order polynomial: second and third powers of X and Y and multiples of X and Y (XY, X^2Y and XY^2). Instead of using the original X and Y variables (with all values positive), you will use the centred versions (Xc and Yc) as well as the centred squares of these coordinates (Xc^2 and Yc^2). This is recommended to reduce the high correlation (and hence dependency) among non-centered polynomial terms.[3] Polynomial terms will be used as explanatory variables in a constrained ordination, but you will not accept blindly all of them, but rather find their subset that describes well the change in oribatid community composition across the sampled area. To be able to define a trend-surface polynomial, you must use not only the Xc, Yc, Xc^2, and Yc^2 variables present in your data, but also define the other terms. They are formed with the help of the Analysis Setup Wizard as interactions and they can be defined only when the QuickWizard mode is switched off, so this must be done first (see the box Quick Wizard and Slow Wizard in Section 2.3).

To reproduce the example *Oribatids-trend-surface* analysis, click the *New* button, uncheck the *Environment* data table on the first page of the New Analysis Wizard, and choose the *Interactive-forward-selection* analysis template on its third page. At the second page of the Analysis Setup Wizard (named *Response and Explanatory Data (1)*), check the *Define interactions of explanatory variables ("spatial coordinates")* box. At the third page (*Selection of explanatory variables*), move the x and y variables to the left-hand list, keeping the other four selected. At the fourth page (*Interactions of explanatory variables*), the polynomial terms must be defined.[4] After you define them, the list of *Resulting interaction terms* (in the lower part of the wizard page) should look as shown in Figure 19–2.

The other choices shown in the setup wizard pages should be left as suggested. When the analysis is executed, the *Forward Selection Step* dialog box appears. Its contents

---

[3] See Legendre and Legendre (2012), p. 569. While Canoco centres and standardises all explanatory variables, this comes too late here – only after the interaction terms were computed. Also note that the centring cannot achieve full independency of the polynomial terms – for that, the orthogonalisation of all terms would be needed.

[4] For example, to create the third power of the X coordinate, you must select Xc in the left-hand list and Xc^2 in the right-hand list and click the *Create* button. To create X^2Y term, you select the Xc^2 variable in the left-hand list and Yc in the right-hand list and then click the *Create* button.

and use were already discussed elsewhere (see Section 5.6), but we note here that it is appropriate to limit the type I error inflation by selecting, at the dialog box bottom, some *P values correction* method, e.g. the *False discovery rate*. Before you select any of the offered candidate terms, the top list shows the simple (independent) effects of each of the nine polynomial terms and it is quite clear that the 'Y-only' terms have the strongest relation with community composition, while the three 'X-only' terms do not have any significant relation. Once you start selecting the terms, however, some of the terms including horizontal direction (X) become significant as well. With no a priori expectation, you can perform the selection by choosing the topmost candidate term of each step until its *P(adj)* value exceeds the 0.05 threshold, selecting Y, Y^3, Y^2, XY, and X terms, but with a very limited contribution of the last two (3.4% out of the total variation in oribatid data).

Reviewing the biplot diagram (*Graph 1* in the example analysis), you can see that nine of the ten best-fitting oribatid taxa decrease their abundance with increasing Y coordinates. But in general, the biplot diagram is not the best way of presenting results for this kind of analysis.[5] Instead, you can again summarise the pattern of (constrained) case scores in the XY coordinate space (see *Graph 2* in the example analysis), using the case scores on the first constrained axis of the present analysis. The first axis accounts for almost 75% of the total variation explained by the five constrained axes (i.e. by the simplified polynomial) and so it characterises quite reliably the fitted trend-surface polynomial.

Note, however, that your present approach is able to reveal only far-reaching trends that take the form of a monotonous change, or of a smooth non-linear change with one maximum ('peak') or one minimum ('valley') with a second-order polynomial, or of a change containing one maximum and one minimum with a third-order polynomial. You can neither reveal other types of spatial heterogeneity pattern (e.g. smaller-scale patchiness) commonly found in the field, nor can you describe and compare spatial heterogeneity occurring at multiple spatial scales. To do so, you can use, instead of a trend-surface polynomial, the eigenfunction spatial predictors, discussed already in Section 10.2 and illustrated in the remaining sections of this case study.

19.3 All-scale spatial variation of community and environment

Canoco 5 supports multiple variants of PCNM (principal coordinates of neighbour matrices) analyses,[6] but here we focus on simpler ones, directly supported by analysis templates. Before opening the New Analysis Wizard, make sure the QuickWizard mode

---

[5] Included multiple polynomial terms express the effects of just two directions, so it is not possible to reliably judge from the arrows for individual terms the actual change of species abundance in space.

[6] We use the still prevailing name PCNM, although the method has been improved in the meantime (Dray et al. 2006) and also its scope was extended. The the Canoco 5 implementation covers primarily what Legendre and Legendre (2012) call dbMEM (distance-based Moran's eigenvector maps). With the possibility of importing a distance matrix into Canoco data table, the methods of general MEM, including binary MEM, are (at least partly) supported as well, as long as the user is able to quantify the connection network using a matrix with connectivity measures (either distances or similarities).

is switched on (you have possibly switched it off in the preceding section), so that you do not need to go through all the setup pages of this quite complex analysis. To re-create the example *Oribatids–PCNM* analysis, keep all three data tables chosen in the wizard and on its last page choose the *Var-part-PCNM* template in the *Variation Partitioning Analyses* folder. Note that there are multiple instances of this template, because the Canoco Adviser cannot understand the meaning of variables in the two tables with predictors and you must select the template described as *oribatid taxa ~ space(spatial coordinates) + environmental variables*, as the table with *spatial coordinates* contains the X and Y variables (here you can again use the non-centred X and Y, because they are used only to compute the distances among the points).

As you have chosen an analysis template without *–FS* at the end of its name, the forward selection will be performed only for the spatial eigenfunction predictors, but not for the environmental variables. This decision might not be appropriate for other data sets, but in this data set all the recorded variables contribute to some extent to an effective prediction of oribatid community composition, and so we can simplify the analysis.

The first page shown in the setup wizard allows you to choose whether the variation partitioning is calculated using the adjusted explained variation values and we recommend that you keep the box checked (see Section 5.7 and Peres-Neto et al. 2006). The fifth page of the setup wizard (when run in QuickWizard mode) named *PCO Options (3)* allows you to specify how the spatial eigenfunction predictors are computed and their subset to be used in the analyses. Recommended choices for your analysis are shown in Figure 19–3.

The cut-off threshold choice in the middle-right area of the wizard page corresponds to recommended defaults of PCNM (see e.g. Legendre & Legendre 2012, p. 863). The choice at the page bottom implies that all the spatial eigenfunctions with positive eigenvalues (corresponding to positive spatial correlations at progressively smaller scales) will enter the forward selection. When leaving the page, you are warned that the global permutation test on the joint effect of all computed spatial eigenfunctions is skipped, because it is not very informative: there are too many predictors (covering all spatial scales that can be represented given the sampling design properties) and the constrained ordination model is therefore too complex.

When executing the analysis, you must select a subset of the computed spatial eigenfunctions (labelled as *PCO.n*) that predict best the oribatid community. Only four of them (with relatively low numbers and hence corresponding to gross spatial patterns[7]) seem to have significant simple (independent) effects (when using false discovery rate estimates for the selection), but as you start the forward selection, more of them are chosen, effectively covering the first eleven predictors plus PCO.16 and PCO.20. After you choose PCO.20, a warning dialog informs you that the (adjusted) percentage of explained variation exceeds the value estimated in the preceding analytical step where all spatial predictors were used, and the selection is suggested to stop (see Blanchet et al. 2008 for a detailed description of this strategy).

---

[7] The spatial eigenfunctions in this analysis describe also the overall trend along the Y axis that you have found earlier. Legendre and Legendre (2012, p. 868) recommended removing such a trend separately, not through the spatial eigenfuctions. We will demonstrate how to do it in the following section.

Separating effects of space and environment

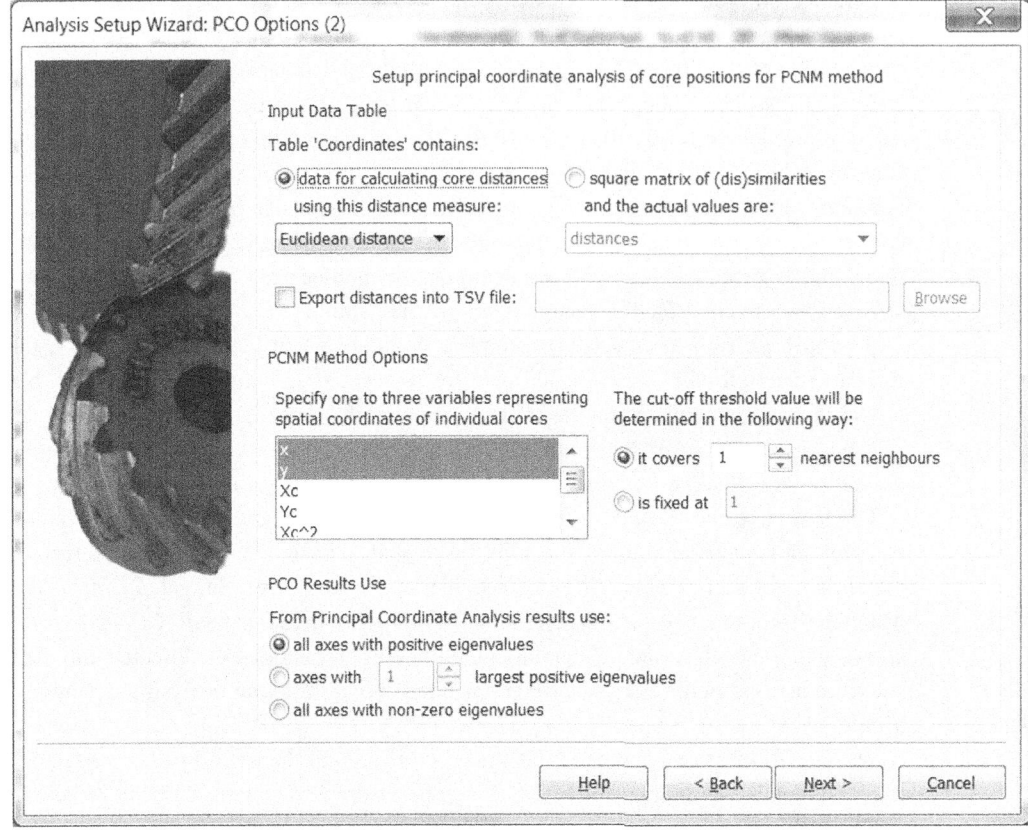

Figure 19-3 Setup wizard page where PCNM options are chosen.

After the analysis is executed, the Graph Wizard offers four different biplots representing different fractions of the oribatid community variation that follow from the variation partitioning model separating the effect of spatial variation and the space-independent environmental effects. We suggest that you keep the second and the fourth suggested biplot selected (corresponding to *Graph 1* and *Graph 2* in the example analysis). You will note that the content of *Graph 1* is identical with *Graph 1* of the *Oribatids-and-environment* analysis. This is indeed expected, because you show here the total effect of environmental variables, without taking the spatial variation into account. In contrast, *Graph 2* shows the conditional effects of environmental variables, after removing all the variability in oribatid community composition that could be (alternatively) explained by the spatial eigenfunctions. Consequently, the explanatory power of the shrub cover (None, Few, Many) or of the ground relief form (raised vs flat) greatly diminishes, while the effects of water contents and substrate density, as well as the community difference in the places with bare peat, remain similarly strong.

Back in Chapter 10 (Section 10.2), we have mentioned the question about the extent, to which the environmental factors are spatially structured. While your preceding analysis provides an indirect hint about the spatial variation of environmental properties (through a large overlap between the variation explained by the space and by the environment), you will now approach this question more directly. To do so, you will predict the environmental data using spatial eigenfunctions and test their predictive ability. The stepwise selection done in the *Oribatids-PCNM* analysis does not necessarily provide a relevant choice of the spatial eigenfunctions, because it was based on predicting oribatids composition. So while you can use the eigenfunctions computed earlier, you must perform their selection independently here. Another important feature of this new analysis is that the environmental measurements will act as response variables, but they are not measured on comparable scales. Consequently, a linear ordination with both centring and standardisation of response variables must be used here.

As you want to use already computed spatial eigenfunctions as explanatory variables, you cannot create the analysis with the New Analysis Wizard, as it works only with data present in project tables.[8] Instead, you must create the new analysis using the *Analysis | Add new analysis | Customized* menu command. Before entering the Analysis Setup Wizard, you must choose the analysis name (the analysis in the sample project uses the name *Environment-PCNM*), and then you go through the full sequence of wizard pages (QuickWizard mode is ignored here). Please note that the customised analyses in Canoco 5 have more features than we demonstrate here.[9]

In the second page of the setup wizard, select the *Environmental variables in 'Environment' table* as the response data and then click the *Advanced role* button in the lower area (labelled *Explanatory Data*). In the *Advanced Role Assignment* dialog, click the *New* button, select the *Oribatids-PCNM | PCO* item in the *Use score results from* list and choose the *PCO* scores* in the *Select score type* list before closing the dialog with the *OK* button. Back in the setup wizard, in the *Ordination Options* page the RDA is preselected, as well as the *Center and standardize* choice for the environmental variables. Finally, in the *Testing and Stepwise selection* wizard page, change the *All constrained axes test* choice at its top to *Forward selection of expl. variables*.

During the analysis execution, you can proceed with the forward selection in exactly the same way as before, selecting the PCO scores based on the false discovery rate criterion until the *P(adj)* value of the topmost score exceeds the 0.05 threshold. This time, you will select eight spatial eigenfunction predictors (1, 2, 4, 6, 8, 10, 19, and 20) and these account for 36% (or 27.6%, if you focus on adjusted variance values) of the total variation in environmental data.

Canoco 5 does not offer any graphing advice for customised analyses, but you can easily create an informative biplot with the *Graph | Biplots | Environmental variables + explanatory variables* menu command. It is interesting to see that the variation along the first (most important) constrained axis is dominated by the PCO.2 predictor and that particularly the water contents and the cover of shrubs correlate with it, while the second

---

[8] You could of course copy the corresponding scores from the second (PCO) step of the *Oribatids-PCNM* analysis to the Clipboard (see section 7.10.2.2 in the Canoco 5 manual) and then create a new data table in the project and paste the scores into it. This is, however, too complex a procedure for the intended brief use of these data.

[9] See Canoco 5 manual, section 4.4, particularly the subsection 4.4.4.

axis is correlated with the PCO.1 predictor (but also with others) and the substrate density, as well as the presence of various cover types are correlated with this axis. You can also create an attribute plot in the XY space, similar to the one created in the *Oribatids-and-environment* analysis, as illustrated by *Graph 3*, where the constrained case scores on both the first and second RDA axis were visualised.

When comparing the pattern with the *Graph 3* in the *Oribatids-and-environment* analysis, one can see clearly that while both the oribatid community and the properties of the environment are strongly spatially structured, their spatial variation differs to some extent. If you now want to focus on the part of spatial structuring in the oribatid community data that cannot be interpreted as a result of spatial structuring of the environmental variables (see footnote 4 in Chapter 10), you must use the PCO.n eigenfunctions selected here as a priori covariates – in the same way the X and Y coordinates are used to detrend the data in the analysis described in the following section – as a part of a new PCNM variation partitioning analysis (but this advanced type of analysis is not illustrated here or in the example project).

19.4 Variation partitioning with spatial predictors

PCNM used in the *Oribatids-PCNM* analysis that you have created in the preceding section, is implemented as a part of an analysis template that performs variation partitioning. You can therefore compare the size of unique effects of the spatial eigenfunctions and of the measured environmental variables, as well as to quantify the overlap in their explanatory ability. We have not discussed the variation partitioning results (summarised by the *PCNM* page in the analysis notebook) in the preceding section. We will do it now, but using a new analysis, named *Oribatids-var-part* in the sample project, which further refines the approach taken in *Oribatids-PCNM*.

Namely, you will follow the advice given in Legendre and Legendre (2012, p. 868) that it is better not to let the spatial eigenfunctions model global trends in data, as they can be modelled much more effectively using a linear term. You will use here only the Y variable (the non-centred version is sufficient), because you have already learned that the Y direction (unlike X) represents an important spatial trend in the oribatid community data. Please note that using the Y coordinates as a covariate and using them at the same time (together with X) to calculate the truncated distance matrix to generate spatial eigenfunctions is a perfectly valid approach. With Y used as a covariate, you remove the linear trend from further consideration and the spatial eigenfunctions then model more complex spatial patterns.

To re-create the example *Oribatids-var-part* analysis, you must use the New Analysis Wizard, with all three data tables selected, and choose in the third wizard page the *Var-part-PCNM-FS-covariates* analysis template, namely its instance with the *oribatid taxa ~ space(spatial coordinates) + environmental variables | spatial coordinates* formula. The a priori covariates are in this analysis used also for the stepwise selection of environmental variables: remember that you remove the global trend and hence some of the originally effective variables might no longer hold significant explanatory power.

In the second setup wizard page (assuming the QuickWizard mode is active), remove *x* from the list of selected covariates and on the fifth page (*PCO Options (3)*) choose *x* and *y* variables in the list.

Compared with the earlier variation partitioning analysis, the stepwise selection of environmental descriptors omits the *Substratum.density* and you also select a smaller number of spatial predictors: 1, 3–7, 10, 11, and 20. This time we recommend that you do not accept any of the graph suggestions made by the Graph Wizard, clicking the *Cancel* button when they are offered.

When reading the results of any type of variation partitioning in Canoco 5, it is best to start at the bottom of the page (see Figure 19–4) to make sure you understand the formal labels used for the distinguished fractions. In your analysis, *a* represents the unique effect of environmental variables, after accounting for the effect of spatial eigenfunctions (and for the linear trend in the Y direction); *b* is the unique effect of spatial eigenfunctions, partialling-out the effect of environmental variables and the linear trend; and *c* is the shared portion of the variation in oribatid community that can be accounted for either by the environmental variables or by the spatial eigenfunctions (but with the linear trend again excluded). When you jump now to the table at the page top, you can immediately see that while both unique effect types are substantial in size, the overlap is even larger than either of the two unique effect fractions.

In this case study, you were lucky that in the forward selection procedure, similar numbers of environmental variables and of spatial eigenfunctions were chosen (10 and 9, respectively). In other cases this might not happen and the percentages displayed in the third and fourth columns of the *Variation Explained* table might be tricky to compare. Perhaps more informative is – in such a case – to compare the amount of variation explained in each group of variables per single degree of freedom (i.e. per one constrained axis), as shown in the *Mean Square* column. In your analysis, however, the outcome is more or less similar, indicating a slightly larger unique effect of space.

19.5 Visualising spatial variation

Interpretation of the spatial eigenfunctions chosen within your analysis is inherently difficult. We have only a vague notion that the larger is its order, the finer spatial scale the eigenfunction represents. In addition, contiguous groups of eigenfunctions are often chosen (as in your three example analyses) and one should be careful when considering interpretation of individual eigenfunctions. For example in the *Oribatids-var-part* analysis, this might be perhaps done for the PCO.20 spatial eigenfunction and it is plotted in *Graph 1* using the method discussed earlier, namely by visualising a fitted loess smoother model. *Graph 1* was created by choosing the *Analysis 'Oribatids-var-part' / Step PCO / PCO* scores / Axis 20* as the plotted attribute. For finer-scale spatial eigenfunctions such as this one, one can expect that they represent more localised and suddenly changing patterns. This was reflected when choosing loess model parameters, namely the use of a local second-order polynomial and the span value 0.67.

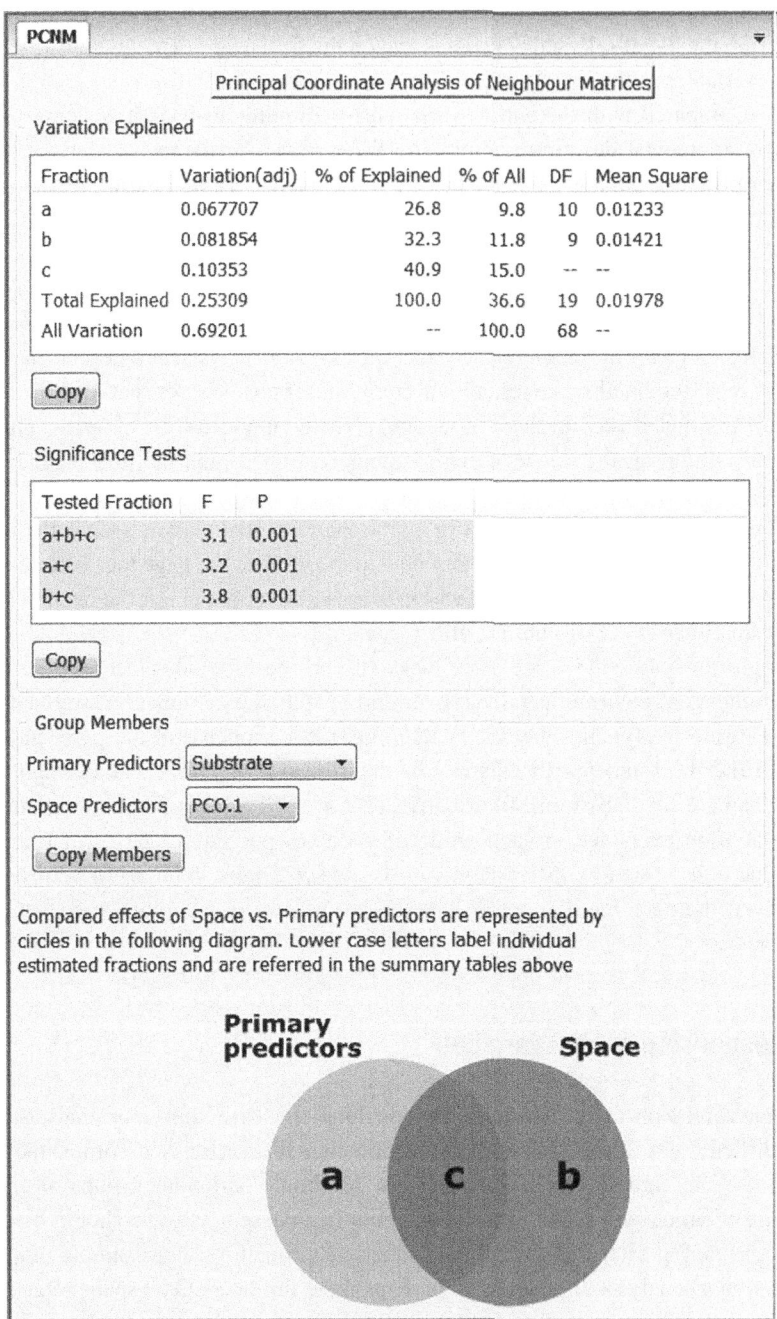

Figure 19–4 Page in the *Oribatids-var-part* analysis notebook summarising variation partitioning results.

19.5 Visualising spatial variation

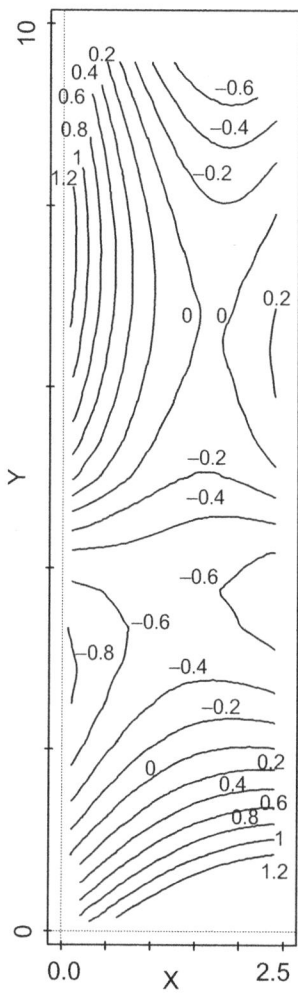

Figure 19-5 Visualisation of the case scores constrained by selected spatial eigenfunctions of the *Oribatids-var-part* analysis.

Instead of displaying a contour plot for a fitted loess (or GAM) model, you can create a colour attribute plot simply by keeping the *Symbol plot* option selected in the *Create XY(Z) Plot* dialog and selecting the plotted attribute in the *Colour based on* list in the lower right area of the dialog. Such a plot may provide substantial insight particularly for data sampled over a regular spatial grid or along one or a few transects.

As an alternative to visualising individual spatial eigenfunctions, you can use the constrained case scores from a PCNM analysis step that represent the effects of selected spatial eigenfunctions. This is illustrated in *Graph 2* (reproduced in Figure 19-5), where the constrained case scores on axis 1 of the Spatial Effects step are plotted.

Yet another possibility would be to split the selected spatial eigenfunctions into two or three subgroups corresponding to different spatial scales and fit their effects (using constrained ordination) separately (see also Legendre & Legendre 2012, pp. 870–872). There should be no overlap in the explained variation among such sub–groups, because individual spatial eigenfunctions are mutually uncorrelated (orthogonal). This implies that you can interpret each such subgroup separately. The definition of subgroups is arbitrary and sometimes difficult to achieve. For the results of the *Oribatids-PCNM* analysis, perhaps the PCO.1 to PCO.11 eigenfunctions would be one group, representing gross-to-medium scale spatial variation, while PCO.16 and PCO.20 would form a group of eigenfunctions reflecting small-scale variation. But in the *Oribatids-var-part* analysis, only the PCO.20 eigenfunction stands apart from the almost contiguous range of the remaining selected eigenfunctions.

20 Case study 9: Performing linear regression with redundancy analysis

In this chapter, we demonstrate how to perform linear regression in Canoco 5, using redundancy analysis (RDA), as well as how to fit a (generalized) linear model in this program. We hope that this case study will help the reader understand better the meaning of the case scores and the difference between constrained and unconstrained axes in a constrained ordination. In addition, the approach shown here allows one to abandon the parametric tests of significance, used in the context of classical regression, and therefore to use these methods under conditions where classical assumptions about the distribution of modelled variables cannot be met.

20.1 Data

The data for linear regression are taken from the first case study (Chapter 12), where the composition of bird assemblages was related to habitat properties. In this chapter, you will focus on predicting abundance of a single bird species – ring ouzel (*Turdus torquatus*) using site altitude. These data are stored in two sheets of the *chap20.xlsx* file: sheet *Regr-R* contains the two variables (response variable *TurdTorq* and predictor *Altit*) ready for import into a data frame of the R program or into any other general statistical package; sheet *Regr-Canoco* contains the two variables separated for the import into the Canoco 5 project. This is because in Canoco, the 'response data' and 'explanatory data' cannot be in the same data table. This project was already created for you and it is stored in the *chap20.c5p* file.

20.2 Linear regression using program R

When you import data from the *Regr-R* sheet, you can fit a simple linear regression using the *lm* function. The variable *TurdTorq* will be used as a response variable and *Altit* as a predictor.

```
lm.20 <- lm(TurdTorq~Altit,data=ch20)
summary(lm.20)
...
```

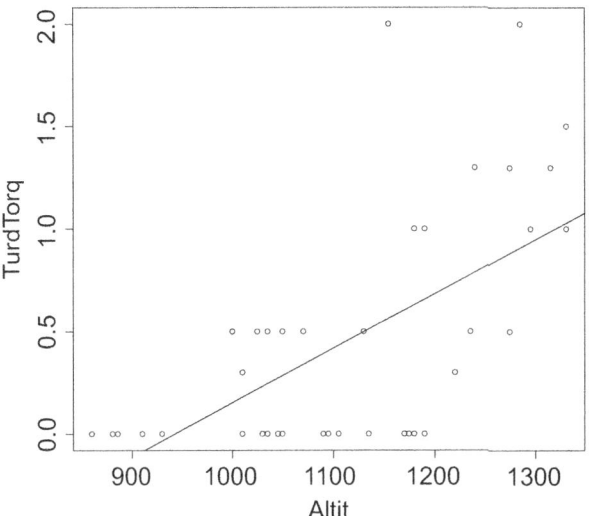

Figure 20-1 Fitted regression line with original data.

```
Coefficients:
              Estimate Std. Error t value Pr(>|t|)
(Intercept) -2.5010062  0.5811748  -4.303 0.000102 ***
Altit        0.0026524  0.0005221   5.080 8.67e-06 ***
...
Residual standard error: 0.4543 on 41 degrees of freedom
Multiple R-squared:  0.3863,    Adjusted R-squared:  0.3713
F-statistic: 25.81 on 1 and 41 DF,  p-value: 8.671e-06
```

It is worth noting that the fitted linear regression model can be expressed as TurdTorq = $-2.50101 + 0.00265 *$ Altit + residual. To create a plot displaying the fitted line (see Figure 20-1), you can use the following two commands in R.

```
plot(TurdTorq~Altit, data=ch20)
abline(lm.20, lwd=2)
```

The R^2 value in the output of the function *summary* (shown above) tells you that ca 38.6% of the variability in the variable *TurdTorq* was explained by the regression (i.e. by the altitude). The same number can be obtained using the analysis of variance table (shown below), by dividing the sum of squares explained in regression by the total sum of squares (5.327 / 13.7907 = 0.38627), where the total sum of squares is equal to (5.3270 + 8.4637), i.e. the sum of explained and residual sums of squares:

```
anova(lm.20)
Analysis of Variance Table
```

20.2 Linear regression using program R

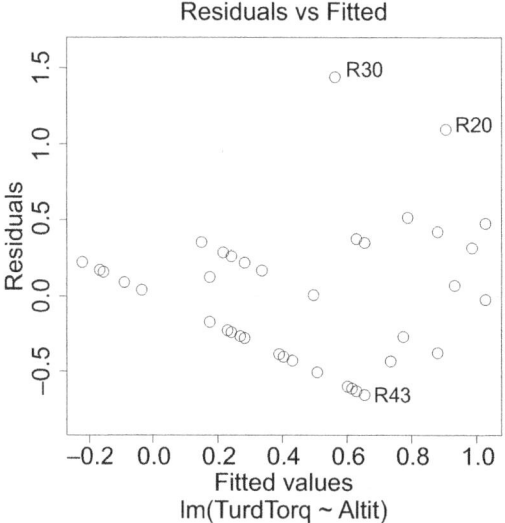

Figure 20–2 Regression residuals plotted against fitted values.

```
Response: TurdTorq
            Df Sum Sq Mean Sq F value    Pr(>F)
Altit        1 5.3270  5.3270  25.805 8.671e-06 ***
Residuals   41 8.4637  0.2064
...
```

As expected, the significance level in the analysis-of-variance table's F-test ($8.67.10^{-6}$) is equal to the one in the t-test of the regression coefficient for *Altit*. This is because you have only one predictor (one independent variable) in this model.

When fitting a regression model, you should also check whether the assumptions about the data are met, to validate the approach. You can do this with some of the **regression diagnostic** plots, such as the one in Figure 20–2, created using the following command:

```
plot(lm.20,which=1,add.smooth=F)
```

This plot of residuals against the fitted values of *TurdTorq* may help reveal systematic variation in the values of the response variable (i.e. in the abundance of *Turdus torquatus*) not described properly by the chosen regression model.[1] There are perhaps more effective varieties of diagnostic plots, but this one will prove useful when interpreting the results you obtain later with RDA.

---

[1] No such patterns can be seen in Figure 20–2, except the 'banding' caused by the limitation of response variable values to (a few) whole numbers, with the fitted values varying continuously.

Summary of Results

Method: RDA

Total variation is 13.79070, explanatory variables account for 38.6%
(adjusted explained variation is 37.1%)

Figure 20-3 Upper part of the *Summary* page of the analysis notebook.

20.3 Linear regression with redundancy analysis

You will start here with the *chap20.c5p* project file or you can create your own project with two data tables: *Response* with a single variable *TurdTorq* (based on columns A and B of the *Regr-Canoco* sheet) and *Predictor* with a single variable *Altit* (based on columns D and E of the *Regr-Canoco* sheet).

Create a new analysis using the New Analysis Wizard (starting with the *New* button below the list of analyses), confirm the *Response* data table as focal, and in the third wizard page select the *Constrained* analysis template from the *Standard Analyses* folder. When the Analysis Setup Wizard window opens (we assume that you use the QuickWizard mode), keep *Altit* selected as the explanatory variable. In the following wizard page (*Ordination Options*), you will find that you cannot change the choice of ordination method (RDA). This is because the *Response* data table was created as a non-compositional table, so that the unimodal methods cannot be applied to it. Another reason, mentioned in the explanation field below the method selection area, is that the unimodal methods cannot be used for data tables with just one response variable. Keep the other options as suggested in the page and make no changes on the following page either, except increasing the *Number of permutations* from 499 to 999.

After clicking the *Finish* button at the last page of the Analysis Setup Wizard, the analysis is performed and the Graph Wizard window is shown, but close it with the *Cancel* button and look at the created analysis notebook. In the upper part of the *Summary of Results* area (shown in Figure 20-3), you can see that the total variation is equal to 13.7907 and hence identical to the value implied by the ANOVA table of the regression model (as the sum of the two sums-of-squares given there). Similarly, the value 38.6% corresponds (after dividing it by 100) to the coefficient of determination (R^2) reported in R.

Perhaps more interesting is to look at the *Summary Table* part of the *Summary* page, where the first eigenvalue (0.3863) again shows the R^2 value. This is because the eigenvalues normally report, in linear ordination methods, the fraction of the total variation explained by particular ordination axis, and in your case, with a single explanatory (and single response) variable, there is only one constrained axis, namely the Axis 1 of RDA. Also, there is just one other axis (Axis 2) that summarises the remaining variation not explained by the predictor(s) – here by the altitude. So you can say that the first axis represents the fitted values (the systematic part of the model), while the second axis represents the regression residuals (unexplained variation, the stochastic part of the model).

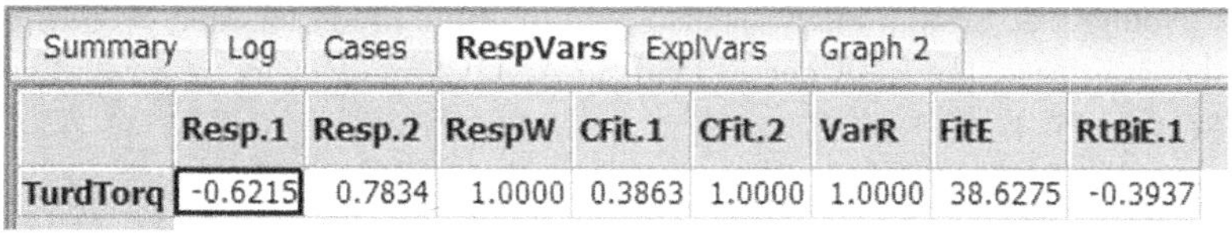

20.3 Linear regression with redundancy analysis

	Resp.1	Resp.2	RespW	CFit.1	CFit.2	VarR	FitE	RtBiE.1
TurdTorq	-0.6215	0.7834	1.0000	0.3863	1.0000	1.0000	38.6275	-0.3937

Figure 20–4 Scores of response variables in the analysis notebook shown in non-brief mode.

You can see it more easily by creating a scatter of case symbols in Canoco 5 (use the *Graph | Scatterplots | Landscape quadrats* menu command), which yields a pattern of points identical with Figure 20–2, which was created as a diagnostic plot in R.

These findings can be generalised to constrained (canonical) and unconstrained (non-canonical) axes in all constrained ordinations: the former ones represent the fitted variation, while the latter ones represent the residual (unexplained) variation.

In the lower right area of the *Summary* page, the results of the Monte Carlo permutation test are reported and it is interesting to note that the pseudo-F statistic is exactly equal to the F value reported by R. The resulting P-value is higher than the *p-level* reported in R, because it was limited by the number of performed permutations [(0+1)/(999+1) = 0.001).

You will now look into the RDA scores produced by Canoco 5 for your analysis and stored in the analysis notebook. Most likely, you will not see them there yet, because after installation, Canoco 5 shows the analysis notebooks in so-called brief mode, which hides the details of limited value for the users in their everyday work. See Section 2.3, near its end, for the instructions on how to change the notebook mode. Click then on the *RespVars* tab at the top of the analysis notebook to see the page illustrated in Figure 20–4.[2]

Only one response variable is listed there, as expected, and its fit on the first ordination axis (*CFit.1*) is – unsurprisingly – again equal to the coefficient of determination from the earlier linear regression model. Also note that the *TurdTorq* score on the first (constrained) axis has a negative value (−0.6215). The page named *ExplVars* in the analysis notebook gives (in the *RegrE.1* column) the standardised regression coefficient for the *Altit* predictor (as −1.0) and after multiplying it with the *Resp.1* value, you obtain 0.6215. This is the regression coefficient for *Altit*, but after the variable was standardised to zero mean and unit variance. To obtain the 'normal' (non-standardised) regression coefficient, you must de-standardise this value, by multiplying it with *TAU* (the response variable standard deviation) and dividing by the standard deviation of the predictor: 0.6215 * 0.566316 / 132.6978 = 0.00265, which is the same value as in the output of the *summary* function in R, shown earlier. The *TAU* value can be found in the *Summary* page, after you click the *Details* button, while the standard deviation of the predictor is shown in the *ExplVars* page, in the *SDE* column. The same page also reports a t-value for the regression coefficient (column *TValE.1*), identical with that shown by R.

---

[2] If you see different values in the first two columns, click the sixth button from the left in the Canoco 5 toolbar (its tooltip says *Edit scaling options* ...) and select the *Focus on response variable correlations* value in the topmost field.

You can do similar analyses also with multiple predictors and even using factors. In the latter case, the RDA corresponds to ANOVA models, as we have already shown in Section 15–4.

20.4 Fitting generalized linear models in Canoco

Here we will show how to fit a standard linear model in Canoco 5. More precisely, Canoco 5 offers an analysis template for generalized linear models (GLM), but when you set up the model with the Gaussian distribution and identity link function, the fitted model is identical with a standard linear regression (see Section 8.3).

To create a GLM analysis, start with the *New* button again and choose the *Generalized-linear-model* analysis template in the *Specialized Analyses* folder.[3] Confirm the identity of response and explanatory variable in the first two pages of the Analysis Setup Wizard, as well as the permutation test settings in the *Testing and Stepwise selection* page. This is the first place where the offered implementation of GLM in Canoco 5 differs from the general statistical programs: you can – beside the standard parametric tests – also perform permutation testing, including the advanced permutation schemes available in Canoco 5 for constrained ordination methods. The next page (titled *GLM Options*) is a page specifically for this analysis template, where the type of response distribution, as well as additional model options, can be specified. The default settings (*Gaussian* and *Model has an intercept term*) are appropriate for your task. After you click *Finish*, the analysis is executed and its notebook is shown.

Its first page is again a custom page, with a standard report on fitted linear model, as well as its estimated parameters (regression coefficients). The F-based test reported there is a parametric test, identical to that provided by R. But if you switch to the *Summary* page, you can see (in its lower part) the results of a permutation test, coherent (but not necessarily identical) with those shown on the *Linear Model* page. For this analysis, you can create the diagnostic plots directly (not as ordination plots), by selecting the *Graph | Attribute plots | XY(Z) diagram* menu command and then choosing X and Y variables in the two upper lists, from the *Analysis 'Generalized-linear-model'* folder (as illustrated by *Graph 1* in the *Generalized-linear-model* analysis of the example project).

---

[3] Note that this template has two instances, but only one makes sense for our data, namely the one with (*bird species ~ landscape parameters*) formula following its name.

Appendix A
Glossary

Attribute plot	Scatter diagram usually based on the **case scores** and displaying one (or two) particular attribute(s) of cases (e.g. values of selected **response variable**, of selected **explanatory variable**, diversity of species composition, etc.), see Section 11.3. Attribute plots can be also created for attributes of individual response variables or other types of variables.
Biplot	**Ordination diagram** displaying (at least) two different types of objects (e.g. **response variables** and **cases** or **cases** and **explanatory variables**) that can be jointly interpreted using the biplot rule (see Section 11.1).
Canonical axis, Constrained axis	Axis of an **ordination** space which is constrained by its definition as an error-free linear multiple regression model using selected **explanatory variables**.
Case	One sampling unit (object, site, record, one row in data table) in an analysis.
Centring	Change of values of a variable by subtracting the arithmetic mean of the variable. After centring, the variable has zero mean.
Constrained ordination	**Ordination** containing at least one **canonical (constrained) axis**, summarising the part of total variation in **response variables** that is related to the values of **explanatory variables**.
Covariates	Also called *covariables* in older versions of Canoco. They are predictor variables with their effects upon the response variables acknowledged, but not of primary interest.
Degrees of freedom	Measure of complexity ('number of parameters') of a statistical model. DFs are also used to express the amount of information left in a statistical sample after information has been extracted with a model.
Dummy variable	A variable with 0 and 1 values (most often), coding a particular level of a **factor** (nominal) variable. This is a user-friendly name for a particular type of factor contrasts.
Eigenvalue	A statistic measuring the importance of an **ordination** axis and representing the amount of variability in the **response variables** explained by the axis. This is a specific interpretation of this term, which does have more general uses in the literature.
Explanatory variables	Variables used to explain directly (in **constrained ordination**) the variability in a table with **response variables**.

(continued)

Factor, Nominal variable	Variable which has a non-quantitative character and indicates mutually exclusive classes (types) to which individual **cases** belong.
Gradient analysis	A statistical method which attempts to explain the values of **response variables** (e.g. the abundance of species in a community) using continuous (quantitative) **explanatory variables**. In the typical context, explanatory variables are supposed to describe the variation of the environment, but they do not need to be directly measured (e.g. in **unconstrained ordination**). **Ordination** methods are a specific group of gradient analysis methods (see Chapter 4).
Joint plot	**Ordination diagram** displaying (at least) two different types of objects (e.g. **response variables** and **cases**) that are jointly interpreted by the centroid principle (see Section 11.2) and not by the biplot rule.
Linear ordination	**Ordination** method based on a **response model**, in which straight lines are supposed to best describe the change of the **response variable** values along the ordination axis (after possible **transformation** and/or **standardisation** and **centring**).
Ordination	Multivariate statistical method that summarises total variation (**unconstrained ordination**) or a specific part of it (**constrained ordination** or **partial ordination**) using a set of ordination axes.
Ordination diagram	Two-dimensional scatterplot displaying, using symbols and/or arrows, **scores** of objects of one or multiple types (e.g. **response variables**, **cases**, **explanatory** or **supplementary variables**), enabling easy interpretation of the multivariate data, summarised by an **ordination**. An important property of the ordination diagram is that its axes cannot be arbitrarily rescaled in respect of one to another.
Partial ordination	**Constrained** or **unconstrained ordination** using **covariates**. Consequently it summarises only a part of the total variation in **response variables**, as the effect of **covariates** is 'partialled-out' ('factored-out').
Permutation test	Test of a null hypothesis, in which the computed value of a test statistic is compared with a constructed estimate of its distribution under the validity of the tested null hypothesis. This distribution is estimated by randomly permuting ('shuffling') the sampling units (**cases**) in accordance with the particular null hypothesis. The extent of randomness of the performed permutations can be restricted to reflect partial dependence among (internal structure of) the individual **cases** (see Chapter 5).
QuickWizard mode	When Canoco 5 works in this mode, the Analysis Setup Wizard does not show pages displaying either rarely used options or no modifiable options. See Section 2.3 for information on how to switch this mode on and off.
Relevé	Traditional term from vegetation science. It refers to a **case** describing vegetation composition (usually percentage cover of individual plant species or estimates of this cover on a semi-quantitative scale – e.g. the Braun-Blanquet scale) recorded in a plot.

Response model	Refers to a fitted curve describing the assumed shape of the change of values of a **response variable** (a population size or percentage cover of a biological species, for example) with the values of **explanatory variables** (including here even hypothetical gradients, such as ordination axes from **unconstrained ordination**).
Response variables	Variables whose values are explained by statistical models (including **constrained** and **unconstrained ordination** or various regression methods). A typical example from the community ecology is a set of species of a biotic community.
Scores	Values representing coordinates of various types of entities (**cases**, **response variables**, **explanatory** or **supplementary variables**) in the ordination space, constructed by an **ordination**. They are used to plot **ordination diagrams** and represent a functional relation of **cases** and of various types of variables to the ordination axes.
SD units	Quantify the extent of variation along ordination axes constructed by **unimodal ordination** methods. Correspond to weighted standard deviation units.
Standardisation	Change of values of a variable by division by its standard deviation or some other measure, so as to ensure that after standardisation, the variables are more comparable, for example to have all unit variation in the case of division by the standard deviation. Standardisation is often done in combination with **centring**.
Supplementary response variables or cases	The **response variables** or **cases** that were not used during the calculations performed by an ordination. Based on the ordination results, however, we can project such variables or cases passively into the ordination space. Therefore, these entities do not contribute to the meaning of the **ordination** results, but their meaning is interpreted using those results.
Supplementary variables	These are additional (predictor) variables projected into the **unconstrained ordination** space (rarely into **constrained ordination** space) to help to interpret the meaning of ordination axes.
Transformation	A change of values of a variable, using a parametric (functional) 'recipe'. In this book, we use this term when the individual entries in a data table are transformed with an identical recipe, not varying across **cases** or variables. Most often, monotonous transformations are used (which do not change the ordering of transformed values), such as log- or square-root transformations.
Triplot	An **ordination diagram** containing three types of objects (**response variables**, **cases**, and **explanatory or supplementary variables**) where all three possible pairs can be interpreted using the biplot rule (see also the definition of the **biplot**).
Unconstrained ordination	**Ordination** with no **canonical** (**constrained**) **axes**. All its ordination axes represent hypothetical predictors (gradients) maximising the amount of variation explained in the **response variable** values.
Unimodal ordination, Weighted averaging methods	**Ordination** based on a **response model**, where a symmetrical, unimodal (bell-shaped) curve is assumed to best describe the change of the response variable values along the ordination axes.

Appendix B
Sample data sets and projects

The sample data sets (in Excel file format, both as the new *xlsx* and the old *xls*), the Canoco 5 project files, and a few additional files discussed in this book are available from our website:

http://regent.prf.jcu.cz/maed2/

You can download the sample files from there and find additional information, including errata and additional notes about the topics discussed in this book. The sample files are available in compressed ZIP files, either separately for each chapter or in one large ZIP archive, containing all the files.

See also Appendix C for information on how to obtain a trial version of Canoco 5 that you can use to repeat all the steps described in the case studies, using the sample data sets.

Appendix C
Access to Canoco and overview of other software

The use of any multivariate statistical method even for a small data set requires a computer program to perform the analysis. Most of the known statistical methods are implemented in multiple statistical programs. In this book, we demonstrate how to use mainly the ordination methods with possibly the most widely employed package, Canoco. We also show how to use the methods not available in Canoco (e.g. clustering) with a general, freely available statistical software R (R Core Team, 2013) – see Appendix D for guidance on work with R.

The following paragraph describes how to obtain a trial version of the Canoco program, which you can use to work through the tutorials provided in this book, using the sample data sets (see Appendix B for information on how to download them). We also provide an overview of other available software in a tabular form and show there both freely available as well as commercial software. The attention is focused on specialised software packages, targeting ecologists (biologists), so we do not cover general statistical packages like Statistica, S-Plus, SAS, or GENSTAT.

> Canoco 5 is a commercial software requiring a valid license for its use. But we have reached agreement with its distributor (Microcomputer Power, Ithaca, NY, USA), who will provide you on request with a trial version of the software, which will be functional for a **minimum** of one month. You can use it to try the sample analyses discussed in this book, using the data and Canoco projects provided on our website (see Appendix B). To contact Dr Richard Furnas from Microcomputer Power, write to the following email address: trial@microcomputerpower.com

Table C–1 lists the most important packages, which can be used for some or many parts of the multivariate analysis of ecological data described (or at least mentioned) in this book, and their functionality is compared. We do not own copies of all the listed programs, so the information is often based on the data excerpted from their web pages. For all of them, however, their authors or distributors reviewed and corrected the provided information. The list of features is actual as of November 2013 and might not be fully correct. The table is followed by footnotes explaining the meaning of individual rows. Then we provide individual paragraphs for each of the listed programs, with the web address for the distributing company.

Table C–1 Comparison of supported multivariate methods and their utilities for a selection of widely used programs. Explanations of comments are shown below the table.

	ade4	Canoco	CAP/ECOM	PAST	PATN	PC-ORD	Primer/P+	vegan
version examined	1.5–2	5	4/2.1	3.0	3	6	6	2.0.7
distribution[1]	F	C	C	F	C	C	C	F
Mac[2]	✔	✘	✘	✘	✘	✘	✘	✔
PCA	✔	✔	✔	✔	✘	✔	✔	✔
RDA	✔	✔	✔	✘	✘	✔	✔[13]	✔
CA	✔	✔	✔	✔	✘	✔	✘	✔
DCA	✘	✔	✔	✔	✘	✔	✘	✔
CCA	✔	✔	✔	✔	✘	✔	✘	✔
permutation tests[3]	rand	full	rand	rand	rand	rand	full[13]	full
stepwise selection[4]	✘	✔	✘	✘	✘	✘	✔[13]	✔
partial ordination[5]	✔	✔	✘	✘	✘	✘	✘	✔
variation partitioning[6]	✘	✔	✘	✘	✘	✘	✔[13]	✔
spatial correlation[7]	✔	✔	✘	✘	✘	✘	✘	✔
functional traits[8]	✔	✔	✘	✘	✘	✘	✘	✘[11]
discriminant analysis	✔	✔	✔	✔	✘	✘	✘	✘[11]
PCoA	✔	✔	✔	✔	✘	✔	✔	✔
distance-based RDA[9]	✘	✔	✘	✘	✘	✘	✔	✔
NMDS	✘	✔	✔	✔	✔	✔	✔	✔
co-inertia analysis	✔	✘[12]	✘	✘	✘	✘	✘	✘
Mantel test	✔	✘	✘	✔	✘	✔	✔	✔
PERMANOVA	✘	✘	✘	✔	✘	✔	✔	✔
ANOSIM/MRPP	✘	✘	✔	✔	✔	✔	✔	✔
clustering[10]	✘[11]	✘	✔	✔	✔	✔	✔	✘[11]
TWINSPAN	✘	✘	✔	✘	✘	✔	✘	✘
multiple regression	✘[11]	✔	✔	✔	✘	✔	✔[13]	✘[11]
GLM/GAM	✘[11]	✔	✘	GLM	✘	✘	✘	✘[11]

[1] F – free, C – commercial;
[2] ✔: runs natively on Mac OS platform; ✘: the program can be likely executed under emulator software;
[3] *rand*: only the completely random swapping of units supported, *full*: full range of restricted permutations, as described in Chapter 5 (split-plot design, transects/time-series, grids);
[4] stepwise selection of a subset of explanatory variables in constrained ordination;
[5] uses the concept of covariates in unconstrained/constrained ordination;
[6] direct support for variation partitioning beyond partial ordination;
[7] supports taking spatial correlation into account when testing and/or modelling it using spatial eigenfunctions;
[8] support for relating species-composition – environment – species-trait data tables in a single analysis;
[9] or CAP method;
[10] supports hierarchical agglomerative clustering; other methods often supported as well;
[11] *ade4* and *vegan* are optional packages in R software: they do not provide this functionality themselves, but it is provided by other R packages, when installed;
[12] co-inertia analysis is implemented but without proper permutation test;
[13] these features are available in the context of distance-based methods only, with full range of restricted permutation types available in PERMANOVA+ add-on;

ade4	is an optional package for R program, which can be added to R installation (on Microsoft Windows) using the *Packages	Install package(s)* menu command or directly downloaded from http://cran.r-project.org/web/packages/ade4
Canoco	program is distributed by Microcomputer Power, USA and you can order it through their www.microcomputerpower.com site. Additional information can be obtained at www.canoco5.com and http://www.canoco.com
CAP and ECOM	are two programs developed and distributed by Pisces Conservation, Ltd., United Kingdom, with complementary functionality (e.g. unconstrained ordination and TWINSPAN methods are available in CAP and constrained ordination methods in ECOM), so we treat them together. Both programs can be purchased through www.pisces-conservation.com
PAST	is freely available software with focus on analyses in the domain of paleontology. It can be obtained from the University of Oslo, Norway, website at http://folk.uio.no/ohammer/past/
PATN	is distributed by Blatant Fabrications Pty Ltd., Australia, and focuses on distance-matrix based multivariate method. The website is www.patn.com.au
PC-ORD	is distributed by MjM Software, USA, and covers full range of unconstrained and constrained ordination, classification methods, as well as additional ones not discussed in this book. The web site is www.pcord.com
Primer and PERMANOVA+	are developed and distributed by Primer-E Ltd., United Kingdom, and similarly to PATN focus on distance-matrix based multivariate methods. The website is www.primer-e.com
vegan	is an optional package for R program, which can be added to R installation (on Microsoft Windows) using the *Packages	Install package(s)* menu command or directly downloaded from http://cran.r-project.org/web/packages/vegan

Appendix D
Working with R

The R software (R Core Team, 2013) is a widely used, powerful statistical environment implementing most of the known statistical methods both in the univariate and multivariate realm. Given its widespread use, there are plentiful brief introductions as well as detailed treatments of the work with this software, and we have no ambition to add to this already bewildering range. For a quick start, however, just to be able to execute in R the tasks illustrated in this book, we provide a simple tutorial on R that you can download from the following URL address:

http://regent.prf.jcu.cz/maed2/RCC.pdf

The above document also contains up-to-date links to more extensive online resources as well as recommendations for referential textbooks that cover either general R use or its use for a particular group of statistical methods.

References

Anderson, M. J. (2001) A new method for non-parametric multivariate analysis of variance. *Austral Ecology*, **26**: 32–46.

Anderson, M. J. & ter Braak, C. J. F. (2003) Permutation tests for multi-factorial analysis of variance. *Journal of Statistical Computation and Simulation*, **73**: 85–113.

Anderson, M. J. & Willis, T. J. (2003) Canonical analysis of principal coordinates: a useful method of constrained ordination for ecology. *Ecology*, **84**: 511–525.

Batterbee, R. W. (1984) Diatom analysis and the acidification of lakes. *Philosophical Transactions of the Royal Society of London, Series B*, **305**: 451–477.

Benjamini, Y. & Hochberg, Y. (1995) Controlling the false discovery rate – a practical and powerful approach to multiple testing. *Journal of the Royal Statistical Society, Series B*, **57**: 289–300.

Benot, M. L., Mony, C., Lepš, J., Penet, L., & Bonis, A. (2013) Are clonal traits and their response to defoliation good predictors of grazing resistance? *Botany*, **91**: 62–68.

Birks, H. J. B. (2012) Overview of numerical methods in paleolimnology. In: Birks, H. J. B, Lother, A. F., Juggins, S., & Smol, J. P. (eds.) *Tracking Environmental Change Using Lake Sediments, Volume 5: Data Handling and Numerical Techniques*. Dordrecht: Springer, pp. 19–92.

Blanchet, F. G., Legendre, P., & Borcard, D. (2008) Forward selection of explanatory variables. *Ecology*, **89**: 2623–2632.

Bonin, A., Ehrich, D., & Manel, S. (2007) Statistical analysis of amplified fragment length polymorphism data: a toolbox for molecular ecologists and evolutionists. *Molecular Ecology*, **16**: 3737–3758.

Borcard, D. & Legendre, P. (2002) All-scale spatial analysis of ecological data by means of principal coordinates of neighbour matrices. *Ecological Modelling*, **153**: 51–68.

Borcard, D., Legendre, P., Avois-Jacquet, C., & Tuomisto, H. (2004) Dissecting the spatial structure of ecological data at multiple scales. *Ecology*, **85**: 1826–1832.

Borcard, D., Legendre, P., & Drapeau, P. (1992) Partialling out the spatial component of ecological variation. *Ecology*, **73**: 1045–1055.

Bray, R. J. & Curtis, J. T. (1957) An ordination of the upland forest communities of southern Wisconsin. *Ecological Monographs*, **27**: 325–349.

Cabin, R. J. & Mitchell, R. J. (2000) To Bonferroni or not to Bonferroni: when and how are the questions. *Bulletin of the Ecological Society of America*, **81**: 246–248.

Chambers, J. M. & Hastie, T. J. (1992) *Statistical Models in S*. Pacific Grove: Wadsworth & Brooks.

Chao, A., Chazdon, R. L., Colwell, R. K., & Shen, T. J. (2005) A new statistical approach for assessing compositional similarity based on incidence and abundance data. *Ecology Letters*, **8**: 148–159.

Chao, A., Chiu, C. H., & Hsieh, T. C. (2012) Proposing a resolution to debates on diversity partitioning. *Ecology*, **93**: 2037–2051.

Clarke, K. R. (1993) Non-parametric multivariate analysis of changes in community structure. *Australian Journal of Ecology*, **18**: 117–143.

Cleveland, W. S. & Devlin, S. J. (1988) Locally-weighted regression: an approach to regression analysis by local fitting. *Journal of the American Statistical Association*, **83**: 597–610.

Colwell, R. K. (2011) *EstimateS, version 8.2: Statistical Estimation of Species Richness and Shared Species from Samples (Software and User's Guide)*. Freeware for Windows and Mac OS. University of Connecticut.

Cox, T. F. & Cox, M. A. A. (1994) *Multidimensional Scaling*. London: Chapman & Hall.

De'ath, G. (2002) Multivariate regression trees: a new technique for constrained classification analysis. *Ecology*, **83**: 1103–1117.

de Bello, F., Carmona, C. P., Mason, N. W. H., Sebastia, M. T., & Lepš, J. (2013) Which trait dissimilarity for functional diversity: trait means or trait overlap? *Journal of Vegetation Science*, **24**: 807–819.

de Bello, F., Lepš, J., & Sebastià, M. T. (2005) Predictive value of plant traits to grazing along a climatic gradient in the Mediterranean. *Journal of Applied Ecology*, **42**: 824–833.

de Bello, F., Lepš, J., & Sebastià, M. T. (2006) Variations in species and functional diversity along climatic and grazing gradients. *Ecography*, **29**: 801–810.

de Bello, F., Thuiller, W., Lepš, J. et al. (2009) Partitioning of functional diversity reveals the scale and extent of trait convergence and divergence. *Journal of Vegetation Science*, **20**: 475–486.

Dem, F. F., Stewart, A. J. A., Gibson, A., Weiblen, G. D., & Novotný, V. (2013) Low host specificity in species-rich assemblages of xylem- and phloem-feeding herbivores (Auchenorrhyncha) in a New Guinea lowland rain forest. *Journal of Tropical Ecology*, **29**: 467–476.

Desdevises, Y., Legendre, P., Azouzi, L., & Morand, S. (2003) Quantifying phylogenetically structured environmental variation. *Evolution*, **57**: 2647–2652.

Diamond, J. (1986) Overview: Laboratory experiments, field experiments, and natural experiments. In: Diamond, J., & Case, T. J. (eds.) *Community Ecology*. New York: Harper & Row, pp. 3–22.

Doledec, S., Chessel, D., ter Braak, C. J. F., & Champely, S. (1996) Matching species traits to environmental variables: a new three-table ordination method. *Environmental and Ecological Statistics*, **3**: 143–166.

Dray, S., Chessel, D., & Thioulouse, J. (2003) Co-inertia analysis and the linking of ecological data tables. *Ecology*, **84**: 3078–3089.

Dray, S. & Legendre, P. (2008) Testing the species traits environment relationships: the fourth-corner problem revisited. *Ecology*, **89**: 3400–3412.

Dray, S., Legendre, P., & Peres-Neto, P. R. (2006) Spatial modelling: a comprehensive framework for principal coordinate analysis of neighbour matrices (PCNM). *Ecological Modelling*, **196**: 483–493.

Ellenberg, H. (1991) Zeigerwerte von Pflanzen in Mitteleuropa. *Scripta Geobotanica*, **18**: 1–248.

Eubank, R. L. (1988) *Smoothing Splines and Parametric Regression*. New York: Marcel Dekker.

Ezekiel, M. (1930) *Methods of Correlation Analysis*. New York: John Wiley and Sons.

Flinders, C. A. & Magoulick, D. D. (2002) Partitioning variance in lotic crayfish community structure based on a spatial scale hierarchy. *unpublished manuscript*.

Garnier, E., Cortez, J., Billès, G. et al. (2004) Plant functional markers capture ecosystem properties during secondary succession. *Ecology*, **85**: 2630–2637.

Götzenberger, L., de Bello, F., Brathen, K. A. et al. (2012) Ecological assembly rules in plant communities – approaches, patterns and prospects. *Biological Reviews*, **97**: 111–127.

Gower, J. C. (1966) Some distance properties of latent root and vector methods used in multivariate analysis. *Biometrika*, **53**: 325–338.

Gower, J. C. & Legendre, P. (1986) Metric and Euclidean properties of dissimilarity coefficients. *Journal of Classification*, **3**: 5–48.

Graffelman, J. & Van Eeuwijk, F. (2005) Calibration of multivariate scatter plots for exploratory analysis of relations within and between sets of variables in genomic research. *Biometrical Journal*, **47**: 863–879.

Grassle, J. F. & Smith, W. (1976) A similarity measure sensitive to the contribution of rare species and its use in investigation of variation in marine benthic communities. *Oecologia*, **25**: 13–22.

Green, R. H. (1979) *Sampling Design and Statistical Methods for Environmental Biologists*. New York: John Wiley and Sons.

Greenacre, M. (2007) *Correspondence Analysis in Practice, 2nd Edn*. London: Chapman and Hall/CRC.

Grime, J. P. (1998) Benefits of plant diversity to ecosystems: immediate, filter and founder effects. *Journal of Ecology*, **86**: 902–910.

Grime, J. P., Hodgson, J. G., & Hunt, R. (1988) *Comparative Plant Ecology*. London: Unwin Hyman.

Grubb, P. J. (1977) The maintenance of species-richness in plant communities: the importance of the regeneration niche. *Biological Reviews*, **52**: 107–145.

Hájek, M., Hekera, P., & Hájková, P. (2002) Spring fen vegetation and water chemistry in the Western Carpathian flysch zone. *Folia Geobotanica*, **37**: 205–224.

Hallgren, E., Palmer, M. W., & Milberg, P. (1999) Data diving with cross-validation: an investigation of broad-scale gradients in Swedish weed communities. *Journal of Ecology*, **87**: 1037–1051.

Harvey, P. H. & Pagel, M. D. (1991) *The Comparative Method in Evolutionary Biology*. Oxford: Oxford University Press, 239 pp.

Hastie, T. J. & Tibshirani, R. J. (1990) *Generalized Additive Models*. London: Chapman and Hall.

Hastie, T., Tibshirani, R., & Friedman, J. (2002) *The Elements of Statistical Learning. Data Mining, Inference, and Prediction*. New York: Springer-Verlag.

Hennekens, S. M. & Schaminee, J. H. J. (2001) TURBOVEG, a comprehensive data base management system for vegetation data. *Journal of Vegetation Science*, **12**: 589–591.

Hill, M. O. (1973) Diversity and evenness: a unifying notation and its consequences. *Ecology*, **54**: 427–432.

Hill, M. O. (1979) TWINSPAN. A FORTRAN program for arranging multivariate data in an ordered two-way table by classification of the individuals and attributes. Ithaca: Ecology and Systematics, Cornell University.

Hill, M. O. & Gauch, H. G. (1980) Detrended correspondence analysis, an improved ordination technique. *Vegetatio*, **42**: 47–58.

Hill, M. O. & Šmilauer, P. (2005) *TWINSPAN for Windows Version 2.3*. Huntingdon & Ceske Budejovice: Centre for Ecology and Hydrology & University of South Bohemia.

Holm, S. (1979) A simple sequentially rejective multiple test procedure. *Scandinavian Journal of Statistic*, **6**: 65–70.

Hotelling, H. (1933) Analysis of a complex of statistical variables into principal components. *Journal of Educational Psychology*, **24**: 417–441, 498–520.

Hurlbert, S. H. (1984) Pseudoreplication and the design of ecological field experiments. *Ecological Monographs*, **54**: 187–211.

Hutchinson, G. E. (1957) Concluding remarks. *Cold Spring Harbor Symposia on Quantitative Biology*, **22**: 415–427.

Ives, A. R. & Helmus, M. R. (2011) Generalized linear mixed models for phylogenetic analyses of community structure. *Ecological Monographs*, **81**: 511–525.

Jaccard, P. (1901) Etude comparative de la distribution florale dans une portion des Alpes et du Jura. *Bulletin de la Société Vaudoise des Sciences Naturelles*, **37**: 547–579.

Jackson, D. A. (1993) Stopping rules in principal components analysis: a comparison of heuristical and statistical approaches. *Ecology*, **74**: 2204–2214.

Jamil, T., Ozinga, W. A., Kleyer, M., & ter Braak, C. J. F. (2013) Selecting traits that explain species-environment relationships: a generalized linear mixed model approach. *Journal of Vegetation Science*, **24**: 988–1000.

Jongman, R. H. G., Ter Braak, C. J. F., & Van Tongeren, O. F. R. (1987) *Data Analysis in Community and Landscape Ecology*. Wageningen, The Netherlands: Pudoc. Reissued in 1995 by Cambridge: Cambridge University Press.

Jost, L. (2006) Entropy and diversity. *Oikos*, **113**: 363–369.

Jost, L. (2007) Partitioning diversity into independent alpha and beta components. *Ecology*, **88**: 2427–2439.

Kattge, J., Díaz, S., Lavorel, S. et al. (2011) TRY – a global database of plant traits. *Global Change Biology*, **17**: 2905–2935.

Kissling, W. D., Baker, W. J., Balslev, H. et al. (2012) Quaternary and pre-Quaternary historical legacise in the global distribution of a major tropical plant lineage. *Global Ecology and Biogeography*, **21**: 909–921.

Kleyer, M, Bekker, R. M., Knevel, I. C. et al. (2008) The LEDA Traitbase: a database of life history traits of the Northwest European flora. *Journal of Ecology*, **96**: 1266–1274.

Kleyer, M., Dray, S., De Bello, F. et al. (2012) Assessing species and community functional responses to environmental gradients: which multivariate methods? *Journal of Vegetation Science*, **23**: 805–821.

Knox, R. G. (1989) Effects of detrending and rescaling on correspondence analysis: solution stability and accuracy. *Vegetatio*, **83**: 129–136.

Kovář, P. & Lepš, J. (1986) Ruderal communities of the railway station Ceska Trebova (Eastern Bohemia, Czechoslovakia) – remarks on the application of classical and numerical methods of classification. *Preslia*, **58**: 141–163.

Kruskal, J. B. (1964) Nonmetric multidimensional scaling: a numerical method. *Psychometrika*, **29**: 115–129.

Kühn, I., Durka, W., & Klotz, S. (2004) BiolFlor – a new plant-trait database as a tool for plant invasion ecology. *Diversity and Distributions*, **10**: 363–365.

Lande, R. (1996) Statistics and partitioning of species diversity and similarity among multiple communities. *Oikos*, **76**: 5–13.

Legendre, P. (1993) Spatial autocorrelation: trouble or new paradigm? *Ecology*, **74**: 1659–1673.

Legendre, P. (2007) Studying beta diversity: ecological variation partitioning by multiple regression and canonical analysis. *Journal of Plant Ecology*, **1**: 3–8.

Legendre, P. & Anderson, M. J. (1999) Distance-based redundancy analysis: testing multi-species responses in multi-factorial ecological experiments. *Ecological Monographs*, **69**: 1–24.

Legendre, P., Borcard, D., & Peres-Neto, P. R. (2005) Analyzing beta diversity: partitioning the spatial variation of community composition data. *Ecological Monographs*, **75**: 435–450.

Legendre, P. & Fortin, M. J. (2010) Comparsion of the Mantel test and alternative approaches for detecting complex multivariate relationships in the spatial analysis of genetic data. *Molecular Ecology Resources*, **10**: 831–844.

Legendre, P. & Gallagher, E. D. (2001) Ecologically meaningful transformations for ordination of species data. *Oecologia*, **129**: 271–280.

Legendre, P. & Legendre, L. (2012) *Numerical Ecology*, Third English Edn. Amsterdam: Elsevier B. V.

Legendre, P., Oksanen, J., & ter Braak, C. J. F. (2011) Testing the significance of canonical axes in redundancy analysis. *Methods in Ecology and Evolution*, **2**: 269–277.

Lepš, J. (1999) Nutrient status, disturbance and competition: an experimental test of relationship in a wet meadow. *Journal of Vegetation Science*, **10**: 219–230.

Lepš, J. & Buriánek, V. (1990) Pattern of interspecific associations in old field succession. In: Krahulec F., Agnew A. D. Q., Agnew, S., & Willems, J. H. (eds.): *Spatial Processes in Plant Communities*. SPB Publishers, pp. 13–22.

Lepš, J., de Bello, F., Lavorel, S., & Berman, S. (2006) Quantifying and interpreting functional diversity of natural communities: practical considerations matter. *Preslia*, **78**: 481–501.

Lepš, J., de Bello, F., Šmilauer, P., & Doležal, J. (2011) Community trait response to environment: disentangling species turnover vs intraspecific trait variability effects. *Ecography*, **34**: 856–863.

Lepš, J. & Hadincová, V. (1992) How reliable are our vegetation analyses? *Journal of Vegetation Science*, **3**: 119–124.

Lepš, J., Novotný, V., & Basset, Y. (2001) Habitat and successional status of plants in relation to the communities of their leaf-chewing herbivores in Papua New Guinea. *Journal of Ecology*, **89**: 186–199.

Lepš, J., Prach, K., & Slavíková, J. (1985) Vegetation analysis along the elevation gradient in the Nízké Tatry Mountains (Central Slovakia). *Preslia*, **57**: 299–312.

Lepš, J. & Šmilauer, P. (2007) Subjectively sampled vegetation data: don't throw out the baby with the bath water. *Folia Geobotanica*, **42**: 169–178.

Lindsey, J. K. (1993) *Models for Repeated Measurement*. Oxford: Oxford University Press.

Little, R. J. A. & Rubin, D. B. (1987) *Statistical Analysis with Missing Data*. New York: John Wiley and Sons.

Ludwig, J. A. & Reynolds, J. F. (1988) *Statistical Ecology*. New York: John Wiley and Sons.

Maccherini, S. & Santi, E. (2012) Long-term experimental restoration in a calcareous grassland: identifying the most effective restoration strategies. *Biological Conservation*, **146**: 123–135.

McCullagh, P. & Nelder, J. A. (1989) *Generalized Linear Models*. Second edn. London: Chapman and Hall.

McCune, B. & Mefford, M. J. (1999) PC-ORD. *Multivariate Analysis of Ecological Data, Version 4*. Gleneden Beach: MjM Software Design.

Mielke, P. W. & Berry, K. J. (2007) *Permutation Methods: A Distance Function Approach*. Second edition, Springer.

Moravec, J. (1973) The determination of the minimal area of phytocenoses. *Folia Geobotanica Phytotaxonomica*, **8**: 429–434.

Morisita, M. (1959) Measuring of interspecific association and similarity between communities. *Memoirs Fac. Sci. Kyushu Univ., Ser E.*, **3**: 65–80.

Mouillot, D., Stubbs, W., Faure, M. et al. (2005) Niche overlap estimates based on quantitative functional traits: a new family of non-parametric indices. *Oecologia*, **145**: 345–353.

Mueller-Dombois, D. & Ellenberg, H. (1974) *Aims and Methods of Vegetation Ecology*. New York: John Wiley and Sons.

Nakagawa, S. (2004) A farewell to Bonferroni: the problems of low statistical power and publication bias. *Behavioral Ecology*, **15**: 1044–1045.

Novotný, V. & Basset, Y. (2000) Rare species in communities of tropical insect herbivores: pondering the mystery of singletons. *Oikos*, **89**: 564–572.

Okland, R. H. (1999) On the variation explained by ordination and constrained ordination axes. *Journal of Vegetation Science*, **10**: 131–136.

Oksanen, L. (2001) Logic of experiments in ecology: is pseudoreplication a pseudoissue? *Oikos*, **94**: 27–38.

Orloci, L. (1967) An agglomerative method for classification of plant communities. *Journal of Ecology*, **55**: 193–205.

Orloci, L. (1978) *Multivariate Analysis in Vegetation Research*. Second edn. The Hague: W. Junk B. V.

Pakeman, R. J., Garnier, E., Lavorel, S. et al. (2008) Impact of abundance weighting on the response of seed traits to climate and land use. *Journal of Ecology*, **96**: 355–366.

Pakeman, R. J. & Quested, H. M. (2007) Sampling plant functional traits: what proportion of the species need to be measured? *Applied Vegetation Science*, **10**: 93–98.

Pélissier, R. & Couteron, P. (2007) An operational, additive framework for species diversity partitioning and beta-diversity analysis. *Journal of Ecology*, **95**: 294–300.

Pélissier, R., Couteron, P., Dray, S., & Sabatier, D. (2003) Consistency between ordination techniques and diversity measurements: two strategies for species occurrence data. *Ecology*, **84**: 242–251.

Peres-Neto, P. R. & Legendre, P. (2010) Estimating and controlling for spatial structure in the study of ecological communities. *Global Ecology and Biogeography*, **19**: 174–184.

Peres-Neto, P. R., Legendre, P., Dray, S., & Borcard, D. (2006) Variation partitioning of species data matrices: estimation and comparison of fractions. *Ecology*, **87**: 2614–2625.

Pielou, E. C. (1977) *Mathematical Ecology*. New York: John Wiley and Sons.

Prentice, H. C. & Cramer, W. (1990) The plant community as a niche bioassay: environmental correlates of local variation in *Gypsohila fastigiata*. *Journal of Ecology*, **78**: 313–325.

Pyšek, P. & Lepš, J. (1991) Response of a weed community to nitrogen fertilization: a multivariate analysis. *Journal of Vegetation Science*, **2**: 237–244.

R Core Team (2013) *R: A Language and Environment for Statistical Computing*. R Foundation for Statistical Computing, Vienna, Austria. www.r-project.org/

Rao, C. R. (1982) Diversity and dissimilarity coefficients – a unified approach. *Theoretical Population Biology*, **21**: 24–43.

Rao, C. R. (1995) A review of canonical coordinates and an alternative to correspondence analysis using Hellinger distance. *Qüestiió (Quaderns d'Estadística i Investigació Operativa)*, **19**: 23–63.

Reckhow, K. H. (1990) Bayesian inference in non-replicated ecological studies. *Ecology*, **71**: 2053–2059.

Rice, W. R. (1989) Analyzing tables of statistical tests. *Evolution*, **43**: 223–225.

Ricotta, C. & Moretti, M. (2011) CWM and Rao's quadratic diversity: a unified framework for functional ecology. *Oecologia*, **167**: 181–188.

Robinson, P. M. (1973) Generalized canonical analysis for time series. *Journal of Multivariate Analysis*, **3**: 141–160.

Schönemann, P. H. (1966) A generalized solution of the orthogonal Procrustes problem. *Psychometrika*, **31**: 1–10.

Shan, H., Kattge, J., Reich, P. et al. (2012) Gap filling in the plant kingdom: trait prediction using hierarchical probabilistic matrix factorization. Edinburgh: *Proceedings of the 29th International Conference on Machine Learning (ICML-12)*, pp. 1303–1310.

Sharma, S., Legendre, P., de Caceres, M., & Boisclair, D. (2011) The role of environmental and spatial processes in structuring native and non-native fish communities across thousands of lakes. *Ecography*, **34**: 762–771.

Shepard, R. N. (1962) The analysis of proximities: multidimensional scaling with an unknown distance function. *Psychometrika*, **27**: 125–139.

Siepelski, A. M. & McPeek, M. A. (2013) Niche versus neutrality in structuring the beta diversity of damselfly assemblages. *Freshwater Biology*, **58**: 758–768.

Silvertown, J. W. & Lovett Doust, J. (1993) *Introduction to Plant Population Biology.* Third edn. Oxford: Blackwell Scientific Publications.

Sneath, P. H. A. (1966) A comparison of different clustering methods as applied to randomly-spaced points. *Classification Society Bulletin*, **1**: 2–18.

Sokal, R. R. & Michener, C. D. (1958) A statistical method for evaluating systematic relationships. *University of Kansas Science Bulletin*, **38**: 1409–1438.

Sokal, R. R. & Rohlf, F. J. (1995) *Biometry.* Third edn., New York: W. H. Freeman.

Sørensen, T. (1948) A method of establishing groups of equal amplitude in plant sociology based on similarity of species contents and its application to analysis of the vegetation on Danish commons. *Biologiska Skrifter (Copenhagen)*, **5**: 1–34.

Spitzer, K., Novotný, V., Tonner, M., & Lepš, J. (1993) Habitat preferences distribution and seasonality of the butterflies (Lepidoptera, Papilionoidea) in a montane tropical rain forest of Vietnam. *Journal of Biogeography*, **20**: 109–121.

Špačková, I., Kotorová, I., & Lepš, J. (1998) Sensitivity of seedling recruitment to moss, litter and dominant removal in an oligotrophic wet meadow. *Folia Geobotanica*, **33**: 17–30.

Stewart-Oaten, A., Murdoch, W. W., & Parker, K. P. (1986) Environmental impact assessment: 'pseudoreplication' in time? *Ecology*, **67**: 929–940.

ter Braak, C. J. F. (1994) Canonical community ordination. Part I: Basic theory and linear methods. *Ecoscience*, **1**: 127–140.

ter Braak, C. J. F., Cormont, A., & Dray, S. (2012) Improved testing of species traits-environment relationships in the fourth corner problem. *Ecology*, **93**: 1525–1526.

ter Braak, C. J. F. & Prentice, I. C. (1988) A theory of gradient analysis. *Advances in Ecological Research*, **18**: 93–138.

ter Braak, C. J. F. & Schaffers, A. P. (2004) Co-correspondence analysis: a new ordination method to relate two community compositions. *Ecology*, **85**: 834–846.

ter Braak, C. J. F. & Šmilauer, P. (2012) *Canoco Reference Manual and User's Guide: Software for Ordination (Version 5.0).* Ithaca: Microcomputer Power.

ter Braak, C. J. F. & Verdonschot, P. F. M. (1995) Canonical correspondence analysis and related multivariate methods in aquatic ecology. *Aquatic Sciences*, **57**: 255–289.

Trekels, H., Van de Meutter, F., & Stoks, R. (2011) Habitat isolation shapes the recovery of aquatic insect communities from a pesticide pulse. *Journal of Applied Ecology*, **48**: 1480–1489.

Underwood, A. J. (1997) *Experiments in Ecology.* Cambridge: Cambridge University Press.

Van den Brink, P. J. & ter Braak, C. J. F. (1998) Multivariate analysis of stress in experimental ecosystems by Principal Response Curves and similarity analysis. *Aquatic Ecology*, **32**: 163–178.

Van den Brink, P. J. & ter Braak, C. J. F. (1999) Principal Response Curves: Analysis of timedependent multivariate responses of a biological community to stress. *Environmental Toxicology and Chemistry*, **18**: 138–148.

Van der Maarel, E. (1979) Transformation of cover-abundance values in phytosociology and its effect on community similarity. *Vegetatio*, **38**: 97–114.

Van der Maarel, E. & Franklin, J. (2013) Vegetation ecology: Historical notes and outline. In: van der Maarel E. & Franklin J. (eds.): *Vegetation Ecology*, 2nd edn. Chichester: John Wiley and Sons, pp. 1–27.

Veen, G. F. & Olff, H. (2011) Interactive effects of soil-dwelling ants, ant mounds and simulated grazing on local plant community composition. *Basic and Applied Ecology*, **12**: 703–712.

Verberk, W. C. E. P., van Noordwijk, C. G. E., & Hildrew, A. G. (2013) Delivering on a promise: integrating species traits to transform descriptive community ecology into a predictive science. *Freshwater Science*, **32**: 531–547.

Verhoeven, K. J. F., Simonsen, K. L., & McIntyre, L. M. (2005) Implementing false discovery rate control: increasing your power. *Oikos*, **108**: 643–647.

Von Ende, C. N. (1993) Repeated measures analysis: growth and other time-dependent measures. In: Scheiner S. M. & Gurevitch J. (eds.): *Design and Analysis of Ecological Experiments*. New York: Chapman and Hall, pp. 113–117.

Vos, P., Hogers, R., Bleeker, M. et al. (1995) AFLP: a new technique for DNA fingerprinting. *Nucleic Acids Research*, **23**: 4407–4414.

Wagner, H. H., Wildi, O., & Ewald, K. C. (2000) Additive partitioning of plant species diversity in an agricultural mosaic landscape. *Landscape Ecology*, **15**: 219–227.

Wartenberg, D., Ferson, S., & Rohlf, F. J. (1987) Putting things in order: a critique of detrended correspondence analysis. *American Naturalist*, **129**: 434–448.

Wessels, K. J., van Jaarsveld, A. S., Grimbeek, J. D., & van der Linde, M. J. (1998) An evaluation of the gradsect biological survey method. *Biodiversity and Conservation*, **7**: 1093–1121.

Westoby, M. (1998) A leaf-height-seed (LHS) plant ecology strategy scheme. *Plant and Soil*, **199**: 213–227.

Whittaker, R. H. (1960) Vegetation of Siskiyou Mountains, Oregon and California. *Ecological Monographs*, **30**: 279–338.

Whittaker, R. H. (1967) Gradient analysis of vegetation. *Biological Review (Camb.)*, **42**: 207–264.

Whittaker, R. H. (1975) *Communities and Ecosystems*. New York: MacMillan.

Williams, W. T. & Lambert, J. M. (1959) Multivariate methods in plant ecology. I. Association analysis in plant communities. *Journal of Ecology*, **49**: 717–729.

Wright, S. P. (1992) Adjusted P-values for simultaneous inference. *Biometrics*, **48**: 1005–1013.

Index to useful tasks in Canoco 5

Here we index brief recipes describing useful minor tasks in the Canoco 5 software. These recipes are dispersed over the book, mostly in its tutorial parts. We include here neither general tasks (e.g. 'How to create a new analysis?') nor the tasks specific to a particular chapter or section (e.g. 'How to perform stepwise selection?'). Instead, we index only the generally useful titbits of working with Canoco. Each task is briefly characterised and the page (and footnote number, when the description appears there) is given. All these tasks are described in more detail in the Canoco 5 manual.

analysis on a subset of cases, defining 281
averaged data rows, creating project with 81n9, 323

best-fitting response variables (e.g. species), choosing for plot 213, 217
black and white diagrams, creating 277
brief and non-brief analysis notebook, showing 31

calibration axis for variable arrow, adding to diagram 188
CANOCO 4.x data formats (outdated), exporting 123n3
case scores derived from response variable scores, using in constrained ordination diagrams 187n4, 236
classification of cases or variables, defining 304, 311
compositional versus general data table, specifying 37

direction of ordination axes, mirroring (flipping) 218
dummy variables and factors, changing one into another 24n7

envelopes (enclosing hulls), plotting 312
equilibrium contribution circle in PCA/RDA plots, adding 214

factor variable, changing reference level 276n15
fuzzy variables, defining 24

graph legend, adding 312
graph legend, modifying contents 277n17
group of variables (e.g. species) or cases, defining 148, 281

interactions of explanatory variables, creating in data table 275n14, 284
interactions of explanatory variables, defining in setup wizard 272, 327

labels of ordination diagram axes, adding 36

opaque background of labels, setting 213
ordination scores, using in other analyses 293

pie plot in ordination space, creating 204

QuickWizard mode, switching off and on 26

recreating graph 213
relative scaling of different scores in biplots and triplots, adjusting 316
response variables (e.g. species) with highest fit, choosing for plot 213, 218
response variables (e.g. species) with highest weight, choosing for plot 229

scaling of ordination diagram scores, changing 184
series (successional vectors), creating 284
species response curves diagram, creating 149
symbol type, varying by classes 311

transformation of variables, in XY(Z) plots 257n13
transformation of variables, setting 239, 310
t-value biplot, creating 290

variable change across ordination space, colour-coded plot 335
variable change across ordination space, contour plot 292
variables or cases in ordination diagram, suppressing 303

Subject index

AIC. *See* regression, Akaike information criterion
analysis of variance 116, 260, 265
   repeated measurements 270
arch effect 28
attribute plot 34, 202, 235, 249, 292, 326

biplot rule 63, 187, 192, 196, 197
Braun-Blanquet scale 8, 226, 246

CA. *See* correspondence analysis
calibration 51, 56
Canoco
   analysis 15, 24–32
   analysis notebook 15, 31
   Analysis Setup Wizard 17, 26, 79, 104, 209
   data import 17, 208, 259
   data tables 12, 15, 19, 22, 37
   Graph Wizard 17, 33–35
   Log page 273
   project 15
   QuickWizard 26
   score pages 291, 341
   Summary page 31, 211, 216, 219, 227, 230, 251, 263, 340
canonical axes. *See* constrained axes
canonical correspondence analysis 62, 173, 230
cases 4
   classification in graphs 202–204, 304, 311
   scores 58, 61, 62, 187, 189, 191, 196, 235
   series in graphs 204, 284
   standardisation by 10, 30, 97, 262, 265
CCA. *See* canonical correspondence analysis
centring. *See* data, centring
centroid principle 196
classification 112
   agglomerative 120
   divisive 121
   K-means clustering 114
   TWINSPAN 121–127, 237
co-correspondence analysis 14, 69, 240–242
co-inertia analysis 69
constrained axes 61, 62, 63, 212
correspondence analysis 58, 62

covariables. *See* covariates
covariates 5, 51, 219, 238, 270
CVA. *See* linear discriminant analysis

data
   centring 9. *See also* response variables, centring
   compositional 5, 10, 37
   standardisation 9. *See also* response variables, standardisation
   transforming 9, 96, 310
db-RDA. *See* principal coordinates analysis, constrained
DCA. *See* detrended correspondence analysis
design
   BACI 45, 47, 267
   completely randomised 39
   factorial 44
   hierarchical 44, 79, 301, 309. *See also* design, split-plot
   Latin square 41
   nested. *See* design, hierarchical
   non-manipulative studies 48, 208, 239
   pseudoreplication 42
   randomised complete blocks 40, 258
   repeated measures 45, 81, 267
   split-plot 44, 267
detrended correspondence analysis 61, 112, 227
detrending 28–29
discriminant analysis. *See* linear discriminant analysis
dissimilarity coefficients. *See* similarity/dissimilarity
distance measures. *See* similarity/dissimilarity
diversity
   alpha 34, 177–182, 221–225, 235
   beta 177–182, 221–225
   functional 102, 159, 290
   gamma 177–182, 221–225
   measures 180
downweighting 30

eigenvalue 60
environmental variables. *See* explanatory variables
Excel data 17

Subject index

explanatory variables 5
　biplot scores 63, 191
　conditional effects 85
　dummy 24, 37
　factors 6, 23, 87, 192, 194, 198, 269
　interactions 268, 327
　nominal. *See* explanatory variables, factors
　scores 190, 191
　simple effects 85, 232
　stepwise selection 83–85, 299, 314. *See also* permutation tests, stepwise selection

false discovery rate. *See* permutation tests, false discovery rate
forward selection. *See* explanatory variables, stepwise selection
functional traits. *See* species traits
fuzzy coding 24, 193

GAM. *See* regression, generalized additive models
GLM. *See* regression, generalized linear models
gradient analysis
　direct. *See* ordination, constrained
　indirect. *See* ordination, unconstrained
gradients 1
　length 27, 197

indicator values 56
inertia 88

linear discriminant analysis 173–174, 237, 313–316
loess. *See* regression, loess smoother

missing values 12, 153–154
model of species response 51
　linear 53, 58, 187
　unimodal 53, 58, 112, 150, 195
Moran eigenvector maps. *See* PCNM

NMDS. *See* non-metric multidimensional scaling
non-metric multidimensional scaling 107–108

ordination 1, 51
　constrained 3, 13, 60, 66, 215, 230, 249, 270
　diagnostics 201
　diagrams 2, 36, 64, 207, 212–215, 228, 229–230, 234, 249, 264, 273, 280, 316
　hybrid 61, 75
　interpretable axes 211
　linear methods 28, 30, 186
　partial 6, 51, 219, 238, 262, 270
　scaling scores 38, 184–186, 218, 263
　selecting type 27, 53, 111, 227, 301, 321
　unconstrained 13, 60, 66, 209, 227
　unimodal methods 28, 61, 195, 225

PCA. *See* principal components analysis
PCNM 170–173, 324–336
PCoA. *See* principal coordinates analysis
permutation tests 66
　Bonferroni correction 86, 91, 259
　data in space and time 76
　design-based versus model-based 75, 262
　F statistic. *See* permutation tests, pseudo-F statistic
　false discovery rate 86, 91, 232
　Holm correction 91
　Mantel test 108–111
　number of permutations 68, 74
　of individual constrained axes 74–75, 231
　PERMANOVA 111
　pseudo-F statistic 73
　significance 67, 74, 81, 87, 259
　split-plot design 79, 81, 176, 273, 305
　stepwise selection 87
　use 75–81, 216, 219, 230, 249, 262, 272, 281, 305
PRC. *See* principal response curves
principal components analysis 58, 62, 210, 302, 310
principal coordinates analysis 103–106, 171, 290, 317–319
　constrained 106, 320–321
principal response curves 168, 279
Procrustes rotation 242–245

R use 114, 116, 118, 120, 247, 260, 261, 270, 337
RDA. *See* redundancy analysis
redundancy analysis 62, 215, 218, 249, 262, 265, 268
regression 51, 129, 247
　Akaike information criterion 146
　classification and regression trees 139, 296
　generalized additive models 137, 146, 294
　generalized linear models 133, 142, 342
　linear versus RDA 294, 337–339
　loess smoother 135, 141
　mixed-effect models 138
　parsimony 131, 146
　species response curves 112, 150
response variables 4
　centring 30, 210
　fit 202, 213, 217, 234, 277
　frequency 259, 294
　prediction 183
　scores 58, 187, 189, 196, 197
　standardisation 30, 210
　weight 228, 234
reverse analysis 70

SD units. *See* turnover units
similarity/dissimilarity 92
　chi-square distance 99, 196, 197

similarity/dissimilarity (*cont.*)
   Euclidean distance 97, 189
   Gower distance 100, 290
   Hellinger distance 98
   Jaccard coefficient 93
   Morisita index 101
   NESS 101
   ordination methods 103–111
   Pearson correlation coefficient 95, 100, 189
   percentage similarity 99
   simple matching coefficient 94, 318
   Sørensen coefficient 93
   V and Q coefficients 95
spatial data analysis. *See* PCNM
species traits 102, 151–166, 286
   plot-based approach 155, 158–162, 287–293
   species-based approach 155, 162–166, 293–299
standardisation. *See* data, standardisation
stepwise selection. *See* explanatory variables, stepwise selection

supplementary variables 6, 209, 227

time series. *See* design, repeated measures
traits. *See* species traits
transformation 96. *See also* data, transforming
   log 7, 226, 239
   ordinal 8, 226
   square-root 8
TURBOVEG 226
turnover units 27, 54, 197
t-value biplot 205, 290
   Van Dobben circles 207
TWINSPAN. *See* classification, TWINSPAN

variation partitioning 88–90, 252–255, 332–333
   hierarchical 169–177, 300–308

weighted averaging 53, 56
whole-plot. *See* permutation tests, split-plot design

Made in the USA
Monee, IL
03 May 2026

49437982R00208